Cooperative Control of Multi-agent Systems

Guanglei Zhao · Hailong Cui · Changchun Hua ·
Shuang Liu

Cooperative Control of Multi-agent Systems

A Hybrid System Approach

Springer

Guanglei Zhao
Yanshan University
Qinhuangdao, Hebei, China

Hailong Cui
Yanshan University
Qinhuangdao, Hebei, China

Changchun Hua
Yanshan University
Qinhuangdao, Hebei, China

Shuang Liu
Yanshan University
Qinhuangdao, Hebei, China

ISBN 978-981-97-0967-0 ISBN 978-981-97-0968-7 (eBook)
https://doi.org/10.1007/978-981-97-0968-7

This Springer imprint is published by the registered company Springer Nature Singapore Pte Ltd.
The registered company address is: 152 Beach Road, #21-01/04 Gateway East, Singapore 189721, Singapore

Preface

Cooperative control of multi-agent systems (MASs) has extensive application scenarios in numerous engineering systems such as unmanned aerial vehicles, secondary control of microgrid, and cooperative firefighting with intelligent robots that makes it become a promising research field in the past decade. Benefitting from the distribution property of cooperative control, many complicated and burdensome tasks can be easily accomplished by MASs through local cooperations, as the global common objective is able to be achieved by using only the local information with both the computation burden and communication burden are efficiently reduced. However, despite the cooperative control having many advantages, the practical implementation of it still needs to deal with the inherent imperfections induced by networked communication, since it may be subject to various network-induced constraints, such as limited bandwidth, time delays, and packet losses.

This book is devoted to providing a new hybrid system approach to address the cooperative control problem of networked MASs subject to various communication constraints, and the main technical challenge comes in two aspects: (1) how to describe the mathematical model of various uncertain network-induced issues and incorporate them into a hybrid dynamical system; (2) how to achieve the global cooperative control objective under the hybrid systems framework while not overusing the communication resources and guaranteeing resilience to network imperfections. As the mathematical model of the network-induced problems always goes beyond the description ability of either a continuous-time dynamical system or a discrete-time dynamical system, the results in the literature dedicated to address the cooperative control problem with communication constraints cannot describe the problem completely, and thus are slightly conservative. For example, to track the number of successive package losses, a counter variable is needed that would keep invariant for a certain period of time and have an instantaneous change (plus 1 or clear its value) according to if the next data package is lost or not. The counter variable, albeit simple, exhibits characteristics of both continuous-time and discrete-time dynamical systems, making it belong to a hybrid dynamical system in essence. Hence, it is crucial to enhance the modeling framework by combining the modeling tools of continuous-time and discrete-time dynamical systems to better describe the networked MASs' closed-loop behavior in the presence of various communication constraints, and provide a corresponding solution to rigorously stabilize the closed-loop system.

In the introduction chapter of this book, we first reveal the inherent characters and implementation limitations in practical applications of networked MASs, and illustrate that the hybrid system approach is an advanced analysis methodology for networked MASs and deserves further investigation. Then, the rest of this book is divided into three parts: Part I, asynchronous sampled-data consensus control of MASs; Part II, hybrid event-triggered consensus control of MASs; Part III, reset control of MASs.

In Part I, Chaps. 2–5 are contained and the results regarding asynchronous sampled-data consensus control of MASs are presented. As the sampled information between agents is exchanged via a shared network, which may be subject to some networked constraints, the network-induced communication delays and non-identical packet losses are considered in Chaps. 2 and 3, respectively. A unified hybrid model is constructed for both leaderless and leader-following MASs, Lyapunov-based stability analysis results and how to explicitly determine the sampling periods are provided. Under static/switching directed network, the sampled-data-based leader-following consensus problem is addressed in Chap. 4 and a larger upper bound of sampling period can be obtained compared with the existing related work. Besides, suppose that the sampled relative output information is available; the dynamic output feedback cooperative control problem is studied under the hybrid systems framework in Chap. 5.

In Part II, Chaps. 6–8 are contained and the results regarding hybrid event-triggered consensus control of MASs are presented. In Chap. 6, the event-triggered bipartite consensus problem of MASs with structurally balanced signed graphs is studied, and a hybrid event-triggering mechanism is given in decentralized form. Consider that the networked MASs always subject to network-induced constraints such as communication delays and non-identical packet losses, Chaps. 7 and 8 aim at proposing a hybrid event-triggered control method to not only address these problems but also guarantee desired consensus performance.

In Part III, Chaps. 9–11 are contained and the results regarding reset control of MASs are presented. Reset control is an effective hybrid control method, which, by resetting the controller state with appropriate reset conditions, is able to improve transient performance efficiently. In this part, the reset control method is applied to MASs with asynchronous sampling control in Chap. 9, event-triggered control in Chap. 10, and observer-based event-triggered control in Chap. 11.

With respect to the aforementioned important topics, this book mainly introduces recent results regarding cooperative control of MAS under hybrid systems framework. The target audience of this book include academic researchers who are interested in cooperative control of MASs, hybrid dynamical systems, and networked control systems, particularly including (1) researchers in universities and government organizations, who are doing research or are interested in cooperative control of MASs; (2) industry control engineers who design and develop cooperative control systems of multiple mobile robots; (3) senior undergraduate students and postgraduate students in a range of fields such as cooperative control of mobile robots, control engineering, and electrical and electronic engineering.

We acknowledge IEEE for granting us permission to reuse materials from our publications copyrighted by these publishers in this book. In addition, we gratefully acknowledge the National Natural Science Foundation of China under Grants 62073277, 61825304, U22A2050, and 62188101, Natural Science Foundation of Hebei Province under Grant F2022203040, Central Government Guided Local Science and Technology Development Fund Project under Grant 226Z0301G, Hebei Innovation Capability Improvement Plan Project under Grant 22567619H, and Basic Innovation Research Project of Yanshan University under Grant 2022LGZD008, and the Institute of Electrical Engineering in Yanshan University for their support.

Qinhuangdao, China
August 2023

Guanglei Zhao
Hailong Cui
Changchun Hua
Shuang Liu

Contents

Abbreviations and Notations

Abbreviations

DOF	Dynamic output feedback
ETC	Event-triggered control
ETM	Event-triggering mechanism
MAS	Multi-agent system
MIET	Minimum inter-event time
PI	Proportional integral
RM	Reset mechanism

Notations

$\mathbb{R}$	Field of real numbers
$\mathbb{R}_{\geq 0}$	Field of nonnegative reals
$\mathbb{R}^n$	n-dimensional real Euclidean space
$\mathbb{R}^{n \times m}$	Space of $n \times m$ real matrices
$\mathbb{N}$	Set of integer numbers
$\mathbb{N}_{\geq 0}$	Set of nonnegative integers
$\lVert \cdot \rVert$	Euclidean norm
$\langle \cdot \rangle$	The inner product of real vectors
$\otimes$	Kronecker product
I_N	The N-dimensional identity matrix
$\text{diag}\{a_1, \cdots, a_n\}$	A diagonal matrix with diagonal elements or (sub-matrices) a_1 to a_n
1_N (0_N)	The N-dimensional vector with each entry being 1 (0)
$[1, N]$	The set of integers from 1 to N
A^T	Transpose of matrix or vector A
A^{-1}	Inverse of matrix A
$A \geq 0$	A symmetric positive semi-definite matrix A

$A > 0$	A symmetric positive definite matrix A
$A \leq 0$	A symmetric negative semi-definite matrix A
$A < 0$	A symmetric negative definite matrix A
$\lambda_n(A)$	n-th eigenvalue of matrix A in ascending order
$\lambda_{\min}(A)$	Minimum eigenvalue of matrix A
$\lambda_{\max}(A)$	Maximum eigenvalue of matrix A
$\mathrm{sgn}(\cdot)$	The signum function
$\longrightarrow$	Tend to or mapping to
sup	Supremum
inf	Infimum
$*$	Symmetric terms in a symmetric matrix
$\cup$	Union of two sets
$\cap$	Intersection of two sets
$\wedge$	Logical operator 'and'
$\vee$	Logical operator 'or'

Chapter 1
Introduction

1.1 Cooperative Control of MASs

In nature, of particular interest is the cooperative behavior, which helps the animal social groups achieve what an independent individual cannot [1]. For instance, when migrating, the bird flock can substantially reduce flying consumption through remaining the tailing birds within wingtip vortices. Collective motion of fish and quadrupeds is helpful for defending against predators and fast foraging for food, while the cooperation of hunters such as lions, wolves, and jackals would result in more efficient pursuit of prey. As the study develops in depth, the rationale behind animal group behaviors has inspired the study of cooperative control of multi-agent systems (MASs). With emphasis on the cooperative behavior of multiple agents, it shows a number of advantages over a single agent, including robustness to failures, comprehensiveness to system composition. Moreover, the cooperative control theory can be applied to perform complex tasks such as multipoint surveillance, hazardous material handling, and distributed reconfigurable sensor networks [2]. Where the agents can be interpreted according to different control context as autonomous robots, robot manipulators, unmanned aerial vehicles, or wireless sensors [3]. Therefore, in recent years, the study of cooperative control of MASs has attracted lots of attention of researchers from various disciplines of engineering and science.

For the cooperative control of MASs, several types of cooperative behaviors that have been widely studied are listed as follows

- rendezvous. All of the participant agents arrive at a location at a common rendezvous time.
- alignment. All of the participant agents reach an agreement in term of orientations of the local reference frames.
- formation. Each participant agent converges to an objective position and maintains a certain geometric pattern relative to each other or relative to a reference.
- synchronization/consensus. All of the participant agents reach an agreement regarding a common quantity of interest.

G. Zhao et al., *Cooperative Control of Multi-agent Systems*,
https://doi.org/10.1007/978-981-97-0968-7_1

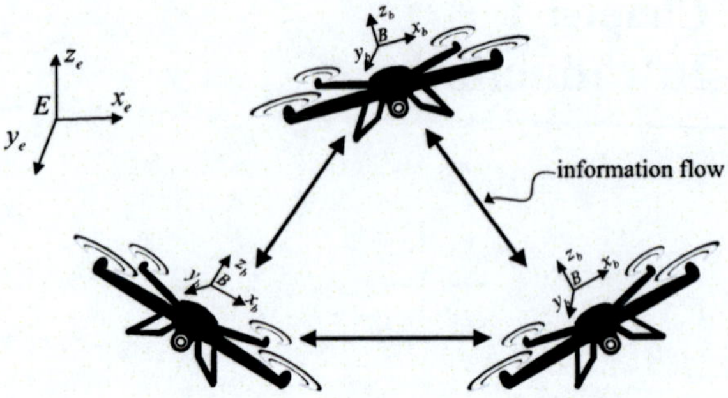

Fig. 1.1 Cooperative control of unmanned aerial vehicles (UAVs). "E" denotes the common earth reference frame, "B" denotes the local body reference frame

Based on the concepts above, it can be found that synchronization/consensus is essentially the most fundamental and typical problem that covers the others as special cases. Note that the definitions of synchronization and consensus are similar in most cases [4, 5] unless otherwise stated. The main difference between them is that the synchronization allows self-dynamics (linear or nonlinear dynamics) so that the final agreement state could be time-varying, while the consensus is widely studied in engineering applications, it usually neglects self-dynamics and the agreement state of MASs is usually constant [6].

Take unmanned aerial vehicles (UAVs), as shown in Fig. 1.1, as an illustrative example, we would like to present some further discussions about the cooperative behaviors. For the rendezvous problem, it can be interpreted as a goal-directed date of a group of UAVs. The target location can be pre-notified to all the individuals, while the rendezvous time is not known a priori and it may be prone to the effects of environment conditions such as nonuniform wind. Thus, it is better to determine the rendezvous time dynamically through team negotiation. To this end, each UAV first creates a state variable denoting the predictive arrival time, and initializes this time state according to its own current position. Then, the consensus control is used for UAVs to negotiate a common rendezvous time by regulating the time states through communicating with their own neighbors. When environment conditions change, UAVs may reset their time states and then repeat the negotiation process [7]. The orientation alignment and formation control of UAVs are often associated. According to whether a common sense of orientation is available for all UAVs or not, there exist two dominate paradigms of formation control, i.e., displacement-based approach [8–10] and distance-based approach [11–13]. To be specific, only when orientation alignment is ensured, will the UAVs be able to measure the relative positions (displacement) with respect to their neighbors in the same reference frame such that the displacement-based consensus control can be developed to maneuver the relative position and drive each individual to converge to desired formation pattern. In this case, some direction sensors such as compass are required to be equipped. On the other hand, without a common sense of orientation, UAVs measure the relative positions with respect to their neighbors only in their own local reference frames, as a result, the distance-based consensus control cannot directly maneuver the relative positions, but in turn to maneuver the norms of the relative positions to achieve desired formation pattern. Although distance and angle measurements sometimes are available for the UAVs at the same time, the transformation problems are complicated [14]. Therefore, both the alignment problem and formation problem are important for the consensus study of UAVs, and their combination makes the consensus control be even challenging. With the common interest possibly being rendezvous time, orientation alignment or formation pattern, the consensus control of UAVs can recover the cases of rendezvous control, alignment control and formation control. As such, consensus control can be viewed as the fundamental technology of cooperative control of MAS. In the sequel of the book we will pay the main attention to the consensus problem of MASs.

It is worth to be noted that the MAS can be viewed as a large-scale system with multiple subsystems, each agent corresponds to a subsystem. Thus, compared with the control of a certain agent, the cooperative control of multiple agents is more difficult and faces more challenges in the sense that the objective of cooperative control is to achieve cooperation operation of MASs, while it is not the case of single system control. In early stage of study, centralized method is widely adopted as it is relatively simple, but it also induces some limitations in terms of computation complexity and communication burden. To relax the limitations, the recent results have devoted to exploring distributive work manner and developing networked communication tools for the cooperative control of MASs. In the following, let us make some discussions to the computation complexity, distributed manner and network communication.

- Computation complexity. For MASs, each subsystem is endowed with its own dynamics and control protocol. To synchronize the states such that the collection appears to be a single entity to better response to external events, the control protocol has to integrate information as much as possible. For instance, the conventional approach to address the consensus problem is centralized control, which collects all agents' state information, makes control decision and returns control commands to each agent. The implementation of a centralized scheme inevitably causes large computation burden to the supervisor or central station, in particular, the computation complexity of the centralized control may become more and more heavy as the number of agents increases. It was pointed out in [15] that centralized control is essentially a straightforward extension of the control of single system, albeit complex. The justification is that the returned command is just the accurate trajectory. In addition, any variation in the collection, such as adding or dropping an agent or a communication link, would result in the throughout redesign of the control protocol. Therefore, there exist some inherent deficiencies in terms of computation complexity, robustness and scalability of the centralized control. To overcome such deficiencies, distributed control is proposed, which generally has less computation complexity due to each agent only collects information from its immediate neighbors.
- Distributed manner. It is well known that distributed control means that each agent only makes use of locally available information, e.g., the information from immediate neighbors and sensor measurements from agent itself. In contrast with cen-

tralized control, less information is required to be transmitted over network, while similar control performance is achieved. However, it doesn't mean that the less information available the better, since too little information may severely degrade the control performance. Thus, in order to achieve desired performance of MASs, the communication links among agents cannot be too little. More specifically, to help describe the information flows among agents, communication graph has been introduced in [16], from which the Laplacian matrix as well as its nonzero smallest eigenvalue can be derived to reflect control performance explicitly. Especially, if it fails to link enough connections among agents such that the non-negative property of Laplacian matrix is guaranteed, the MASs may exhibit undesirable behaviors or even become uncontrollable.

- Networked communication. The information flow from one agent to another plays a crucial role in consensus control of MASs, thus communication is usually indispensable. The currently most prevailing fashion of communication is via wireless network, since the networked communication has vast advantages in the sense of low cost, resources sharing and use flexibility. However, from the viewpoint of practice, there exists no ideal network environment, i.e., network with infinite bandwidth and load capabilities. This implies that the control protocols requiring continuous communication, as widely studied in many existing literature, are rather limited in general. In contrast, discrete communication could be well suitable for practical implementation of cooperative control of MASs. But discrete communication also suffers from inevitable imperfections such as time delays, packet losses, quantization, noise interference, denial of service attacks, topology changes, bandwidth limitations, etc, that usually degrade the control performance of the MASs and even lead to instability (see [17] for an illustrative example).

Now, it is clear that the cooperative control of MASs is closely related to distributed control and network communication. Therefore, in this book, we will consider the distributed consensus problem of networked MASs. This topic has been a hot issue in the literature, however, different from the existing approaches, this book introduces a hybrid system approach to deal with the considered cooperative control problem. Specifically, this book provides a hybrid system approach to study the asynchronous sampled-data consensus control problem, hybrid event-triggered consensus control problem, as well as reset control problem of MASs with network-induced constraints. First, with respect to asynchronous sampled-data consensus control, it is more practical than traditional consensus control methods that need continuous communication between agents, since continuous communication is not applicable for digital communication network. By contrast, sampled-data control can be better applied to network environment with digital sensors. In addition, note that the asynchronous sampling mechanism can cover synchronous sampling as a special case. Therefore, asynchronous sampled-data consensus control of MASs is a significant topic. Second, with respect to event-triggered consensus control, due to the communication resources are usually limited, e.g., the agents have limited energy and the network has limited bandwidth, event-triggered control approach can be used to efficiently save usage of communication resources, that is useful for practical applications.

However, one fundamental problem of event-triggered control, i.e., how to exclude Zeno behavior, has not been well solved, especially in the cases with disturbances or network-induced constraints. Therefore, hybrid event-triggered consensus control, which can guarantee Zeno-freeness, is a significant topic. Third, with respect to reset control, since most of the results related to consensus of MASs focus on steady state performance rather than transient performance, how to design controller to improve transient performance is also important. Therefore, in this book, a reset control method is proposed to improve transient consensus performance, which is also a significant topic.

1.2 Network-Induced Constraints

As mentioned before, the cooperative control of MAS relies on communication, and the networked communication builds a relatively new bridge for the information exchanges among agents. Nevertheless, the public sharing character and the packet-based digital character pose more challenges to the control compared with using the dedicated point-to-point wired communication. For example, the sensors' data and actuators' data have to be quantized and be transmitted only at discrete time instants. This section introduces some network-induced constraints including discrete communication, time delay, packet loss and switching topology, that will be considered in latter sections. Their distinctions and a relatively comprehensive review are given as well.

Discrete communication: There exist two commonly used discrete communication schemes, i.e., sampled-data scheme and event-triggered scheme. The sampled-data scheme determines its next communication purely based on time (system internal clock). Whenever a specified period of time elapsed, the communication occurs. On the other hand, event-triggered scheme determines its next communication based on particular events, e.g., the system state exceeds some thresholds. Both of the two schemes have distinct advantages in their areas, but also suffers from some limitations. The time based sampled-data scheme is always periodically activated in time, so that it is predictable and easy to implement under network environment. However, it may result in redundant communication when no disturbances are acting on the system and the system is operating desirably [18]. The event-triggered scheme is another alternative communication implementation, by which the next information transmission instant can be determined according to the current system states. It generally leads to less communication, but the uncertainty of triggering instants may cause some technical difficulties to the event detections and samplings. In this regard, while each approach has its own pros and cons, the choice mostly boils down to existing development and runtime platforms.

The consensus of MASs with sampled-data control has been extensively studied in the past decade. For the single/double-integrator MASs, [19] addressed the leaderless consensus problem. Reference [20] studied the sampled-data consensus problem under static network case for first-order MASs, and it was subsequently extended to switching network case in [21]. The sampled-data consensus problem was studied in [22] based on sampled position data and in [23] based on impulsive control laws for second-order MASs. For general linear MASs, [24] studied the time-varying topologies, and [25] investigated the robust consensus problem. Consider the sampled-data leader-following consensus problem, [26] proved that if all agents' sampling periods are smaller than a given threshold, the consensus tracking could be achieved by utilizing asynchronous sampled-data consensus control algorithm. Reference [27] proposed weighted consensus protocols for double-integrator MAS, and [28] developed delay-dependent stability conditions for general linear MASs. Reference [29] studied consensus of nonlinear MASs with stochastic sampling. More results with regard to sampled-data cooperative control of MASs can be referred to survey paper [30]. Besides, in the context of event-triggered control (ETC), distributed event-triggered sampling strategies were proposed for second-order MAS in [31] and for general linear MAS in [32]. Reference [33] investigated the rendezvous problem and proposed a combinational measurement approach. In [34], distributed event-triggered impulsive control approach was utilized for leader-following consensus of MAS. Reference [38] proposed a dynamic event-triggering mechanism to further reduce triggering number compared with static event-triggering mechanism, and the dynamic event-triggering mechanism was later applied to set-membership estimation [39], leaderless consensus [40] and leader-following consensus [41]. Decentralized event-triggered consensus protocols were proposed in [35], more results can be found in recent survey papers [36, 37].

Time delay: Time delays for networked communication usually can be classified as network access delays (i.e., the time it takes for information to get access permission of network channel) and the transmission delays (i.e., the time during which information is transmitting on network). They are inevitable, due to all the agents' information transmissions share the common communication channel, while the network resources and bandwidth are generally limited. Besides, if large amount of information occupies communication channel, the caused information congestion would also aggravate the delay phenomena. It is well known that even very small communication delays may destroy or degrade the stability and performance of the closed-loop systems. Therefore, it is an important research topic and has attracted considerable research interest. For the consensus problem of MASs, constant time delays were considered with sampled-data control [42–44], continuous event-triggered control [45, 46], periodic event-triggered control [47]. Time-varying delays were considered in [48–51].

Packet loss: Packet losses typically result from transmission errors in the wireless communication channels induced by limited bandwidth, cyber attack, noise interference, and packet disorder, or from buffer overflows due to congestion. Long transmission delays sometimes result in packet reordering, which essentially amounts to a packet loss if the receiver discards "outdated" arrivals. Some reliable transmission protocols have retransmission mechanism, that is, within maximum tolerance time the information packet can be retransmitted, but will be discarded once exceeding the tolerance. This kind of dropouts is called passive packet loss, and if the number of packets retransmission increases, it will also significantly degrade the system performance. The other kind of packet loss is active packet loss. During transmission process, advanced scheduling protocols can be used to intentionally drop non-real-time packets so that the receiver system always obtains the newest information. Nonetheless, regardless of passive or active manner, there always exists some of important information being dropped, which inevitably results in control performance degradation. Therefore, it is of significance to consider the effect of packet losses for cooperative control of MASs. Reference [24] developed a switched system approach to address the sampled-data based Bernoulli distributed packet losses and deterministic packet losses. References [53, 54] addressed the Bernoulli distributed packet losses by means of the event-triggered consensus strategy and event-triggered model predictive control strategy, respectively. In [52], Markovian packet losses were assumed and corresponding consensus control law was developed. In [55], it was assumed that the MAS include Lipschitz nonlinearities and packet losses, and consensus tracking was achieved with sampled information exchange. According to parametric Riccati equation, consensus protocol with low gain was proposed in [44] for MASs with packet losses and sampling transmission mechanism.

Switching topologies: The communication topology is dedicated to denote what information is available for a certain agent at a given time instant. Therefore, if we want to take into account the communication link failure and link creation, reconfigurable networks and nearest neighbor coupling, the communication topology is no more fixed but time-dependent. Such changes of connections among agents result in switching topologies phenomenon. With respect to this issue, lots of valuable works regarding MASs have been investigated. For instance, under the assumption that the switching topologies satisfy dwell-time condition, i.e., all adjacent switching instants are larger than a positive number, many different consensus control results (e.g., target aggregation, event-triggered control, time-varying formation control, and output-feedback control) were developed for various linear MASs [56–62] and nonlinear MASs [63–67]. Under the assumption that the switching topologies satisfy an improved average dwell-time condition, the case that each graph must be connected was studied in [68–70], and the case that some of the graphs can be unconnected was studied in [71–74].

It is reflected by the aforementioned introduction and discussions that there often exist some inherent imperfections for the networked communication in practice. Thus, the network-induced constraints have to be considered for the MASs to make the proposed consensus control strategy be more practical. In this book, different from other books, we model these network problems into a hybrid dynamical system and propose a hybrid system approach to deal with the consensus problem of closed-loop MASs. Some preliminaries with respect to hybrid systems are given in the following section.

1.3 Hybrid Dynamic Systems

In this section, the basic model of a hybrid system is introduced. Some distinct features of using a hybrid system approach in sampling control, event-triggered control and reset control are discussed and we present three control examples to illustrate the effectiveness of the hybrid model and the hybrid system approach.

1.3.1 Modeling Framework

The basic model of a hybrid dynamical system is given by

$$\mathbb{H}: \begin{cases} \dot{\xi} = \mathscr{F}(\xi), & \text{If } \xi \in \mathbb{F}, \\ \xi^{+} \in \mathscr{J}(\xi), & \text{If } \xi \in \mathbb{J}, \end{cases} \tag{1.1}$$

where $\xi \in \mathbb{X} \subseteq \mathbb{R}^m$ denotes the state vector. $\mathscr{F}(\xi)$ is the flow dynamics, $\mathscr{J}(\xi)$ is the jump dynamics. $\mathbb{F} \subseteq \mathbb{X}$ and $\mathbb{J} \subseteq \mathbb{X}$ are the flow set and jump set, respectively. It's assumed that sets $\mathbb{F}$ and $\mathbb{J}$ are closed sets. Then, the mathematical expression (1.1) suffices to describe various state evolutions that possibly involving both continuous-time and discrete-time dynamical systems by means of differential equation $\dot{\xi} = \mathscr{F}(\xi)$ and difference equation $\xi^{+} = \mathscr{J}(\xi)$. Due to the state vector ξ is not differentiable everywhere, $\dot{\xi}$ actually denotes the generalized directional derivative (i.e., Clarke derivative [75]), and ξ^{+} represents the value of the state after an instantaneous change. A hybrid system in the form of (1.1) can be represented by notation $\mathbb{H} = (\mathbb{F}, \mathscr{F}, \mathbb{J}, \mathscr{J})$ or in short, by $\mathbb{H}$. Here, we define that the initial conditions $\xi(0, 0)$ belong to the set $\mathbb{X}_0 \subseteq \mathbb{F} \cup \mathbb{J} \subseteq \mathbb{X}$, i.e., $\xi(0, 0) \in \mathbb{X}_0$.

With respect to the hybrid model (1.1), some necessary definitions are presented in the following.

Definition 1.1 A compact hybrid time domain is a set $\mathscr{T} = \bigcup_{k=0}^{\Omega-1} [t_k, t_{k+1}] \times \{k\} \subset \mathbb{R}_{\geq 0} \times \mathbb{N}_0$ with some finite sequence of times $0 = t_0 \leq t_1, \ldots, \leq t_{\Omega}$. A hybrid time domain is a set $\mathscr{T} \subset \mathbb{R}_{\geq 0} \times \mathbb{N}_0$ such that $\mathscr{T} \cap ([0, T] \times \{0, 1, \ldots, \Omega\})$ is a compact hybrid time domain for each $(T, \Omega) \in \mathscr{T}$.

Definition 1.2 A hybrid arc is a function ξ defined on its hybrid time domain dom ξ that is locally absolutely continuous in t on the interval $(\text{dom } \xi) \cap (\mathbb{R}_{\geq 0} \times \{k\})$ for every $k \in \mathbb{N}_0$.

Definition 1.3 A time instant $t \in \mathbb{R}_{\geq 0}$ is called a jump time for a hybrid arc ξ if there exists $k \in \mathbb{N}$ such that $(t, k-1) \in \text{dom } \xi$ and $(t, k) \in \text{dom } \xi$.

Definition 1.4 A hybrid arc $\xi \in \mathbb{X}$ is a solution to $\mathbb{H}$, if the initial condition $\xi(0, 0) \in \mathbb{X}_0$ and the following is satisfied:

(1) for all $k \in \mathbb{N}_0$ and for almost all t such that $(t, k) \in \text{dom } \xi$, there exist $\xi(t, k) \in \mathbb{F}$ and $\dot{\xi}(t, k) = \mathscr{F}(\xi(t, k))$;

(2) for all $(t, k) \in \text{dom } \xi$ such that $(t, k+1) \in \text{dom } \xi$, there exist $\xi(t, k) \in \mathbb{J}$ and $\xi(t, k+1) = \mathscr{J}(\xi(t, k))$.

Definition 1.5 Consider the hybrid system $\mathbb{H}$ given by (1.1) defined on $\mathbb{X}$, the set $\{\xi \in \mathbb{X} | \xi = 0_m\}$ is said to be uniformly globally asymptotically stable (UGAS) if there exists a function $\beta \in \mathscr{K}\mathscr{L}$ such that for any initial condition $\xi(0, 0) \in \mathbb{X}_0$, all corresponding solutions ξ satisfy for all $(t, k) \in \operatorname{dom} \xi$

$$\|\xi(t, k)\| \leq \beta(\|\xi(0, 0)\|, t).$$

A function $\alpha\colon \mathbb{R}_{\geq 0} \to \mathbb{R}_{\geq 0}$ is said to be of class $\mathscr{K}$ if it is continuous, strictly increasing and $\alpha(0) = 0$. It's said to be of class $\mathscr{K}_\infty$, if it is of class $\mathscr{K}$ and is unbounded. A continuous function $\beta\colon \mathbb{R}_{\geq 0} \times \mathbb{R}_{\geq 0} \to \mathbb{R}_{\geq 0}$ is said to be of class $\mathscr{K}\mathscr{L}$ if, for each fixed s, the mapping $r \to \beta(r, s)$ belongs to class $\mathscr{K}$, and for each fixed r the mapping $s \to \beta(r, s)$ is decreasing and $\beta(r, s) \to 0$ as $s \to \infty$. A continuous function $\gamma\colon \mathbb{R}_{\geq 0} \times \mathbb{R}_{\geq 0} \times \mathbb{R}_{\geq 0} \to \mathbb{R}_{\geq 0}$ is said to be of class $\mathscr{K}\mathscr{L}\mathscr{L}$ if, for each $r \geq 0$, both $\gamma(\cdot, \cdot, r)$ and $\gamma(\cdot, r, \cdot)$ belong to class $\mathscr{K}\mathscr{L}$.

The hybrid model with the identified form (1.1) consisting of a flow set, flow map, jump set, and jump map, was first appeared in [76] and more formally stated in [77]. Then, it follows a great number of excellent results in aspect of reset systems [78–80], networked control systems [81–86], MASs [87–90]. From the viewpoint of modeling and analyzing, the continuous-time system is with respect to the time variable $t \in \mathbb{R}_{\geq 0}$, while it may result in some limitations when parameterizing a hybrid dynamical system merely by $t \in \mathbb{R}_{\geq 0}$ [91, 92]. Therefore, hybrid time domain given in Definition 1.1 is studied, by which the solution of a hybrid system is parameterized through a finite or infinite sequence of intervals $[t_k, t_{k+1}]$, where $t_k \leq t_{k+1}$, $k = 1, 2, \ldots$, which suggests that the solutions of a hybrid system are with respect to both t, the amount of time elapsed, and k, the number of jumps that have occurred. In other words, the evolution of a hybrid system state $\xi(t, k)$ is actually corresponding to the hybrid time domain $\mathscr{T}$, which is a subset of $\mathbb{R}_{\geq 0} \times \mathbb{N}_0$. Definitions 1.2–1.4 define a solution taking on the hybrid time domain. In Definition 1.5, we show the stability conditions for hybrid system (1.1), which will be used latter. Moreover, for simplicity, the hybrid domain of ξ is not explicitly mentioned unless it is necessary, and it is assumed that each hybrid arc ξ comes from a hybrid time domain $\operatorname{dom} \xi$.

1.3.2 Why Using Hybrid System Approach

Reviewing the hybrid model (1.1), it can be seen that a hybrid dynamical system, or just a hybrid system, is actually a dynamical system that is modeled with a combination of common modeling tools for continuous-time, e.g., differential equations, and discrete-time dynamical systems, e.g., difference equations. Such modeling approach makes the hybrid model (1.1) be efficient to be applied to numerous dynamical systems that are beyond the descriptive ability of either purely continuous-time dynamical systems or purely discrete-time dynamical systems. For example, a continuous-time system controlled by an algorithm involving discrete-time control updates, such as sampled-data control, event-triggered control and reset control. By virtue of this hybrid model (1.1), a hybrid system approach capable of making full use of the discrete-time dynamics is inspired to improve the consensus performance of networked MAS subject to some network issues. This will be the main study topics of this book, and the next subsection will present some illustrative examples associated with hybrid system.

In addition to facilitating modeling, by using the hybrid model (1.1), timer variable and some other continuous-discrete internal variables can be introduced into the closed-loop state ξ, with which, sampling mechanism and event-triggering mechanism can exhibit several distinct features. With sampling mechanism, the timer variable gradually increases, and when reaching or exceeding a threshold, it clears its value. Repeating such a process, equidistant sampling sequence can be obtained immediately. In between any adjacent sampling instants (i.e., sampling interval), the closed-loop state of MAS evolves along the flow dynamics with the control inputs being constant values or time-varying open-loop estimation values, but not the actual values of states. At every sampling instant, the system closed-loop state performs an instantaneous change, at which the control input updates to be new value and the timer variable is reset to 0. Moreover, the MAS essentially involves many sensor measurements and samplings resulting from multiple agents. If all the agents of a MAS are required to sample at the same times (i.e., synchronously), global internal clock is needed, which is impractical in implementation and contradicts to the distribution goal. Therefore, assigning independent sampling periods for multiple agents through introducing more timer variables to achieve asynchronous samples becomes of great interest, that is difficult to be tackled with the approaches in the existing books, but can be easily achieved by utilizing the proposed hybrid system approach in this book.

With event-triggering mechanism, the purpose of timer variable is changed, it turns into recording an inter-event time, rather than a fixed time period required by sampling mechanism, to rule out the nontrivial Zeno behavior (i.e., the phenomena with infinite triggers in finite time period). In general, event-triggering mechanism always leads to less communication and is helpful for saving communication resources. However, one fundamental problem of event-triggered control, i.e., how to exclude Zeno behavior, has not been well solved [93–95], especially in the cases with disturbances or network-induced constraints. Compared with guaranteeing ruling out Zeno behavior by introducing an exponential decay term into triggering condition, the hybrid system approach used in this book can force a strictly positive interval between any two adjacent triggering events. Such a positive interval provided by means of timer variables helps the event-triggered mechanism be more practical for physical implementation, since physical devices always impose a positive minimum time constraint between adjacent triggering instants.

In addition, reset control is also a typical "hybrid" scheme of continuous-time and discrete-time dynamics. In classical continuous-time control, the output of a controller evolves continuously in time, while the output of a reset controller possibly experiences jumps caused by reset conditions that may depend on the value of the controller inputs. The full dynamics of reset control system can be clearly described by the hybrid model (1.1). Then, by using a hybrid system approach, the discontinuous reset states can be incorporated into Lyapunov function to help to design reset conditions and reset map functions. Besides, similar to the case of event-triggering control, timer variables can be used to provides positive time intervals for adjacent reset instants by the hybrid system approach as well. It is shown in [96] that reset control always lead to improved system performance in comparison to (non-reset) classical continuous-time control.

Based on the above observations, we can see that hybrid model can completely describe system dynamics with both flow dynamics and jump dynamics, that is useful for stability analysis and control design, and hybrid system approach has its own advantages in the sense of relaxing limitations of traditional consensus methods. This is the main motivation of this book and why hybrid system is used to deal with the cooperative control problem subject to various network issues.

1.3.3 Control System Examples

In this section, we present three classical examples, i.e., sampled-data control systems, event-triggered control systems and reset control systems, and show the modeling process of them toward hybrid model (1.1). Then, we take sampled-data control as an example to show how to develop a hybrid system approach.

Sampled-data control systems

Consider a continuous-time control system driven by a discrete-time state-feedback controller. The controller works in periodic sample-and-hold manner, which typically takes the following repeating measures: **(i)** measure the state of the system, and use the feedback controller to compute the control value based on the measurements; **(ii)** apply the computed constant control value for certain amount of time. By using the generated piecewise control inputs, the system state behaves (between periodic updates) according to the continuous-time dynamics of the system. At the sampling instant, there exists an abrupt change to the closed-loop system dynamics, which essentially leads to the jump dynamics. Therefore, such a control system can be better modeled by the hybrid model (1.1), where the closed-loop system include both flow dynamics and jump dynamics.

To be convenient, we assume that the control system has the form

$$\dot{x} = Ax + Bu \tag{1.2}$$

where $x \in \mathbb{R}^{n_p}$ is the state of system, $u \in \mathbb{R}^{n_c}$ is the control input. A and B are system matrices with compatible dimensions. If continuous control input $u = Kx$ is used with $K \in \mathbb{R}^{n_c \times n_p}$ being the gain matrix, it leads to an ideal continuous-time system

$$\dot{x} = (A + BK)x, \tag{1.3}$$

which can be asymptotically stable as long as the real part of the eigenvalues of $A + BK$ are negative.

If sample-and-hold control input $u = Kx(t_k)$ is used with t_k, $k = 0, 1, \ldots$, denoting the sampling instants, it is better to model the system with the closed-loop state

$$\xi = \begin{bmatrix} x \\ e_x \\ \tau \end{bmatrix} \in \mathbb{R}^{n_p+n_c+1}, \tag{1.4}$$

where we take the state error $e_x = x(t_k) - x$ into account as the component of closed-loop state. Note that it will be instantly reset to 0 at sampling instant t_k. τ is the timer variable introduced to record a fixed sampling period T. During flow, the system state x evolves along the continuous dynamics $\dot{x} = (A + BK)x + BKe_x$, τ keeps tracking the elapsed time with $\dot{\tau} = 1$, and the state error with $\dot{e}_x = -(A + BK)x - BKe_x$ is piecewise continuous. Thus, to construct hybrid model in the form of (1.1), the flow set and the flow dynamics can be described as

$$\mathbb{F} = \mathbb{R}^{n_p} \times \mathbb{R}^{n_c} \times [0, T), \quad \mathscr{F}(\xi) = \begin{bmatrix} (A + BK)x + BKe_x \\ -(A + BK)x - BKe_x \\ 1 \end{bmatrix}. \tag{1.5}$$

Jumps occur when the timer variable reaches T. The system state x does not change, but the timer variable clears its value, and the state error is updated to be $e_x = 0$. Then, the jump set and jump dynamics can be determined as

$$\mathbb{J} = \mathbb{R}^{n_p} \times \mathbb{R}^{n_c} \times \{T\}, \quad \mathscr{J}(\xi) = \begin{bmatrix} x \\ 0 \\ 0 \end{bmatrix}. \tag{1.6}$$

Event-triggered control systems

Compared with time-based sampled-data control systems, with event-triggering mechanism, the controller is usually less updated as only when some certain prespecified conditions are satisfied, an event occurs. In between adjacent triggering instants, the closed-loop system states also behave in continuous manner while experiencing an instantaneous jump at each triggering instant. Therefore, the event-triggered control has similarities with sampled-data control in many cases. Continuing with the example of system (1.2), the system flow dynamics with event-triggered control keeps the same as that of (1.5), but alternatively, the flow set becomes

$$\mathbb{F} = \left\{(x, e_x, \tau) \in \mathbb{R}^{n_p} \times \mathbb{R}^{n_c} \times \mathbb{R}_{\geq 0} | f(x, e_x) > 0 \vee \tau < \rho\right\}. \tag{1.7}$$

where $\vee$ denotes logical operation 'or' and $\wedge$ denotes logical operation 'and', $f(x, e_x)$ denotes the triggering function and ρ is a positive constant. Letting timer variable τ larger than a positive time ρ means that there always exists a minimal inter-event interval between adjacent events such that Zeno behavior is ruled out naturally. From (1.7), it can be seen that the possible state jumps occur only at the instants when the two conditions are satisfied simultaneously. The jump dynamics is the same as that of (1.6), and the jump set becomes

$$\mathbb{J} = \left\{(x, e_x, \tau) \in \mathbb{R}^{n_p} \times \mathbb{R}^{n_c} \times \mathbb{R}_{\geq 0} | f(x, e_x) \leq 0 \wedge \tau \geq \rho\right\}. \tag{1.8}$$

Reset control systems

In general, the controller output of a continuous-time plant is evolving continuously in time, while the output of a reset controller possibly experiences jumps caused by reset conditions, which makes it different form traditional controllers. The reset action removes some unexpected control input values, such that, in many application scenarios, the reset controllers always lead to improved system performance [96].

Consider a reset control system where the plant state is x_p and the controller state is x_c. The closed-loop state is

$$\xi = \begin{bmatrix} x_p \\ x_c \end{bmatrix} \in \mathbb{R}^{n_p + n_c} \tag{1.9}$$

Since the closed-loop system without reset is linear, the flow dynamics can be described by a linear function

$$\mathscr{F}(\xi) = A_f \xi. \tag{1.10}$$

The state after reset can be denoted by some mapping matrices, so that the jump dynamics can be described by

$$\mathscr{J}(\xi) = A_g \xi. \tag{1.11}$$

For the purpose of removing some unexpected gains, the reset condition can be designed such that two variables x_p and x_c can keep the same sign, and resetting the controller state x_c as long as they have opposite sign. This can be achieved by letting the state ξ satisfy some certain quadratic inequality. Thus, the jump set can be given in the form of

$$\mathbb{J} = \{\xi \in \mathbb{R}^{n_p + n_c} | \xi^T M \xi \leq 0\} \tag{1.12}$$

where M is a symmetric matrix. Then, the flow set can be described as $\mathbb{F} = \mathbb{R}^{n_p + n_c} \setminus \mathbb{J}$.

Besides, similar to the Zeno behavior existing in event-triggered control systems, the reset controller may encounter infinite jumps at a certain time instant. More specifically, such a time instant does not belong to the flow set but belongs to jump set, while the jump dynamics will always map it to jump set again. Thus, at this time instant, the reset controller state is only possible to jump without ever flowing. To address this issue, timer variable τ can be introduced to force a positive time interval ρ in between jumps. In this case, the closed-loop state (1.9) is augmented with timer variable τ and the jump set is updated to be

$$\mathbb{J} = \{(\xi, \tau) \in \mathbb{R}^{n_p+n_c} \times \mathbb{R}_{\geq 0} | \xi^T M \xi \leq 0 \wedge \tau \geq \rho\}. \tag{1.13}$$

The flow set is

$$\mathbb{F} = \{(\xi, \tau) \in \mathbb{R}^{n_p+n_c} \times \mathbb{R}_{\geq 0} | \xi^T M \xi > 0 \vee \tau < \rho\}. \tag{1.14}$$

The jump dynamics is augmented with the equation $\tau^+ = 0$ and the flow dynamics is augmented with the equation $\dot{\tau} = 1$.

According to the afore-presented examples, it is shown that many control systems can be described by a hybrid model, which implies that these control systems can be studied under hybrid system framework. By means of a hybrid system approach, new Lyapunov function design methodology can be provided for the system stability analysis, so it will bring many new and interesting results. Here, we take the sampled-data control system as an example to illustrate the superiority of the hybrid system approach.

For linear system (1.2) with sampled-data control, Eqs. (1.4)–(1.6) have modeled the closed-loop dynamics in hybrid system form. By utilizing hybrid system approach, the stability of original system (1.2) can be achieved as long as (x, e_x) of the closed-loop state achieves UGAS according to Definition 1.5. By introducing a timer variable τ, we construct a Lyapunov function candidate $V = x^T x + \phi(\tau) e_x^T e_x$, it should be pointed out that the state variable e_x is discontinuous, and $\phi(\tau) : \mathbb{R}_{\geq 0} \rightarrow \mathbb{R}_{\geq 0}$ with initial condition $\phi(0) > 0$ is a decreasing function with respect to timer variable τ. Consider the effect of the jump dynamics (1.6) at each sampling instant t_k, i.e., $\xi \in \mathbb{J}$, due to $x^+ = x$, $e_x^+ = 0$ and $\tau^+ = 0$, it has $V^+ - V = -\phi(\tau) e_x^T e_x \leq 0$. Then, taking derivative for V along the flow dynamics (1.5) and using Young's Inequality yield that

$$\begin{aligned} \dot{V} &= 2x^T(A + BK)x + 2x^T BKe_x + \dot{\phi}(\tau) e_x^T e_x + 2\phi(\tau) e_x^T(-(A + BK)x - BKe_x) \\ &\leq -(\sigma - \mu)x^T x + (\dot{\phi}(\tau) + \sigma\phi^2(\tau) + 2\mu\phi(\tau) + \mu) e_x^T e_x \end{aligned} \tag{1.15}$$

where it is supposed that $A + BK = -\sigma I_{n_p}$ and $BK \leq \mu I_{n_p}$ with $\sigma > \mu > 0$, then if the dynamics of $\phi(\tau)$ is carefully designed, it guarantees that $\dot{V} \leq 0$ for any $\xi \in \mathbb{F}$. Without loss of generality, we design $\dot{\phi}(\tau) = -\sigma\phi^2(\tau) - 2\mu\phi(\tau) - \mu$. Then, to guarantee $\phi(\tau) \geq 0$ for the well-posedness of V, the argument τ has a maximal value $\tau_M = (1/\sqrt{\sigma\mu - \mu^2}) \arctan \sqrt{\sigma\mu - \mu^2}\phi(0)/(\mu(\phi(0) + 1))$, which implies that the sampling period T can be set to be any value within the range $(0, \tau_M]$. This shows one advantage of hybrid system approach, that is, the upper bound of sampling period can be explicitly calculated, this is because the state error variable e_x (with jump dynamics) and timer variable τ (with jump dynamics) are introduced into the Lyapunov function construction.

1.4 Algebraic Graph Theory

In MASs, the communication topology among N agents is typically modeled by a graph $\mathscr{G} = (\mathscr{V}, \mathscr{E}, \mathscr{A})$ with a vertex/node set $\mathscr{V} = \{1, 2, \ldots, N\}$, an edge set $\mathscr{E} \subseteq \mathscr{V} \times \mathscr{V}$, and an adjacency matrix $\mathscr{A} = [a_{ij}]_{N\times N}$. A directed edge from node j to node i is denoted by $\mathscr{E}_{ij}$, which means that node i can receive information from node j. For any $i, j \in \mathscr{V}, i \neq j$, $a_{ij} = 1 \Leftrightarrow \mathscr{E}_{ij} \in \mathscr{E}$, otherwise, $a_{ij} = 0$, the diagonal elements $a_{ii} = 0$. $\mathscr{G}$ is called undirected graph if $\mathscr{A}$ is symmetric with $a_{ij} = a_{ji}$ and $\mathscr{G}$ is directed graph if $\mathscr{A}$ is asymmetric. A path from agent i to agent j is a sequence of edges $(i, i_1), (i_1, i_2), \ldots, (i_p, j)$. $\mathscr{G}$ is called connected if for any $i, j \in \mathscr{V}$, there is a path from agent i to agent j. A directed graph is called strongly connected if there exists a directed path between any pair of different agents. The in- and out- neighbor set of node i is defined as $\mathscr{N}_i^{in} = \{j \in \mathscr{V} : \mathscr{E}_{ij} \in \mathscr{E}\}$ and $\mathscr{N}_i^{out} = \{j \in \mathscr{V} : \mathscr{E}_{ji} \in \mathscr{E}\}$ respectively. The matrix $\mathscr{D} = \mathrm{diag}\{d_1, d_2, \ldots, d_N\}$ is called the degree matrix of $\mathscr{G}$, where $d_i = \sum_{j=1}^{N} a_{ij}$. The Laplacian matrix is defined as $\mathscr{L} = \mathscr{D} - \mathscr{A}$.

For the above (follower) node set $\mathscr{V}$, if there is a leader node 0 in addition to the N agents, which can influence the rest followers while not influenced by them, then an augmented topology graph $\bar{\mathscr{G}} = (\bar{\mathscr{V}}, \bar{\mathscr{E}}, \bar{\mathscr{A}})$ is obtained, which consists of $\bar{\mathscr{V}} = \{0\} \cup \mathscr{V}$, $\bar{\mathscr{E}} \supset \mathscr{E}$, $\bar{\mathscr{A}} \supset \mathscr{A}$, where $\bar{\mathscr{A}} \in \mathbb{R}^{(N+1)\times(N+1)}$ and $\mathscr{A}$ is a sub-matrix of $\bar{\mathscr{A}}$. The augmented topology graph $\bar{\mathscr{G}}$ is said to be connected if there exists at least one follower node that can receive information from leader, and topology graph $\mathscr{G}$ is connected. Define a diagonal matrix $\mathscr{B} = \mathrm{diag}\{b_1, b_2, \ldots, b_N\}$ to represent connection matrix between leader and followers, where $b_i = 1$ if agent i, $i \in \mathscr{V}$, can receive information from leader agent, otherwise $b_i = 0$. Then, the augmented topology matrix is marked as $\mathscr{H} = \mathscr{L} + \mathscr{B}$.

In the same topology graph, if there exist cooperative and antagonistic interaction relationships among agents simultaneously, the topology graph is signed, otherwise, unsigned. A signed communication graph is denoted by $\mathscr{G} = (\mathscr{V}, \mathscr{E}, \mathscr{A})$, which is similar to the above normal one, but in the signed weighed adjacency matrix $\mathscr{A} = [a_{ij}]_{N\times N}$, $a_{ij} > 0$ or $a_{ij} < 0$ is used to denote the cooperative or antagonistic interaction relationship between agent i and j. The Laplacian matrix for signed graph $\mathscr{G}$ is $\mathscr{L} = \mathscr{D} - \mathscr{A}$ where $\mathscr{D} = \mathrm{diag}\{d_1, d_2, \ldots, d_N\}$ with $d_i = \sum_{j=1}^{N} |a_{ij}|$. In leader-following case, the augmented topology matrix is $\mathscr{H} = \mathscr{L} + \mathscr{B}$.

If the communication connections in augmented topology graph don't change with the time, then, the augmented topology graph is fixed, otherwise, it is called time-varying (i.e., switching topology). A time-varying graph is denoted by $\bar{\mathscr{G}}_{\sigma(t)} = (\bar{\mathscr{V}}, \bar{\mathscr{E}}_{\sigma(t)}, \bar{\mathscr{A}}_{\sigma(t)})$, where $\sigma(t)$ is a piecewise constant switching signal with $\sigma(t) : \mathbb{R}_{\geq 0} \rightarrow \mathscr{P}$, $\mathscr{P} = \{1, 2, \ldots, n_0\}$. The node set $\bar{\mathscr{V}} = \{0, 1, \ldots, N\}$, the edge set $\bar{\mathscr{E}}_{\sigma(t)} \subseteq \bar{\mathscr{V}} \times \bar{\mathscr{V}}$, the adjacency matrix $\bar{\mathscr{A}}_{\sigma(t)} = [a_{ij}(t)]_{(N+1)\times(N+1)}$. $(i, j) \in \bar{\mathscr{E}}_{\sigma(t)}$ implies that agent i can receive information from agent j at time t. The graph $\bar{\mathscr{G}}_{\sigma(t)}$ is said to have a spanning tree, if there is a root node, and there exists a directed path from the root node to every other node. For any $i, j \in \bar{\mathscr{V}}, i \neq j$, $a_{ij}(t) = 1 \Leftrightarrow (i, j) \in \bar{\mathscr{E}}_{\sigma(t)}$, otherwise, $a_{ij}(t) = 0$, the diagonal elements $a_{ii}(t) = 0$. $\mathscr{N}_i^{in}(t) = \{j \in \bar{\mathscr{V}} : (i, j) \in \bar{\mathscr{E}}_{\sigma(t)}\}$ and $\mathscr{N}_i^{out}(t) = \{j \in \bar{\mathscr{V}} : (j, i) \in \bar{\mathscr{E}}_{\sigma(t)}\}$ denote the in- and out- neighbor set of agent i at time t. Then, the switching topology graph can be represented as $\mathscr{H}_{\sigma(t)} = [d_{ij}(t)] \in \mathbb{R}^{N\times N}$ with $d_{ij}(t) = -a_{ij}(t)$ for $i, j \in [1, N]$ and $i \neq j$, and $d_{ii}(t) = \sum_{j=0}^{N} a_{ij}(t)$ for $i \in [1, N]$.

References

1. Lewis, Frank L., Zhang, H., Hengster-Movric, K., Das, A.: Cooperative Control of Multi-Agent Systems: Optimal and Adaptive Design Approaches. Springer London, London (2014)
2. Bai, H., Arcak, M., Wen, J.: Cooperative Control Design: A Systematic, Passivity-Based Approach. Springer New York, New York, NY (2011)
3. Sun, Z.: Cooperative Coordination and Formation Control for Multi-agent Systems. Springer International Publishing, Cham (2018)
4. Cao, Y., Yu, W., Ren, W., Chen. G,: An overview of recent progress in the study of distributed multi-agent coordination. IEEE Trans. on Ind. Inform. **9**, 427-438 (2013)
5. Wielanda, P., Sepulchre, R., Allgower, F.: An internal model principle is necessary and sufficient for linear output synchronization. Automatica **47**, 1068-1074 (2011)
6. Tang, Y., Qian, F., Gao, H., Kurths, J.: Synchronization in complex networks and its application - A survey of recent advances and challenges. Annual Reviews in Control **37**(2), 184-198 (2014)
7. Ren, W., Beard, R., Atkins, E.: Information consensus in multivehicle cooperative control. IEEE Control Syst. Mag. **27**(2), 71-82 (2007)
8. Lin, Z., Francis, B., Maggiore, M.: Necessary and sufficient graphical conditions for formation control of unicycles. IEEE Trans. Autom. Control **50**(1), 121-127 (2005)
9. Ren, W., Atkins, E.: Distributed multi-vehicle coordinated control via local information exchange. Int. J. Robust Nonlinear Control **17**, 1002-1033 (2007)
10. Dimarogonas, D., Kyriakopoulos K.: A connection between formation inachievability and velocity alignment in kinematic multiagent systems. Automatica **44**(10), 2648-2654 (2008)
11. Krick, L., Broucke, M., Francis, B.: Stabilization of infinitesimally rigid formations of multi-robot networks. Int. J. Control **82**(3),423-439 (2009)
12. Dorfler, F., Francis, B.: Formation control of autonomous robots based on cooperative behavior. Proc. Eur. Control Conf. 2432-2437 (2009)
13. Oh, K., Ahn, H.: Formation control of mobile agents based on inter-agent distance dynamics. Automatica **47**(10), 2306-2312 (2011)
14. Oh, K., Ahn, H.: Formation Control and Network Localization via Orientation Alignment. IEEE Trans. Autom. Control **59**(2), 540-545 (2014)
15. Song, Y, Wang, Y.: Cooperative Control of Nonlinear Networked Systems: Infinite-time and Finite-time Design Methods. Springer International Publishing, Cham (2019)
16. Merris, R.: Laplacian matrices of graphs: A survey. Linear Algebra and its Applications **197-198**, 143-176 (1994)
17. Hespanha, J., Naghshtabrizi, P., Xu, Y.: A Survey of Recent Results in Networked Control Systems. Proc. IEEE **95**(1), 138-162 (2007)

18. Heemels, W., Donkers, M., Teel, A.: Periodic Event-Triggered Control for Linear Systems. IEEE Trans. Autom. Control **58**(4), 847-861 (2013)
19. Xiao, F., Wang L.: Asynchronous consensus in continuous-time multi-agent systems with switching topology and time-varying delays. IEEE Trans. Autom. Control **53**(8), 1804-1816 (2008)
20. Xie, G., Liu, H., Wang, L., Jia, Y.: Consensus in networked multiagent systems via sampled control: fixed topology case. American Control Conference, St. Louis, MO, USA, 3902-3907 (2009)
21. Xie, G., Liu, H., Wang, L., Jia, Y.: Consensus in networked multi-agent systems via sampled control: switching topology case. American Control Conference, St. Louis, MO, USA, 4525-4530 (2009)
22. Yu, W., Zheng X., Chen G., Ren, W., Cao, J.: Second-order consensus in multi-agent dynamical systems with sampled position data. Automatica **47**(7), 1496-1503 (2011)
23. Guan, Z. Liu, Z., Feng, G., Jian, M.: Impulsive consensus algorithms for second-order multi-agent networks with sampled information. Automatica **48**(7), 1397-1404 (2012)
24. Zhang, W., Tang, Y., Han, Q., Liu, Y.: Sampled-Data Consensus of Linear Time-Varying Multiagent Networks With Time-Varying Topologies. IEEE Trans. Cybern. **52**(1), 128-137 (2022)
25. Xiao, F., Shi, Y., Ren, W.: Robustness Analysis of Asynchronous Sampled-Data Multiagent Networks With Time-Varying Delays. IEEE Trans. Autom. Control **63**(7), 2145-2152 (2018)
26. Liu, W., Huang, J.: An updated version of "Leader-following consensus for linear multi-agent systems via asynchronous sampled-data control". IEEE Trans. Autom. Control **65**(7), 3215-3222 (2020)
27. Tang, Z., Huang, T., Shao, J., Hu, J.: Leader-following consensus for multi-agent systems via sampled-data control. IET Contr. Theory Appl. **5**(14), 1658-1665 (2011)
28. Ding, L., Han, Q., Guo, G.: Network-based leader-following consensus for distributed multi-agent systems. Automatica **49**(7), 2281-2286 (2013)
29. He, W., Zhang, B., Han, Q.: Leader-following consensus of nonlinear multiagent systems with stochastic sampling. IEEE Trans. Cybern. **47**(2), 327-338 (2017)
30. Ge, X., Han, Q., Ding, D., Zhang, X., Ning, B.: A survey on recent advances in distributed sampled-data cooperative control of multi-agent systems. Neurocomputing **275**(5), 1684-1701 (2018)
31. Li, H., Liao, X., Huang, T., Zhu, W.: Event-Triggering Sampling Based Leader-Following Consensus in Second-Order Multi-Agent Systems. IEEE Trans. Autom. Control **60**(7), 1998-2004 (2015)
32. Hu, W., Liu, L., Feng, G.: Consensus of Linear Multi-Agent Systems by Distributed Event-Triggered Strategy. IEEE Trans. Cybern. **46**(1), 148-157 (2016)
33. Fan, Y., Feng, G., Wang, Y., Song, C.: Distributed event-triggered control of multi-agent systems with combinational measurements. Automatica **49**(2), 671-675 (2013)
34. Qin, J., Fu, W., Gao, H., Zheng, W.: Distributed K-means algorithm and fuzzy C-means algorithm for sensor networks based on multiagent consensus theory. IEEE Trans. Cybern. **47**(3), 772-783 (2017)
35. Yang, D., Ren, W., Liu, X., Chen, W.: Decentralized event-triggered consensus for linear multi-agent systems under general directed graphs. Automatica **69**, 242-249 (2016)
36. Ding, L., Han, Q., Ge, X., Zhang, X.: An Overview of Recent Advances in Event-Triggered Consensus of Multiagent Systems. IEEE Trans. Cybern. **48**(4), 1110-1123 (2018)
37. Qin, J., Ma, Q., Shi, Y., Wang, L.: Recent Advances in Consensus of Multi-Agent Systems: A Brief Survey. IEEE Trans. Ind. Electron **64**(6), 4972-4983 (2017)
38. Girard, A.: Dynamic triggering mechanisms for event-triggered control. IEEE Trans. Autom. Control **60**(7), 1992-1997 (2015)
39. Ge, X., Han, Q., Wang, Z.: A Dynamic Event-Triggered Transmission Scheme for Distributed Set-Membership Estimation Over Wireless Sensor Networks. IEEE Trans. Cybern. **49**(1), 171-183 (2019)
40. Hu, W., Yang, C., Huang, T., Gui, W.: A Distributed Dynamic Event-Triggered Control Approach to Consensus of Linear Multiagent Systems With Directed Networks. IEEE Trans. Cybern. **50**(2), 869-874 (2020)

41. Du, S., Liu, T., Ho, D.: Dynamic event-triggered control for leader-following consensus of multiagent systems. IEEE Trans. Syst. Man, Cybern: Syst. **23**(6), 602-619 (2018)
42. Jiang, W., Chen, Y., Charalambous, T.: Consensus of General Linear Multi-Agent Systems With Heterogeneous Input and Communication Delays. IEEE Control Syst. Lett. **5**(3), 851-856 (2020)
43. Su, H., Zhang, J., Chen, X.: A Stochastic Sampling Mechanism for Time-Varying Formation of Multiagent Systems With Multiple Leaders and Communication Delays. IEEE Trans. Neural Netw. Learn. Syst. **30**(12), 3699-3707 (2019)
44. Xing, M., Deng F., Hu, Z.: Sampled-Data Consensus for Multiagent Systems With Time Delays and Packet Losses. IEEE Trans. Syst., Man, Cybern: Syst. **50**(1), 203-210 (2020)
45. Miao, G., Cao, J., Alsaedi, A., Alsaadi, F.: Event-triggered containment control for multi-agent systems with constant time delays. J. Franklin Inst. **354**(15), 6956-6977 (2017)
46. Zhu, W., Jiang, Z.: Event-based leader-following consensus of multi-agent systems with input time delay. IEEE Trans. Autom. Control **60**(5), 1362-1367 (2015)
47. Duan, G., Xiao, F., Wang, L.: Asynchronous Periodic Edge-Event Triggered Control for Double-Integrator Networks With Communication Time Delays. IEEE Trans. Cybern. **48**(2), 675-688 (2018)
48. Qian, W., Gao, Y., Yang, Y.: Global Consensus of Multiagent Systems With Internal Delays and Communication Delays. IEEE Trans. Syst., Man, Cybern: Syst. **49**(10), 203-210 (2019)
49. Garcia, E., Cao, Y., Casbeer, D.: Periodic Event-Triggered Synchronization of Linear Multi-Agent Systems With Communication Delays. IEEE Trans. Autom. Control **62**(1), 366-371 (2017)
50. Deng, C., Che, W., Wu, Z.: A Dynamic Periodic Event-Triggered Approach to Consensus of Heterogeneous Linear Multiagent Systems With Time-Varying Communication Delays. IEEE Trans. Cybern. **51**(4), 1812-1821 (2021)
51. Chen, M., Yan, H., Zhang, H., Chen, S., Li, Z.: Event-Triggered Consensus of Multiagent Systems With Time-Varying Communication Delay. IEEE Trans. Syst., Man, Cybern: Syst. **52**(5), 2706-2720 (2021)
52. Xu, L., Mo, Y., Xie, L.: Distributed Consensus Over Markovian Packet Loss Channels. IEEE Trans. Autom. Control **65**(1), 279-286 (2020)
53. Wang, F., Wen, G., Peng, Z., Huang, T., Yu, Y: Event-Triggered Consensus of General Linear Multiagent Systems With Data Sampling and Random Packet Losses. IEEE Trans. Syst., Man, Cybern: Syst. **51**(2), 1313-1321 (2020)
54. H. Yang, H., Zhao, H., Xia, Y., Zhang, J.: Event-Triggered Active MPC for Nonlinear Multiagent Systems With Packet Losses. IEEE Trans. Cybern. **51**(6), 3093-3102 (2021)
55. Jiang, X., Xia, X., Feng, G., Jiang, Z.: Consensus Tracking of Data-Sampled Nonlinear Multi-Agent Systems With Packet Loss and Communication Delay. IEEE Trans. Netw. Sci. Eng. **8**(1), 126-137 (2021)
56. Hong, Y., Gao, L., Cheng, D., Hu, J.: Lyapunov-Based Approach to Multiagent Systems With Switching Jointly Connected Interconnection. IEEE Trans. Autom. Control **52**(5), 943-948 (2007)
57. Shi, G., Hong, Y.: Global target aggregation and state agreement of nonlinear multi-agent systems with switching topologies. Automatica, **45**(5), 1165-1175 (2009)
58. Cheng, B., Wang, X., Li, Z.: Event-Triggered Consensus of Homogeneous and Heterogeneous Multiagent systems With Jointly Connected Switching Topologies. IEEE Trans. Cybern. **49**(12), 4421-4430 (2019)
59. Wang, X., Zhu, J., Feng, J.: A New Characteristic of Switching Topology and Synchronization of Linear Multiagent Systems. IEEE Trans. Autom. Control **64**(7), 2697-2711 (2019)
60. Dong, X., Hu, G.: Time-varying formation control for general linear multi-agent systems with switching directed topologies. Automatica **73**, 47-55 (2016)
61. Back, J., Kim, J.: Output Feedback Practical Coordinated Tracking of Uncertain Heterogeneous Multi-Agent Systems Under Switching Network Topology. IEEE Trans. Autom. Control **62**(12), 6399-6406 (2017)

62. Su, S., Lin, Z.: Distributed Consensus Control of Multi-Agent Systems With Higher Order Agent Dynamics and Dynamically Changing Directed Interaction Topologies. IEEE Trans. Autom. Control **61**(2), 515-519 (2016)
63. Liu, K., Duan, P., Duan, Z., Cai, H., Lu, J.: Leader-following consensus of multi-agent systems with switching networks and event-triggered control. IEEE Trans. Circuits Syst. I-Regul. Pap. **65**(5), 1696-1706 (2018)
64. You, X., Hua, C., Peng, D., Guan, X.: Leader-following consensus for multiagent systems subject to actuator saturation with switching topologies and time-varying delays. IET Contr. Theory Appl. **10**(2), 144-150 (2016)
65. Rezaee, H., Abdollahi, F.: Consensus problem in high-order multiagent systems with lipschitz nonlinearities and jointly connected topologies. IEEE Trans. on Syst. Man, and Cybern.: Syst. **47**(5), 741-748 (2017)
66. Meng, H., Chen Z., Middleton, R.: Consensus of multiagents in switching networks using input-to-state stability of switched systems. IEEE Trans. on Autom. Control **63**(11), 3964-3971 (2018)
67. Liu, W., Huang, J.: Adaptive leader-following consensus for a class of higher-order nonlinear multiagent systems with directed switching networks. Automatica **79**(5), 84-92 (2017)
68. Casadei, G., Marconi, KL., Isidori, A.: About synchronization of homogeneous nonlinear agents over switching networks. 53rd IEEE Conference on Decision and Control, 0191-2216 (2014)
69. Wen, G., Yu, W., Xia, Y., Yu, X., Hu, J.: Distributed tracking of nonlinear multiagent systems under directed switching topology: an observer-based protocol. IEEE Trans. on Syst. Man, Cybern.: Syst. **47**(5), 869-881 (2017)
70. Wang, P., Wen, G., Yu, X., Yu, W., Lv, Y.: Consensus disturbance rejection for linear multiagent systems with directed switching communication topologies. IEEE Trans. Neural Netw. Learn. Syst. **7**(1), 254-265 (2020)
71. Zheng, D., Zhang H., Zheng, Q.: Consensus analysis of multiagent systems under switching topologies by a topology-dependent average dwell time approach. IET Contr. Theory Appl. **11**(13), 429-438 (2017)
72. Casadei, G., Marconi, L., Isidori, A.: About disconnected topologies and synchronization of homogeneous nonlinear agents over switching networks. Int. J. Robust Nonlinear Control **28**(3), 901-917 (2018)
73. Wen, G., Zheng, W.: A Stochastic Sampling Mechanism for Time-Varying Formation of Multiagent Systems With Multiple Leaders and Communication Delays. IEEE Trans. Autom. Control **64**(9), 3796-3803 (2019)
74. Li, K., Hua, C., You, X., Guan, X.: Distributed Consensus Control for Nonlinear Multiagent Systems Under Directed Graphs of Dynamic Frequency Switches. IEEE Trans. Autom. Control **66**(2), 841-848 (2021)
75. Clarke, F.: Optimization and nonsmooth analysis, SIAM, Philadelphia, PA (1990)
76. Goebel, R., Hespanha, J., Teel, A., Cai, C., Sanfelice, R.: Hybrid systems: generalized solutions and robust stability. In IFAC Symposium on Nonliear Control Systems, Stuttgart (2004)
77. Goebel, R., Teel, A.: Solutions to hybrid inclusions via set and graphical convergence with stability theory applications. Automatica **42**(4), 573-587 (2006)
78. Baños, A., Barreiro, A.: Reset control systems: the zero-crossing resetting law. Nonlinear Analysis: Hybrid Systems **46**, 101259 (2022)
79. Loon, S., Gruntjens, K., Heertjes, M., Wouw, N., Heemels, W.: Frequency-domain tools for stability analysis of reset control systems. Automatica **82**, 101-108 (2017)
80. Beerens, R., Bisoffi, A., Zaccarian, L., Heemels, W., Nijmeijer, H., Wouw, N.: Reset integral control for improved settling of PID-based motion systems with friction. Automatica **107**, 483-492 (2019)
81. Postoyan, R., Nesic, D.: A Framework for the Observer Design for Networked Control Systems. IEEE Trans. Autom. Control **57**(5), 1309-1314 (2012)
82. Postoyan, R., Wouw, N., Nesic, D., Heemels, W.: Emulation-based tracking solutions for nonlinear networked control systems. Proceedings of the 51th IEEE Conference on Decision and Control, Maui, HI, USA, 740-745 (2012)

83. Postoyan, R., Wouw, N., Nesic, D., Heemels, W.: Tracking Control for Nonlinear Networked Control Systems. IEEE Trans. Autom. Control **59**(6), 153-1554 (2014)
84. Heijmans, S., Postoyan, R., Nesic, D., Heemels W.: Computing Minimal and Maximal Allowable Transmission Intervals for Networked Control Systems Using the Hybrid Systems Approach. IEEE Control. Syst. Lett. **1**(1), 56-61 (2017)
85. Wang, W., Postoyan, R., Nesic, D., Heemels, W.: Periodic Event-Triggered Control for Nonlinear Networked Control Systems. IEEE Trans. Autom. Control **65**(2), 620-635 (2020)
86. Heijmans, S., Postoyan, R., Nesic, D., Heemels W.: An Average Allowable Transmission Interval Condition for the Stability of Networked Control Systems. IEEE Trans. Autom. Control **66**(6), 2526-2541 (2021)
87. Dolk, V., Postoyan, R., Heemels, W.: Event-Triggered Consensus for Multi-Agent Systems with Guaranteed Robust Positive Minimum Inter-Event Times. 58th IEEE Conference on Decision and Control, Nice, France 2604-2609 (2019)
88. Scheres, K., Dolk, V., Chong, M., Postoyan, R., Heemels, W.: Distributed Periodic Event-triggered Control of Nonlinear Multi-Agent Systems. IFAC-PapersOnLine **55**(13), 168-173 (2022)
89. Yang, J., Yu, H., Chen, T.: Cooperative Output Regulation With Asynchronous Transmissions and Time-Varying Delays. IEEE Trans. Autom. Control **67**(3), 1438-1445 (2022)
90. Yu, H., Chen, T.: A new Zeno-free event-triggered scheme for robust distributed optimal coordination. Automatica **129**, 109639 (2021)
91. Deshpande, A.: Control of hybrid systems. PhD thesis, University of California, Berkeley (1994)
92. Lygeros, J.: Hierarchical, hybrid control of large scale systems. PhD thesis, University of California, Berkeley (1996)
93. Xu, W., He, W., Ho, D., Kurths, J.: Fully distributed observerbased consensus protocol: Adaptive dynamic event-triggered schemes. Automatica **139**, 110188 (2022)
94. Zhang, H., Cai, Y., Wang, Y., Su, H.: Adaptive bipartite event triggered output consensus of heterogeneous linear multiagent systems under fixed and switching topologies. IEEE Trans. Neural Netw. Learn. Syst. **31**(11), 4816-4830 (2020)
95. Qian, Y., Liu, L., Feng, G.: Distributed dynamic event-triggered control for cooperative output regulation of linear multiagent systems. IEEE Trans. Cybern. **50**(7), 3023-3032 (2020)
96. Meng, X., Xie, L., Soh, Y.: Reset control for synchronization of multi-agent systems. Automatica **104**, 189-195 (2019)

Part I
Asynchronous Sampled-Data Cooperative Control

Chapter 2
Sampled-Data Consensus Control of MAS with Time Delay

2.1 Introduction

As mentioned in Chap. 1, sampled-data based control is one of the useful and effective strategies to avoid continuous communication. The implementation of sampling mechanisms can be divided into synchronous sampling and asynchronous sampling. The former one requires that all the agents in a MAS share global and synchronous clock, and have the same sampling period [1–4], while the latter is not needed. In general, the asynchronous sampling is less restrictive compared with synchronous sampling in practical applications, but still has not been sufficiently studied especially when there exist time-varying communication delays. Notice that communication delays are usually inevitable during information exchanges among agents via shared network and it is well known that time delays may degrade system control performance or even result in instability. Thus, of particular concern is to suppress the effect of time delays caused for sampled-data based control, for example, constant time delays were considered in [3, 5] and time-varying delays were considered in [1, 4, 6, 7]. However, one common feature of the above-mentioned results is that the delays are assumed to be homogeneous rather than heterogeneous,[1] which is rather restrictive in physical systems. This is mainly because the communication delays are usually different for different communication channels, as communication delay may not only include transmission delay, but also include sending delay and access delay, both of which are induced by context processing of each agent.

In addition, the generalized concept of cooperative control of MAS sometimes not only contains cooperative relationship between agents, but also there may exist antagonistic relationship between agents. That is, similar as human society, in order to work together to complete a task, in a group of agents, maybe some agents work cooperatively, but the rest work in antagonistic manner, for example in the multi-target surround tracking application scenarios. To describe this category of applica-

[1] Heterogeneous (homogeneous) delay means that the transmitted data from each agent experiences different (same) time delays from (as) other agents.

G. Zhao et al., *Cooperative Control of Multi-agent Systems*,
https://doi.org/10.1007/978-981-97-0968-7_2

tions, the concept of 'bipartite consensus' was introduced in the literature, and structurally balanced signed graph is used to describe the communication connection relationship among cooperative and antagonistic agents. In this chapter, we introduce a hybrid system approach to the sampled-data leaderless and leader-following bipartite consensus control problem of MAS with heterogeneous and time-varying communication delays. Distributed asynchronous sampled-data bipartite consensus protocols are proposed based on estimators. By introducing appropriate intermediate variables and internal auxiliary variables, a unified hybrid model, consisting of flow dynamics and jump dynamics, is constructed to describe the closed-loop dynamics of both leaderless and leader-following MAS. It is shown that the leaderless and leader-following bipartite consensus is equivalent to stability of a hybrid system. The stability results are developed with some specific linear matrix inequalities and the explicit maximum allowable sampling period (MASP) and maximum allowable delay (MAD) are presented for each agent based on given parameters.

From this chapter, readers can learn how to construct hybrid model (with both flow dynamics and jump dynamics) for MAS under sampling communication network environment, such that the closed-loop dynamics can be completely described. Moreover, a novel Lyapunov function with jump dynamics is constructed for the purpose of stability analysis and explicit stability conditions and parameter design procedure are provided.

2.2 Networked MAS with Time Delay

This section formulates the bipartite consensus problem for MAS with or without leaders subject to time delays. It is noteworthy that the concept of "bipartite consensus" is introduced to identify the cooperative and antagonistic relationship in MAS, which may coexist in some natural or social systems, that is why we introduce the results with respect to bipartite consensus of MAS. In addition, note that the connection relationship between agents, formulated by graph theory, is important for achieving the objective of consensus control of MAS, the signed graph is used to describe the cooperative and antagonistic relationship between agents.

It is well known that there exist two classes of MAS according to whether there exists leader or not, i.e., leaderless MAS and leader-following MAS. In this chapter, we consider both classes of MAS and study the bipartite consensus problem of MAS with time delay under hybrid systems framework. First, we give the agent's system model considered in this chapter, each follower agent i's ($i \in \mathscr{V}$) dynamics is described by

$$\dot{x}_i(t) = Ax_i(t) + Bu_i(t), \tag{2.1}$$

where $x_i(t) \in \mathbb{R}^n$ and $u_i(t) \in \mathbb{R}^m$ denote the state and control input of agent i respectively. Suppose that the pair (A, B) is stabilizable. When there exists leader agent, usually, we use the subscript 0 to denote that it is a leader. The leader agent 0's dynamics is described by

$$\dot{x}_0(t) = Ax_0(t), \tag{2.2}$$

where $x_0(t) \in \mathbb{R}^n$ denotes the state.

Some mathematical symbols and preliminaries on communication graph have been introduced in Chap. 1, in this chapter, we do not introduce them repeatedly, readers can found them in Sect. 1.4 if necessary. For proceeding this chapter, we make the following standard Assumptions on the communication graphs.

Assumption 2.1 The signed graph $\mathscr{G}$ is undirected, connected and structurally balanced.

Assumption 2.2 In leader-following case, at least one follower can receive information from the leader, i.e., there exists a spanning tree in the graph $\mathscr{G}$ with the leader agent 0 as the root.

Definition 2.1 A graph $\mathscr{G}$ is said to be *structurally balanced*, if the node set $\mathscr{V}$ can be divided into two subsets $\mathscr{V}_1$ and $\mathscr{V}_2$, with $\mathscr{V}_1 \cup \mathscr{V}_2 = \mathscr{V}$ and $\mathscr{V}_1 \cap \mathscr{V}_2 = \emptyset$, such that $a_{ij} \geq 0$ for $i, j \in \mathscr{V}_l (l \in \{1, 2\})$ and $a_{ij} \leq 0$ for $i \in \mathscr{V}_l$, $j \in \mathscr{V}_q$, $l \neq q (l, q \in \{1, 2\})$.

Then, by means of the *gauge transformation* technique proposed in [13], a change of orthant order in $\mathbb{R}^n$ can be performed via a matrix $M = \operatorname{diag}\{m_1, m_2, \ldots, m_N\}$, with $m_i \in \{\pm 1\}$, such that the modulus of the entries of matrix $\mathscr{L}$ and $\mathscr{H}$ keep unchanged, but the sign pattern can be modified. Regarding the Laplacian matrix $\mathscr{L}$, the gauge transformation matrix M, and the graph related matrices $\mathscr{A}$ and $\mathscr{H}$, the following lemma holds.

Lemma 2.1 ([12]) *If the connected signed graph $\mathscr{G}$ is structurally balanced, the entries of the matrix $\mathscr{A}_M = M\mathscr{A}M$ are all nonnegative, the matrix $\mathscr{L}_M = M\mathscr{L}M$ is positive semi-definite, and the matrix $\mathscr{H}_M = M\mathscr{H}M$ is positive definite. Moreover, if a vector x satisfies $\mathbf{1}^T x = 0$, we have*

$$\min_{x \neq 0} \frac{x^T \hat{\mathscr{L}} x}{x^T x} = a(\hat{\mathscr{L}})$$

where $\hat{\mathscr{L}} = \mathscr{L}_M$ or $\hat{\mathscr{L}} = \mathscr{H}_M$, $a(\hat{\mathscr{L}})$ denotes the minimum positive eigenvalue of $\hat{\mathscr{L}}$.

For the considered MAS with dynamics (2.1) and (2.2), the objective of bipartite consensus is to make some agents' states tend to the same values, i.e., these agents work cooperatively, while, the rest agents' states tend to the same modulus but with opposite signs. To achieve such objective, we need to formulate the bipartite consensus in rigorous mathematical form: (1). the *leaderless bipartite consensus* is said to be achieved if the conditions $\lim_{t \to \infty} \left\| x_i(t) - m_i m_j x_j(t) \right\| = 0$ hold for $i, j \in \mathscr{V}$.

(2). the *leader-following bipartite consensus* is said to be achieved if the conditions $\lim_{t\to\infty} \|x_i(t) - m_i x_0(t)\| \doteq 0$ hold for $i \in \mathscr{V}$. Then, based on the above mathematical expressions, we need to design control law $u_i(t)$ in appropriate form such that the leaderless/leader-following bipartite consensus is achieved.

Note that the MAS usually works under wireless network environment, that is, for the purpose of consensus control, the information transmission among agents via wireless network, is necessary. However, as is well known that, under network environment, the information transmitted from a certain agent to its neighbors is packet-based sampled-data, since continuous measurement and transmission of information is usually not available in practical applications. Besides, due to the limitation of network bandwidth and processing ability of microprocessors, the time delay usually exists, i.e., the information transmitted from one agent to its neighbors cannot arrive the destination instantly, this phenomenon generally exists in application scenarios and it may cause performance degradation or even destabilization of closed-loop system. Thus, with sampled and delayed information, how to design distributed consensus control law such that the objective of bipartite consensus is achieved, and how to select design parameters appropriately should be studied. This chapter presents a hybrid system approach to solve this problem.

For a certain agent i, because neighbor agent j's state $x_j(t)$ is not continuously available as discussed in the last paragraph, in order to design control law, we need to define a variable to approximately estimate neighbor agent j's state information. Specifically, we define $\hat{x}_{ji}(t)$ to denote the estimation of neighbor agent j's state, whose dynamics is designed as

$$\begin{aligned} \dot{\hat{x}}_{ij}(t) &= \begin{cases} A\hat{x}_{ij}(t), i \neq j, t \in (t^i_{k_i} + d^i_{k_i}, t^i_{k_i+1} + d^i_{k_i+1}] \\ A\hat{x}_{ii}(t), i = j, t \in (t^i_{k_i}, t^i_{k_i+1}] \end{cases} \\ \hat{x}_{ij}(t^+) &= \begin{cases} F(d^i_{k_i}) x_i(t - d^i_{k_i}), i \neq j, t = t^i_{k_i} + d^i_{k_i} \\ x_i(t), i = j, t = t^i_{k_i} \end{cases} \end{aligned} \tag{2.3}$$

where $t^i_{k_i}$ denotes agent i's sampling instant, $d^i_{k_i}$ denotes communication delay. Let h_i denotes the sampling period of agent i, without loss of generality, we assume that $0 \leq d^i_{k_i} \leq \bar{d}_i \leq h_i \leq \bar{h}_i$, where $\bar{d}_i$ and $\bar{h}_i$ denote the upper bound of communication delays and sampling period respectively. $F(d^i_{k_i}) = \mathrm{expm}(Ad^i_{k_i})$ with $\mathrm{expm}(\cdot)$ denoting matrix exponential operation.

Then, for the purpose of bipartite consensus control of MAS, according to whether there exists leader agent or not, we design the bipartite consensus control laws

$$u_i(t) = K \sum_{i \in \mathscr{N}_i} a_{ij}(\hat{x}_{ji}(t) - \mathrm{sgn}(a_{ij})\hat{x}_{ii}(t)), \tag{2.4}$$

and

$$u_i(t) = K \left(\sum_{i \in \mathscr{N}_i} a_{ij}(\hat{x}_{ji}(t) - \mathrm{sgn}(a_{ij})\hat{x}_{ii}(t)) - b_i(\hat{x}_{ii}(t) - m_i x_0(t)) \right). \tag{2.5}$$

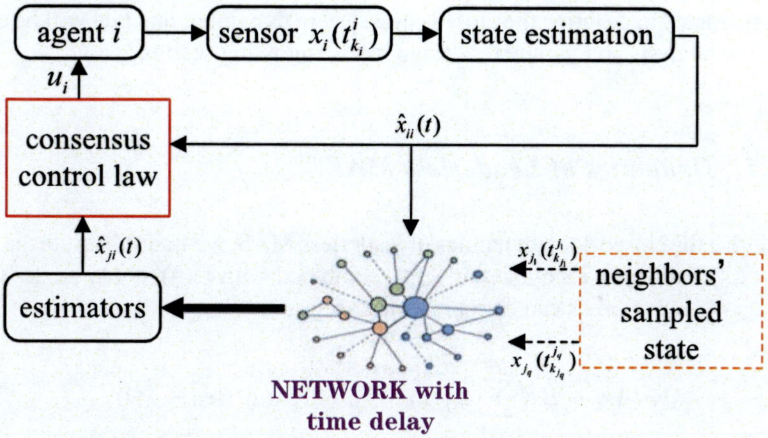

Fig. 2.1 Schematic of control structure

for leaderless MAS and leader-following MAS respectively, where $\text{sgn}(\cdot)$ denotes the sign function. We can see that, in contrast with leaderless MAS, one extra term regarding the error between estimated state and leader agent's state is added in the leader-following case, the role of which is used to make the following agents' states to track the leader agent's state. How to design the control gain K will be given later. In addition, from (2.4), the sampled and delayed information of agent i and its neighbors are used to construct the control law, which is implementable in physical. In addition, based on (2.3), it implies that the delays $d^i_{k_i}$ are needed to be known when its neighbors update $\hat{x}_{ij}(t)$ at $t^i_{k_i} + d^i_{k_i}$, this can be achieved using timestamp when agent i sending data packet. The schematic of control structure of MAS with sampled and delayed information is given in Fig. 2.1.

2.3 Closed-Loop System Model

Based on the agents' model and control law in last section, this section aims to derive the closed-loop system dynamics of the MAS, and by exploiting the feature of closed-loop dynamics of both leaderless and leader-following MAS, a unified hybrid model with both flow dynamics and jump dynamics will be constructed. This hybrid model can completely describe the closed-loop dynamics of MAS no matter there exist leader or not, and it helps to study the leaderless/leader-following bipartite consensus control problem in a unified framework. Next, we first derive the closed-loop dynamics of MAS by defining appropriate closed-loop states, then, by exploring the common feature of system dynamics in leaderless case and leader-following case, a unified hybrid model, with both flow dynamics and jump dynamics, is constructed to describe the closed-loop system dynamics, and that will be used for stability analysis and parameter design in the following section.

2.3.1 Dynamics of Leaderless MAS

To derive the closed-loop dynamics of leaderless MAS, we first substitute the control law (2.4) into agent i's dynamics (2.1), combined with (2.3), it can be derived that the agent i's dynamics can be represented as

$$
\begin{aligned}
\dot{x}_i &= Ax_i + BK\sum_{i=1}^{N} a_{ij}(x_j + e_{ji} - \operatorname{sgn}(a_{ij})(e_{ii} + x_i)) \\
&= Ax_i - BK(\mathscr{L}_r^i \otimes I_n)x - BKd_i e_{ii} + BK(\mathscr{A}_r^i \otimes I_n)e_b,
\end{aligned}
$$

where, for simplicity, $\mathscr{L}_r^i$ and $\mathscr{A}_r^i$ denote the ith row of matrices $\mathscr{L}$ and $\mathscr{A}$ respectively. The error variables e_{ii} and e_{ij} are defined as

$$
\begin{aligned}
e_{ii} &= \hat{x}_{ii} - x_i, \\
e_{ij} &= \hat{x}_{ij} - x_i.
\end{aligned} \tag{2.6}
$$

Next, define the closed-loop variables

$$
\begin{aligned}
x &= \operatorname{col}(x_1, x_2, \ldots, x_N), \\
e_a &= \operatorname{col}(e_{11}, e_{22}, \ldots, e_{NN}) = \operatorname{col}(e_{a1}, e_{a2}, \ldots, e_{aN}), \\
e_b &= \operatorname{col}(e_{1J}, e_{2J}, \ldots, e_{NJ}) = \operatorname{col}(e_{b1}, e_{b2}, \ldots, e_{bN}),
\end{aligned} \tag{2.7}
$$

with $e_{iJ} = e_{bi} = e_{ij}|_{j\in\mathscr{N}_i}$, which implies that the estimation $\hat{x}_{ij}$, $j \in \mathscr{N}_i$, as seen by agent i's neighbors are same, this is reasonable in practice since the communication delays from agent i to its neighbors is mainly caused by agent i's internal processing time rather than transmission delays.

It follows from (2.6) and (2.7) that

$$
\dot{x} = (I_N \otimes A)x - (\mathscr{L} \otimes BK)x - (\mathscr{D} \otimes BK)e_a + (\mathscr{A} \otimes BK)e_b \tag{2.8}
$$

However, note that consensus of MAS usually does not imply the convergence of x, thus, take x as closed-loop state for stability analysis is not appropriate. For the purpose of stability analysis and achieving the objective of bipartite consensus, we need to introduce new closed-loop variables such that the convergence of closed-loop variables implies bipartite consensus of MAS, i.e., the closed-loop system is stable. First, introduce an intermediate variable $z_i = m_i x_i$, and define $z = \operatorname{col}(z_1, z_2, \ldots, z_N)$, i.e., $z = (M \otimes I_n)x$. In leaderless MAS case, define combined consensus error $\delta_i = \sum_{i=1}^{N}(z_j - z_i)$, with $\delta = \operatorname{col}(\delta_1, \delta_2, \ldots, \delta_N)$, it follows that $\delta = -(\mathscr{L}_M \otimes I_n)z$. Besides, define $\bar{e}_{ai} = m_i e_{ai}$, $\bar{e}_{bi} = m_i e_{bi}$, that leads to $\bar{e}_a = (M \otimes I_n)e_a$ and $\bar{e}_b = (M \otimes I_n)e_b$. Then, according to (2.8), (2.7) and the above definitions, it can be derived that

$$\begin{aligned}\dot{\delta} &= (I_N \otimes A - \mathscr{L}_M \otimes BK)\delta + (\mathscr{L}_M \mathscr{D} \otimes BK)\bar{e}_a - (\mathscr{L}_M \mathscr{A}_M \otimes BK)\bar{e}_b \\ \dot{\bar{e}}_a &= (I_N \otimes A + \mathscr{D} \otimes BK)\bar{e}_a - (I_N \otimes BK)\delta - (\mathscr{A}_M \otimes BK)\bar{e}_b \\ \dot{\bar{e}}_b &= (I_N \otimes A - \mathscr{A}_M \otimes BK)\bar{e}_b - (I_N \otimes BK)\delta + (\mathscr{D} \otimes BK)\bar{e}_a.\end{aligned} \tag{2.9}$$

Equation (2.9) gives the flow dynamics of closed-loop variables in leaderless case. It is worth to be noted that, differently from most of the existing results, the variables, such as $\bar{e}_a, \bar{e}_b$, with jump dynamics are also included in the closed-loop system, which is useful to describe the complete dynamics of closed-loop system, but also bring difficulties for stability analysis and control design. Due to whether the element $\bar{e}_{ai}$ in $\bar{e}_a$ has jump dynamics is only related to agent i, next, we describe the jump dynamics of $\bar{e}_{ai}$. From (2.6) and note that there exist time delay for information transmission, the jump dynamics of e_{ii} and e_{ij} can be denoted as $e_{ii}^+(t_{k_i}^i) = 0$ and $e_{ij}^+(t_{k_i}^i + d_{k_i}^i) = F(d_{k_i}^i)x_i(t_{k_i}^i) - x_i(t_{k_i}^i + d_{k_i}^i) = -F(d_{k_i}^i)e_{ij}(t_{k_i}^i) + e_{ij}(t_{k_i}^i + d_{k_i}^i)$. Then, based on the definition of $\bar{e}_{ai}, \bar{e}_{bi}$, it can be further derived that

$$\bar{e}_{ai}^+(t_{k_i}^i) = 0, \quad \bar{e}_{bi}^+(t_{k_i}^i + d_{k_i}^i) = -F(d_{k_i}^i)\bar{e}_{bi}(t_{k_i}^i) + \bar{e}_{bi}(t_{k_i}^i + d_{k_i}^i), \tag{2.10}$$

for $i \in \mathscr{V}$. Equation (2.10) describes the jump dynamics of the elements of closed-loop variables. The jump dynamics can not only describe the closed-loop system dynamics completely, but also is useful for stability analysis and parameter design, that will be shown later.

2.3.2 Dynamics of Leader-Following MAS

Similar as in the leaderless case, we aim to describe the closed-loop leader-following MAS with both flow dynamics and jump dynamics. Because the control objective is not the same as in leaderless case, that is, when there exists a leader, the control objective is to make each agent's state to track the leader's state bipartitely. To achieve the control objective, define $\delta_{mi} = x_i - m_i x_0$, then, δ_{mi} $(i \in \mathscr{V})$ converges to zero is equivalent to leader-following bipartite consensus. Based on (2.6), we have $\hat{x}_{ii} = e_{ii} + \delta_{mi} + m_i x_0$ and $\hat{x}_{ji} = e_{ji} + \delta_{mj} + m_j x_0$. It follows that the leader-following bipartite consensus control law (2.5) can be represented as $u_i = -K((\mathscr{L}_r^i \otimes I_n)\delta_m + b_i\delta_{mi} + d_i e_{ii} - (\mathscr{A}_r^i \otimes I_n)e_b)$, where $\delta_m = \mathrm{col}(\delta_{m1}, \delta_{m2}, \ldots, \delta_{mN})$. Define $\delta = (M \otimes I_n)\delta_m$, $\bar{e}_a = (M \otimes I_n)e_a$ and $\bar{e}_b = (M \otimes I_n)e_b$, and take $(\delta, \bar{e}_a, \bar{e}_b)$ as closed-loop states. Then, using the definitions in (2.7) and following similar derivation process as in Sect. 2.3.1, the flow dynamics of closed-loop MAS can be represented as

$$\begin{aligned}\dot{\delta} &= (I_N \otimes A - \mathscr{H}_M \otimes BK)\delta - (\mathscr{D} \otimes BK)\bar{e}_a + (\mathscr{A}_M \otimes BK)\bar{e}_b,\\ \dot{\bar{e}}_a &= (I_N \otimes A + \mathscr{D} \otimes BK)\bar{e}_a + (\mathscr{H}_M \otimes BK)\delta - (\mathscr{A}_M \otimes BK)\bar{e}_b,\\ \dot{\bar{e}}_b &= (I_N \otimes A - \mathscr{A}_M \otimes BK)\bar{e}_b + (\mathscr{H}_M \otimes BK)\delta + (\mathscr{D} \otimes BK)\bar{e}_a.\end{aligned} \tag{2.11}$$

The jump dynamics of $\bar{e}_{ai}$ and $\bar{e}_{bi}$ have the same form as (2.10). Thus, the leader-following closed-loop MAS with bipartite consensus protocol (2.5) can be described by (2.11) and (2.10).

From (2.9) to (2.11), it can be seen that the closed-loop system forms of MAS in leaderless and leader-following cases are the same, only the system matrices are different. Therefore, the leaderless and leader-following bipartite consensus problems of MAS with sampled and delayed information can be studied in a unified framework.

2.3.3 Reformulation as Hybrid Model

Observe that the above-mentioned closed-loop dynamics is not yet appropriate for bipartite consensus analysis and calculation of maximum allowable sampling periods. In this subsection, by means of internal auxiliary variables, we will reformulate the closed-loop MAS as a hybrid system form (1.1).

First, an internal timer variable $\tau_i(t) : \mathbb{R}_{\geq 0} \to [0, h_i]$, is introduced in the following, which is used to describe the sampling period of each agent in the hybrid model, its dynamics has the form of $\dot{\tau}_i(t) = 1, t \in (t^i_{k_i}, t^i_{k_i+1}]$ and $\tau_i^+(t) = 0, t = t^i_{k_i}$, which implies that $\tau_i \in [0, h_i]$, with $h_i = t^i_{k_i+1} - t^i_{k_i}$ denoting the sampling period of agent i. In addition, if there is no time delay, the jump dynamics of error variables only occurs at sampling instants. But, the communication delay leads to another error jump dynamics when transmitted information arrives at neighbor agents. Thus, to appropriately describe the error jump dynamics, a binary variable $l_i(t) \in \{0, 1\}$ is introduced to indicate that the next event is *transmission event* ($l_i(t) = 0$) or *update event*[2] ($l_i(t) = 1$). The dynamics of $l_i(t)$ can be described by $\dot{l}_i(t) = 0$, for $t \neq t^i_{k_i}$ and $t \neq t^i_{k_i} + d^i_{k_i}$, $l_i^+(t) = 1, t = t^i_{k_i}$, $l_i^+(t) = 0, t = t^i_{k_i} + d^i_{k_i}$. Besides, we introduce another variable $s_i(t) : \mathbb{R}_{\geq 0} \to \mathbb{R}^n$, which is used to memorize some useful information at sampling instant.

With the above introduced interval variables and the closed-loop variables defined in Sects. 2.3.1 and 2.3.2, in order to construct a hybrid closed-loop system model for stability analysis, define the combined variables as follows

$$\tau = \mathrm{col}(\tau_1, \tau_2, \ldots, \tau_N), l = \mathrm{col}(l_1, l_2, \ldots, l_N), s = \mathrm{col}(s_1, s_2, \ldots, s_N), \xi = \mathrm{col}(\delta, \bar{e}_a, \bar{e}_b, \tau, l, s), \tag{2.12}$$

[2] Transmission event mean that at sampling instant $t^i_{k_i}$, agent i samples its own state and transmits it to its neighbors. Update event means that when transmitted information arrives at its neighbors at $t^i_{k_i} + d^i_{k_i}$, the neighbor agents' controllers will update using the received information.

where ξ denotes the closed-loop state. According to the role and meaning of variables τ_i, l_i, s_i, and note that the variables $\bar{e}_a$ and $\bar{e}_b$ have jump dynamics, therefore, by defining appropriate flow set and jump set, the closed-loop dynamics of MAS can be described in hybrid systems form (1.1). Specifically, according to the definition of ξ, (2.9), and (2.11), the flow dynamics of ξ can be formulated as

$$\dot{\xi} = \mathscr{F}(\xi), \text{ If } \xi \in \mathbb{F}, \tag{2.13}$$

where

$$\mathscr{F}(\xi) = [A_{11}\delta + A_{12}\bar{e}_a + A_{13}\bar{e}_b; A_{21}\delta + A_{22}\bar{e}_a + A_{23}\bar{e}_b; A_{31}\delta + A_{32}\bar{e}_a + A_{33}\bar{e}_b; 1_N; 0_N; 0_{nN}], \tag{2.14}$$

where A_{11}–A_{33} denote system matrices. For leaderless MAS, $A_{11} = I_N \otimes A - \mathscr{L}_M \otimes BK$, $A_{12} = \mathscr{L}_M \mathscr{D} \otimes BK$, $A_{13} = -(\mathscr{L}_M \mathscr{A}_M \otimes BK)$, $A_{21} = -I_N \otimes BK$, $A_{22} = I_N \otimes A + \mathscr{D} \otimes BK$, $A_{23} = \mathscr{A}_M \otimes BK$, $A_{31} = -I_N \otimes BK$, $A_{32} = \mathscr{D} \otimes BK$, $A_{33} = I_N \otimes A - \mathscr{A}_M \otimes BK$. For leader-following MAS, $A_{11} = I_N \otimes A - \mathscr{H}_M \otimes BK$, $A_{12} = -\mathscr{D} \otimes BK$, $A_{13} = -\mathscr{A}_M \otimes BK$, $A_{21} = \mathscr{H}_M \otimes BK$, $A_{22} = I_N \otimes A + \mathscr{D} \otimes BK$, $A_{23} = -\mathscr{A}_M \otimes BK$, $A_{31} = \mathscr{H}_M \otimes BK$, $A_{32} = \mathscr{D} \otimes BK$, $A_{33} = I_N \otimes A - \mathscr{A}_M \otimes BK$.

Note that the jump dynamics of closed-loop system variables only occurs in two cases: (1). at sampling instant of agent i; (2). when the information transmitted from agent i arrives at its neighbors at $t^i_{k_i} + d^i_{k_i}$. At rest time, the closed-loop variables evolve according to flow dynamics. Thus, the state space can be divided into flow set and jump set. In flow set, the system dynamics evolve continuously. In jump set, some of the closed-loop system states will jump. Traditional method regarding cooperative control of MAS cannot describe the jump dynamics appropriately, only flow dynamics is taken into account. Here, according to the values l_i (related to transmission event and update event) and τ_i (related to sampling period), the flow set $\mathbb{F}$ is defined as $\mathbb{F} := \bigcap_{i \in \mathscr{V}} \mathbb{F}_i$, with

$$\mathbb{F}_i := \{\xi \in \mathbb{X} | (l_i = 0 \wedge \tau_i < h_i) \vee (l_i = 1 \wedge \tau_i < d^i_{k_i})\}, \tag{2.15}$$

where $\mathbb{X} := \{(\delta, \bar{e}_a, \bar{e}_b, \tau, l, s) \in \mathbb{R}^{nN} \times \mathbb{R}^{nN} \times \mathbb{R}^{nN} \times \mathbb{R}^N_{\geq 0} \times \{0, 1\}^N \times \mathbb{R}^{nN}\}$.

Consider the jump dynamics of ξ, according to (2.6) and (2.12), the jump dynamics of ξ can be descried by

$$\xi^+ \in \mathscr{J}(\xi) = \bigcup_{i \in \mathscr{V}} \mathscr{J}_i(\xi), \text{ If } \xi \in \mathbb{J}, \tag{2.16}$$

where

$$\mathscr{J}_i(\xi) := \begin{cases} \mathscr{J}_{0i}(\xi), & \text{If } \xi \in \mathbb{J}_i \text{ and } l_i = 0, \\ \mathscr{J}_{1i}(\xi), & \text{If } \xi \in \mathbb{J}_i \text{ and } l_i = 1, \end{cases}$$

The two cases of $\mathscr{J}_i(\xi)$ corresponds to different jump dynamics. They are related to transmission event ($l_i = 0$) and update event ($l_i = 1$), specifically, $l_i = 1$ means that transmission event is activated at sampling instant, agent i needs to broadcast information to its neighbors, and some error variables and l_i itself have jump dynamics. $l_i = 1$ indicates that the transmitted information from agent i arrives at its neighbors, and its neighbors' control law will update according to (2.4) or (2.5). According to these two cases, $\mathscr{J}_{0i}(\xi)$ and $\mathscr{J}_{1i}(\xi)$ can be described as

$$\mathscr{J}_{0i}(\xi) = [\delta; ((I_N - \Lambda_i) \otimes I_n)\bar{e}_a; \bar{e}_b; 0_N; l + \Lambda_i 1_N; -(\Lambda_i \otimes I_n)\bar{e}_b],$$
$$\mathscr{J}_{1i}(\xi) = [\delta; \bar{e}_a; (\Lambda_i \otimes F(d^i_{k_i}))s + \bar{e}_b; 0_N; l - \Lambda_i 1_N; ((I_N - \Lambda_i) \otimes I_n)s].$$

The jump set $\mathbb{J}$ has the form of $\mathbb{J} := \bigcup_{i\in\mathscr{V}} \mathbb{J}_i$ with

$$\mathbb{J}_i := \{\xi \in \mathbb{X} | (l_i = 0 \wedge \tau_i \geq h_i) \vee (l_i = 1 \wedge \tau_i \geq d^i_{k_i})\}. \quad (2.17)$$

The jump dynamics (2.16) describes how ξ changes at sampling instants or at the instants when transmitted information from a certain agent arrives at its neighbors. Now, with appropriate defined internal variables and closed-loop variables, the dynamics of closed-loop MAS (both leaderless and leader-following cases) is reformulated in hybrid systems form. The bipartite consensus of MAS is equivalent to the stability of the hybrid system with flow dynamics (2.13), jump dynamics (2.16), flow set (2.15), and jump set (2.17). The stability property of hybrid system is defined as follows.

Definition 2.2 ([8]) Consider the hybrid system consisting of (2.13) and (2.16), the set $\{\xi \in \mathbb{X} | \delta = 0_{nN}, \bar{e}_a = 0_{nN}, \bar{e}_b = 0_{nN}\}$ is said to be uniformly globally asymptotically stable (UGAS) if there exists a function $\beta \in \mathscr{KL}$, such that for initial condition $\xi(0) \in \mathbb{X}$, the solution ξ satisfies

$$\|(\delta(t), \bar{e}_a(t), \bar{e}_b(t))\| \leq \beta(\|(\delta(0), \bar{e}_a(0), \bar{e}_b(0))\|, t). \quad (2.18)$$

2.4 Stability Analysis and Parameter Design

In this section, we present how to design system parameters and how to design Lyapunov function to theoretically prove that the hybrid system (with flow dynamics (2.13), jump dynamics (2.16), flow set (2.15), and jump set (2.17)) is asymptotically stable, which indicates that the leaderless/leader-following bipartite consensus is achieved. The stability analysis and parameter design are performed under hybrid systems framework, in addition to the stability results, the upper bound of sampling period and maximum allowable delay can be explicitly calculated with given parameters.

2.4.1 Lyapunov Function Design

For stability analysis of control systems, Lyapunov method is usually used. An appropriate Lyapunov function is the key for stability analysis. Different from most of the existing results regarding cooperative control of MAS, a hybrid model is constructed for MAS, which not only includes flow dynamics, but also includes jump dynamics. With this property of hybrid model, for stability analysis, it is not only required that the Lyapunov function is decreasing along closed-loop system's flow dynamics, but also is non-increasing when jump dynamics occurs, this must be considered when constructing Lyapunov function. In this chapter, we design a Lyapunov function in the following form

$$U(\xi) = V(\delta) + \sum_{i=1}^{N} \gamma_{l_i i} \phi_{l_i i}(\tau_i) \tilde{W}_i^2(\bar{e}_{ai}, \bar{e}_{bi}, s_i, l_i).$$

It can be seen that all closed-loop variables are included in the Lyapunov function. In order to guarantee that the Lyapunov function is well-defined, the functions $V(\delta)$, $\phi_{l_i i}(\tau_i)$, $\tilde{W}_i^2(\bar{e}_{ai}, \bar{e}_{bi}, s_i, l_i)$, and parameters $\gamma_{l_i i}$ should satisfy some properties. Next, it gives what properties they should satisfy, such that it can be used for stability analysis.

First, consider the function $\phi_{l_i i}(\tau_i)$ with respect to timer variable τ_i. The first requirement for function $\phi_{l_i i}(\tau_i)$ is that it should be non-negative such that the Lyapunov function is well-defined. The second requirement is that it is decreasing along flow dynamics of τ_i, such that it is useful for stability proof. Thus, to satisfy the above two requirements, design the function $\phi_{l_i i}(\tau_i) : \mathbb{R}_{\geq 0} \to \mathbb{R}_{\geq 0}$ in the following form

$$\dot{\phi}_{l_i i}(\tau_i) = -2L_{l_i i}\phi_{l_i i}(\tau_i) - \kappa_i \gamma_{l_i i}(\phi_{l_i i}^2(\tau_i) + 1), \tag{2.19}$$

where the constant parameters $L_{l_i i} > 0$, $\kappa_i > 0$, and $\gamma_{l_i i} > 0$.

Second, consider the function $\tilde{W}_i^2(\bar{e}_{ai}, \bar{e}_{bi}, s_i, l_i)$. Due to the variables $\bar{e}_{ai}, \bar{e}_{bi}, s_i, l_i$ have jump dynamics at sampling instant or when transmitted information is received by neighbors, the key point to design $\tilde{W}_i^2(\bar{e}_{ai}, \bar{e}_{bi}, s_i, l_i)$ is that it should be non-increasing no matter which variable of $\bar{e}_{ai}, \bar{e}_{bi}, s_i, l_i$ jumps. To satisfy this requirement, suppose that the following assumptions hold.

Assumption 2.3 There exist functions $\tilde{W}_i : \mathbb{R}^n \times \mathbb{R}^n \times \mathbb{R}^n \times \{0, 1\}$, $i \in \mathscr{V}$, with $\tilde{W}_i(\bar{e}_{ai}, \bar{e}_{bi}, s_i, l_i)$ locally Lipschitz for $l_i \in \{0, 1\}$, $\mathscr{K}_\infty$ functions $\underline{\alpha}_{\tilde{W}i}, \bar{\alpha}_{\tilde{W}_i}$, constants $0 < \lambda_i < 1$, such that for all $l_i \in \{0, 1\}$, $s_i \in \mathbb{R}^n$, $\bar{e}_{ai} \in \mathbb{R}^n$, $\bar{e}_{bi} \in \mathbb{R}^n$, the following conditions hold

$$\begin{aligned} &\underline{\alpha}_{\tilde{W}i}(\|(\bar{e}_{ai}, \bar{e}_{bi})\|) \leq \tilde{W}_i(\bar{e}_{ai}, \bar{e}_{bi}, s_i, l_i) \leq \bar{\alpha}_{\tilde{W}i}(\|(\bar{e}_{ai}, \bar{e}_{bi})\|), \\ &\tilde{W}_i(0, \bar{e}_{bi}, -\bar{e}_{bi}, 1) \leq \lambda_i \tilde{W}_i(\bar{e}_{ai}, \bar{e}_{bi}, s_i, 0), \\ &\tilde{W}_i(\bar{e}_{ai}, \bar{e}_{bi} + F(d_{k_i}^i)s_i, 0, 0) \leq \tilde{W}_i(\bar{e}_{ai}, \bar{e}_{bi}, s_i, 1). \end{aligned} \tag{2.20}$$

Assumption 2.4 Given the storage function $\tilde{W}_i(\bar{e}_{ai}, \bar{e}_{bi}, s_i, l_i)$ defined in (2.25), there exist continuous functions $H_{1i}, H_{2i} : \mathbb{R}^{nN} \times \mathbb{R}^{nN} \times \mathbb{R}^{nN} \to \mathbb{R}$, positive constants $L_{l_i i}$, such that for $l_i \in \{0, 1\}$, $s_i \in \mathbb{R}^n$, $\delta \in \mathbb{R}^{nN}$, $\bar{e}_a \in \mathbb{R}^{nN}$, $\bar{e}_b \in \mathbb{R}^{nN}$,

$$\left\langle \frac{\partial \tilde{W}_i}{\partial \bar{e}_{ai}}, f_{ai}(\delta, \bar{e}_a, \bar{e}_b) \right\rangle + \left\langle \frac{\partial \tilde{W}_i}{\partial \bar{e}_{bi}}, f_{bi}(\delta, \bar{e}_a, \bar{e}_b) \right\rangle \le L_{l_i i}\tilde{W}_i + H_{1i}(\delta, \bar{e}_a, \bar{e}_b) + H_{2i}(\delta, \bar{e}_a, \bar{e}_b), \tag{2.21}$$

where $f_{ai}(\delta, \bar{e}_a, \bar{e}_b) = (R_i \otimes I_n)(A_{21}\delta + A_{22}\bar{e}_a + A_{23}\bar{e}_b)$, $f_{bi}(\delta, \bar{e}_a, \bar{e}_b) = (R_i \otimes I_n)(A_{31}\delta + A_{32}\bar{e}_a + A_{33}\bar{e}_b)$, $R_i \in \mathbb{R}^{1\times N}$ denotes a row matrix with ith element being 1 and the rest being zeros.

If the function $\tilde{W}_i^2(\bar{e}_{ai}, \bar{e}_{bi}, s_i, l_i)$ satisfies the above conditions, it is obvious that the function $\tilde{W}_i^2(\bar{e}_{ai}, \bar{e}_{bi}, s_i, l_i)$ is radially unbounded and non-increasing when jump dynamics occur. These conditions will be used to formally prove that the hybrid system is asymptotically stable.

Third, consider the function $V(\delta)$. Due to δ is always continuous and it does not contain jump dynamics, it is only required that $V(\delta)$ is non-negative and radially bounded, and satisfies some properties when evolves along system's flow dynamics. Design $V(\delta)$ such that the conditions in the following assumption hold.

Assumption 2.5 There exists a function $V(\delta) : \mathbb{R}^{nN} \to \mathbb{R}_{\ge 0}$, $\mathcal{K}_\infty$ functions $\underline{\alpha}_V, \bar{\alpha}_V$, continuous functions $H_{1i}, H_{2i} : \mathbb{R}^{nN} \times \mathbb{R}^{nN} \times \mathbb{R}^{nN} \to \mathbb{R}$, positive functions $\rho(\cdot) : \mathbb{R}_{\ge 0} \to \mathbb{R}_{\ge 0}$, constants $0 < \lambda_i < 1$, $0 < \epsilon_i < 1$, $\kappa_i > 0$, $\gamma_{l_i i} > 0$, such that for $l_i \in \{0, 1\}$, $s_i \in \mathbb{R}^n$, $\delta \in \mathbb{R}^{nN}$, $\bar{e}_a \in \mathbb{R}^{nN}$, $\bar{e}_b \in \mathbb{R}^{nN}$,

$$\underline{\alpha}_V(\|\delta\|) \le V(\delta) \le \bar{\alpha}_V(\|\delta\|), \tag{2.22}$$

$$\begin{aligned}
&\langle \nabla V(\delta), A_{11}\delta + A_{12}\bar{e}_a + A_{13}\bar{e}_b \rangle \\
&\le -\rho(\|\delta\|) - \sum_{i=1}^{N} \Big(\frac{1}{\kappa_i}(H_{1i}^2(\delta, \bar{e}_a, \bar{e}_b) + H_{2i}^2(\delta, \bar{e}_a, \bar{e}_b)) - \frac{(1-\epsilon_i)\kappa_i \gamma_{l_i i}^2}{\max\{1, l_i/\lambda_i^2\}}(W_i^2(\bar{e}_{ai}) + W_i^2(\bar{e}_{bi})) \Big),
\end{aligned} \tag{2.23}$$

$$\tilde{W}_i^2 \ge (W_i^2(\bar{e}_{ai}) + W_i^2(\bar{e}_{bi}))/\max\{1, l_i/\lambda_i^2\}, \tag{2.24}$$

where the function $W_i(x) = \|x\|$.

It seems that the conditions in these assumptions are complex and the specific form of functions $V(\delta)$ and $\tilde{W}_i(\bar{e}_{ai}, \bar{e}_{bi}, s_i, l_i)$ are not given. However, it is worth to be noted that some usually used functions in the literature can be selected such that these conditions can be satisfied. For example, one selection method is to choose $W_i(\bar{e}_{ai}) = \|\bar{e}_{ai}\|$ and $W_i(\bar{e}_{bi}) = \|\bar{e}_{bi}\|$, and design $\tilde{W}_i(\bar{e}_{ai}, \bar{e}_{bi}, s_i, l_i)$ as

$$\begin{aligned}\tilde{W}_i(\bar{e}_{ai}, \bar{e}_{bi}, s_i, l_i) &= W_i(\bar{e}_{ai}) + W_i(\bar{e}_{bi}), \text{ If } l_i = 0 \\ \tilde{W}_i(\bar{e}_{ai}, \bar{e}_{bi}, s_i, l_i) &= \max\left\{\lambda_i(W_i(\bar{e}_{ai}) + W_i(\bar{e}_{bi})), W_i(\bar{e}_{ai}) + W_i(\bar{e}_{bi} + F(d^i_{k_i})s_i)\right\}, \text{ If } l_i = 1\end{aligned} \tag{2.25}$$

where $F(d^i_{k_i}) = \text{expm}(Ad^i_{k_i})$. Then, we have the following result.

Proposition 2.1 ([14]) *Consider the hybrid system consisting of* (2.13) *and* (2.16). *Suppose that* $W_i(\bar{e}_{ai}) = \|\bar{e}_{ai}\|$ *and* $W_i(\bar{e}_{bi}) = \|\bar{e}_{bi}\|$, *and* $\tilde{W}_i(\bar{e}_{ai}, \bar{e}_{bi}, s_i, l_i)$ *has the form* (2.25). *If the communication delays* $d^i_{k_i} \le \bar{d}_i$ *with the maximum allowable delay* $\bar{d}_i$ *satisfies*

$$\begin{aligned}\bar{d}_i &= \max\ d_i \\ \text{s.t.}\ \ &\|I - F(d_i)\| \le \lambda_i\end{aligned} \tag{2.26}$$

Then, the function $\tilde{W}_i(\bar{e}_{ai}, \bar{e}_{bi}, s_i, l_i)$ *satisfies the conditions in Assumptions* 2.3 *and* 2.4.

Proof From [9, 10], it is known that the following conditions always hold for delay-free case. For a function defined on $W_i(x) : \mathbb{R}^n \to \mathbb{R}_{\ge 0}$, select the specific form for it as $W_i(x) = \|x\|$, then, the following properties hold true

$$\begin{aligned}\underline{\alpha}_{Wi}\ \|x\| \le W_i(x) \le \bar{\alpha}_{Wi}\ \|x\|, \\ W_i(0) \le \lambda_i W_i(x)\end{aligned} \tag{2.27}$$

where $\underline{\alpha}_{Wi} \le 1$, $\bar{\alpha}_{Wi} \ge 1$, $0 < \lambda_i < 1$. As $W_i(\bar{e}_{ai}) = \|\bar{e}_{ai}\|$ and $W_i(\bar{e}_{bi}) = \|\bar{e}_{bi}\|$, it is not difficult to testify that $\tilde{W}_i(\bar{e}_{ai}, \bar{e}_{bi}, s_i, l_i)$ in (2.25) satisfies the first inequality condition in (2.20) with $\underline{\alpha}_{\tilde{W}_i}(r) = \lambda_i r$, $\bar{\alpha}_{\tilde{W}_i}(r) = r$, $0 < \lambda_i < 1$.

When $\xi \in \mathbb{J}_i\ \wedge\ l_i = 1$, that is, at the instants $t^i_{k_i} + d^i_{k_i}$, it has $l_i(t^i_{k_i} + d^i_{k_i}) = 1$. According to the jump dynamics (2.17), it yields $l_i((t^i_{k_i} + d^i_{k_i})^+) = 0$, $\bar{e}_{ai}((t^i_{k_i} + d^i_{k_i})^+) = \bar{e}_{ai}(t^i_{k_i} + d^i_{k_i})$, $\bar{e}_{bi}((t^i_{k_i} + d^i_{k_i})^+) = \bar{e}_{bi}(t^i_{k_i} + d^i_{k_i}) + F(d^i_{k_i})s_i(t^i_{k_i} + d^i_{k_i})$, $s_i((t^i_{k_i} + d^i_{k_i})^+) = 0$. Then, we have

$$\begin{aligned}&\tilde{W}_i(\bar{e}^+_{ai}, \bar{e}^+_{bi}, s^+_i, l^+_i) \\ &= W_i(\bar{e}_{ai}) + W_i(\bar{e}_{bi} + F(d^i_{k_i})s_i) \\ &= \tilde{W}_i(\bar{e}_{ai}, \bar{e}_{bi} + F(d^i_{k_i})s_i, 0, 0) \\ &\le \tilde{W}_i(\bar{e}_{ai}, \bar{e}_{bi}, s_i, 1)\end{aligned} \tag{2.28}$$

where the last inequality holds naturally based on the definition of $\tilde{W}_i(\bar{e}_{ai}, \bar{e}_{bi}, s_i, l_i)$ in (2.25).

When $\xi \in \mathbb{J}_i\ \wedge\ l_i = 0$, that is, at sampling instants $t^i_{k_i}$, it corresponds $l_i(t^i_{k_i}) = 0$. According to the jump dynamics (2.17), it yields $l_i(t^{i+}_{k_i}) = 1$, $\bar{e}_{ai}(t^{i+}_{k_i}) = 0$, $s_i(t^{i+}_{k_i}) = -\bar{e}_{bi}(t^i_{k_i})$. Since the constructed function $\tilde{W}_i(\bar{e}_{ai}, \bar{e}_{bi}, s_i, l_i)$ in (2.25) has two possible forms for $l_i = 1$, thus, if $\lambda_i(W_i(\bar{e}_{ai}) + W_i(\bar{e}_{bi})) \ge W_i(\bar{e}_{ai}) + W_i(\bar{e}_{bi} + F(d^i_{k_i})s_i)$, we have

$$\begin{aligned}\tilde{W}_i(\bar{e}_{ai}^+, \bar{e}_{bi}^+, s_i^+, l_i^+) &= \lambda_i(W_i(\bar{e}_{ai}^+) + W_i(\bar{e}_{bi}^+)) \\ &= \lambda_i(W_i(0) + W_i(\bar{e}_{bi})) = \tilde{W}_i(0, \bar{e}_{bi}, -\bar{e}_{bi}, 1) \\ &\le \lambda_i(W_i(\bar{e}_{ai}) + W_i(\bar{e}_{bi})) = \lambda_i \tilde{W}_i(\bar{e}_{ai}, \bar{e}_{bi}, s_i, 0).\end{aligned} \tag{2.29}$$

If $\lambda_i(W_i(\bar{e}_{ai}) + W_i(\bar{e}_{bi})) \le W_i(\bar{e}_{ai}) + W_i(\bar{e}_{bi} + F(d_{k_i}^i)s_i)$, we have

$$\begin{aligned}\tilde{W}_i(\bar{e}_{ai}^+, \bar{e}_{bi}^+, s_i^+, l_i^+) &= W_i(\bar{e}_{ai}^+) + W_i(\bar{e}_{bi}^+ + F(d_{k_i}^i)s_i^+) \\ &= W_i(0) + W_i(\bar{e}_{bi} - F(d_{k_i}^i)\bar{e}_{bi}) = \tilde{W}_i(0, \bar{e}_{bi}, -\bar{e}_{bi}, 1) \\ &\le \lambda_i W_i(\bar{e}_{ai}) + \left\|(I - F(d_{k_i}^i))\bar{e}_{bi}\right\| \\ &\le \lambda_i W_i(\bar{e}_{ai}) + \left\|(I - F(d_{k_i}^i))\right\| \|\bar{e}_{bi}\| \\ &\overset{(2.26)}{\le} \lambda_i W_i(\bar{e}_{ai}) + \lambda_i W_i(\bar{e}_{bi}) \\ &= \lambda_i \tilde{W}_i(\bar{e}_{ai}, \bar{e}_{bi}, s_i, 0).\end{aligned} \tag{2.30}$$

where the fifth inequality holds if the communication delays $d_{k_i}^i \le \bar{d}_i$ with the MAD $\bar{d}_i$ satisfies (2.26). For this moment, it can be seen that the constructed function $\tilde{W}_i(\bar{e}_{ai}, \bar{e}_{bi}, s_i, l_i)$ makes Assumption 2.3 hold.

Next, we prove the constructed function $\tilde{W}_i(\bar{e}_{ai}, \bar{e}_{bi}, s_i, l_i)$ makes Assumption 2.4 hold. Notice taht

$$\begin{aligned}\dot{\bar{e}}_{ai} &= (R_i \otimes I_n)(A_{21}\delta + A_{22}\bar{e}_a + A_{23}\bar{e}_b), \\ \dot{\bar{e}}_{bi} &= (R_i \otimes I_n)(A_{31}\delta + A_{32}\bar{e}_a + A_{33}\bar{e}_b),\end{aligned} \tag{2.31}$$

it yields

$$\begin{aligned}\|\dot{\bar{e}}_{ai}\| &\le \underbrace{\left\|(R_i \otimes I_n)(A_{21}\delta + A_{22}\bar{\Lambda}_i\bar{e}_a + A_{23}\bar{e}_b)\right\|}_{H_{1i}(\delta,\bar{e}_a,\bar{e}_b)} + \|(R_i \otimes I_n)A_{22}\| \|\bar{e}_{ai}\| \\ \|\dot{\bar{e}}_{bi}\| &\le \underbrace{\left\|(R_i \otimes I_n)(A_{31}\delta + A_{32}\bar{e}_a + A_{33}\bar{\Lambda}_i\bar{e}_b)\right\|}_{H_{2i}(\delta,\bar{e}_a,\bar{e}_b)} + \|(R_i \otimes I_n)A_{33}\| \|\bar{e}_{bi}\|\end{aligned} \tag{2.32}$$

where $\bar{\Lambda}_i = I_{nN} - \Lambda_i \otimes I_n$.

For the case with $l_i = 0$, based on (2.25), we obtain

$$\begin{aligned}&\left\langle \frac{\partial \tilde{W}_i}{\partial \bar{e}_{ai}}, f_{ai}(\delta, \bar{e}_a, \bar{e}_b)\right\rangle + \left\langle \frac{\partial \tilde{W}_i}{\partial \bar{e}_{bi}}, f_{bi}(\delta, \bar{e}_a, \bar{e}_b)\right\rangle \\ \le &\left\|\frac{\partial \tilde{W}_i}{\partial \bar{e}_{ai}}\right\| \|\dot{\bar{e}}_{ai}\| + \left\|\frac{\partial \tilde{W}_i}{\partial \bar{e}_{bi}}\right\| \|\dot{\bar{e}}_{bi}\| \\ \le &H_{1i}(\delta, \bar{e}_a, \bar{e}_b) + H_{2i}(\delta, \bar{e}_a, \bar{e}_b) + L_{0i}\tilde{W}_i,\end{aligned} \tag{2.33}$$

where $L_{0i} = \max\{\|(R_i \otimes I_n)A_{22}\|, \|(R_i \otimes I_n)A_{33}\|\}$.

For the case with $l_i = 1$, according to (2.25), if $\tilde{W}_i = \lambda_i(W_i(\bar{e}_{ai}) + W_i(\bar{e}_{bi}))$, we obtain

$$\begin{aligned}
&\left\langle \frac{\partial \tilde{W}_i}{\partial \bar{e}_{ai}}, f_{ai}(\delta, \bar{e}_a, \bar{e}_b) \right\rangle + \left\langle \frac{\partial \tilde{W}_i}{\partial \bar{e}_{bi}}, f_{bi}(\delta, \bar{e}_a, \bar{e}_b) \right\rangle \\
\le & H_{1i}(\delta, \bar{e}_a, \bar{e}_b) + H_{2i}(\delta, \bar{e}_a, \bar{e}_b) + \frac{L_{0i}}{\lambda_i} \lambda_i (W_i(\bar{e}_{ai}) + W_i(\bar{e}_{bi})) \\
\le & H_{1i}(\delta, \bar{e}_a, \bar{e}_b) + H_{2i}(\delta, \bar{e}_a, \bar{e}_b) + L_{1i}\tilde{W}_i,
\end{aligned}$$

where $L_{1i} = L_{0i}/\lambda_i$. If $\tilde{W}_i = W_i(\bar{e}_{ai}) + W_i(\bar{e}_{bi} + F(d^i_{k_i})s_i)$,

$$\begin{aligned}
&\left\langle \frac{\partial \tilde{W}_i}{\partial \bar{e}_{ai}}, f_{ai}(\delta, \bar{e}_a, \bar{e}_b) \right\rangle + \left\langle \frac{\partial \tilde{W}_i}{\partial \bar{e}_{bi}}, f_{bi}(\delta, \bar{e}_a, \bar{e}_b) \right\rangle \\
\le & H_{1i}(\delta, \bar{e}_a, \bar{e}_b) + H_{2i}(\delta, \bar{e}_a, \bar{e}_b) + L_{0i}(W_i(\bar{e}_{ai}) + W_i(\bar{e}_{bi})) \\
\le & H_{1i}(\delta, \bar{e}_a, \bar{e}_b) + H_{2i}(\delta, \bar{e}_a, \bar{e}_b) + L_{1i}\tilde{W}_i.
\end{aligned}$$

Based on the above derivations, the constructed function $\tilde{W}_i(\bar{e}_{ai}, \bar{e}_{bi}, s_i, l_i)$ makes Assumption 2.4 hold. This completes the proof. □

In addition, we can design $V(\delta) = \delta^T(I_N \otimes P)\delta$ and design appropriated control gain K such that the conditions in Assumption 2.5 hold. Specifically, we have the following result.

Proposition 2.2 ([14]) *Choose* $V(\delta) = \delta^T(I_N \otimes P)\delta$ *and controller gain* $K = \mu B^T P$, *where* μ *and* P *are obtained by solving the following Riccati equation*

$$PA + A^T P - 2\mu a(\hat{\mathscr{L}})PBB^T P + Q = 0, \tag{2.34}$$

where $Q > 0$ *is positive definite,* $\hat{\mathscr{L}} = \mathscr{L}_M$ *and* $\hat{\mathscr{L}} = \mathscr{H}_M$ *in leaderless and leader-following cases respectively,* $a(\hat{\mathscr{L}})$ *denotes the minimum positive eigenvalue of matrix* $\hat{\mathscr{L}}$.

Suppose that there exist constants $0 < \sigma_0 < 1$, $\gamma_i > 0$, $\kappa_i > 0$, *and* $1 > \epsilon_i > 0$, *such that*

$$\left[\begin{matrix}
I_N \otimes (\sigma_0 - 1)Q + \mathsf{sum}(\kappa_i, \mathsf{H}_i^{\delta,\delta}) & (I_N \otimes P)A_{12} + \mathsf{sum}(\kappa_i, \mathsf{H}_i^{\delta,a}) \\
* & -\sum_{i=1}^{N}(1-\epsilon_i)\kappa_i\gamma_i^2(R_i^T R_i \otimes I_n) + \mathsf{sum}(\kappa_i, \mathsf{H}_i^{a,a}) \\
* & *
\end{matrix}\right.$$

$$\begin{bmatrix} \cdots & (I_N \otimes P)A_{13} + \mathsf{sum}(\kappa_i, \mathsf{H}_i^{\delta,b}) \\ \cdots & \mathsf{sum}(\kappa_i, \mathsf{H}_i^{a,b}) \\ \cdots & -\sum_{i=1}^{N}(1-\epsilon_i)\kappa_i\gamma_i^2(R_i^T R_i \otimes I_n) + \mathsf{sum}(\kappa_i, \mathsf{H}_i^{b,b}) \end{bmatrix} \le 0 \qquad (2.35)$$

where the notation $\mathsf{H}_i^{x_1,x_2}$ *(with* x_i, $i \in \{1,2\}$, *replaced by* δ, a, *or* b*) has the form of* $\mathsf{H}_i^{x_1,x_2} = H_{1i}^{x_1 T} H_{1i}^{x_2} + H_{2i}^{x_1 T} H_{2i}^{x_2}$, *and* $\mathsf{sum}(\kappa_i, \mathsf{H}_i^{x_1,x_2}) = \sum_{i=1}^{N} \kappa_i^{-1} \mathsf{H}_i^{x_1,x_2}$, *and* $H_{1i}^{\delta} = (R_i \otimes I_n)A_{21}$, $H_{1i}^{a} = (R_i \otimes I_n)A_{22}\bar{\Lambda}_i$, $H_{1i}^{b} = (R_i \otimes I_n)A_{23}$, $H_{2i}^{\delta} = (R_i \otimes I_n)A_{31}$, $H_{2i}^{a} = (R_i \otimes I_n)A_{32}$, $H_{2i}^{b} = (R_i \otimes I_n)A_{33}\bar{\Lambda}_i$. *Then, the conditions* (2.22) *and* (2.23) *in Assumption* 2.5 *hold with* $\gamma_{0i} = \gamma_i$ *and* $\gamma_{1i} = \gamma_i/\lambda_i$.

Proof For the function $V(\delta) = \delta^T(I_N \otimes P)\delta$, let $\underline{\alpha}_V(r) = \lambda_{\min}(I_N \otimes P)r$ and $\bar{\alpha}_V(r) = \lambda_{\max}(I_N \otimes P)r$, it can be easily verified that condition (2.22) holds.

Then, we show that the condition (2.23) in Assumption 2.5 holds. Based on flow dynamics (2.13), we can obtain

$$\begin{aligned} \dot{V}(\delta) &= \langle \nabla V(\delta), A_{cl}^{11}\delta + A_{cl}^{12}\bar{e}\rangle \\ &= \delta^T((I_N \otimes P)(I_N \otimes A - \hat{\mathscr{L}} \otimes BK) + (I_N \otimes A - \hat{\mathscr{L}} \otimes BK)^T(I_N \otimes P))\delta \\ &\quad + 2\delta^T(I_N \otimes P)(A_{12}\bar{e}_a + A_{13}\bar{e}_b) \\ &= \delta^T(I_N \otimes (PA + A^T P) - 2\mu\hat{\mathscr{L}} \otimes PBB^T P)\delta + 2\delta^T(I_N \otimes P)(A_{12}\bar{e}_a + A_{13}\bar{e}_b) \\ &\le -\delta^T(I_N \otimes Q)\delta + 2\delta^T(I_N \otimes P)(A_{12}\bar{e}_a + A_{13}\bar{e}_b). \end{aligned}$$

Consider the linear matrix inequality (2.35) in Proposition 2.2, let it right- and left-multiplied by $\mathrm{col}(\delta, \bar{e}_a, \bar{e}_b)$ and its transpose, it can be calculated that

$$\begin{aligned} &\langle \nabla V(\delta), A_{cl}^{11}\delta + A_{cl}^{12}\bar{e}\rangle \\ &= -\delta^T(I_N \otimes \sigma_0 Q)\delta - \sum_{i=1}^{N}\frac{1}{\kappa_i}\left\|H_{1i}^{\delta}\delta + H_{1i}^{a}\bar{e}_a + H_{1i}^{b}\bar{e}_b\right\|^2 \\ &\quad - \sum_{i=1}^{N}\frac{1}{\kappa_i}\left\|H_{2i}^{\delta}\delta + H_{2i}^{a}\bar{e}_a + H_{2i}^{b}\bar{e}_b\right\|^2 + \sum_{i=1}^{N}(1-\epsilon_i)\kappa_i\gamma_i^2\left(\bar{e}_{bi}^T\bar{e}_{bi} + \bar{e}_{bi}^T\bar{e}_{bi}\right) \end{aligned}$$

Since $\gamma_{l,i}$ is a free parameter, let $\gamma_{0i} = \gamma_i$ and $\gamma_{1i} = \gamma_i/\lambda_i$, $\kappa_i\gamma_i^2$ can be represented as $\kappa_i\gamma_i^2 = \frac{\kappa_i\gamma_{l,i}^2}{\max\{1, l_i/\lambda_i^2\}}$, then

$$
\begin{aligned}
&\langle \nabla V(\delta), A_{cl}^{11}\delta + A_{cl}^{12}\bar{e}\rangle \\
&\leq -\rho(\|\delta\|) - \sum_{i=1}^{N} \frac{1}{\kappa_i}\left(H_{1i}^2(\delta, \bar{e}_a, \bar{e}_b) + H_{2i}^2(\delta, \bar{e}_a, \bar{e}_b)\right) \\
&\quad + \sum_{i=1}^{N} \frac{(1-\epsilon_i)\kappa_i \gamma_{l_i i}^2}{\max\{1, l_i/\lambda_i^2\}} \left(W_i^2(\bar{e}_{ai}) + W_i^2(\bar{e}_{bi})\right),
\end{aligned}
$$

which shows that the condition (2.23) is satisfied, and $\rho(r) = \lambda_{\min}(I_N \otimes \sigma_0 Q)r$. this completes the proof. □

Next, we show that condition (2.24) holds with the constructed function $\tilde{W}_i$. First, if $l_i = 0$, from (2.25), $\tilde{W}_i = W_i(\bar{e}_{ai}) + W_i(\bar{e}_{bi})$. Due to $W_i(\bar{e}_{ai}) \geq 0$ and $W_i(\bar{e}_{bi}) \geq 0$, it is naturally that $\tilde{W}_i^2 \geq W_i^2(\bar{e}_{ai}) + W_i^2(\bar{e}_{bi})$, i.e., condition (2.24) holds. Then, if $l_i = 1$, condition (2.24) implies that $\tilde{W}_i^2 \geq \lambda_i^2(W_i^2(\bar{e}_{ai}) + W_i^2(\bar{e}_{bi}))$. From (2.25), when $\tilde{W}_i = \lambda_i(W_i(\bar{e}_{ai}) + W_i(\bar{e}_{bi}))$, it is obvious that condition (2.24) holds. When $\tilde{W}_i = W_i(\bar{e}_{ai}) + W_i(\bar{e}_{bi} + F(d_{k_i}^i)s_i)$, $\tilde{W}_i \geq \lambda_i(W_i(\bar{e}_{ai}) + W_i(\bar{e}_{bi}))$ also indicates that condition (2.24) holds. Thus, we can conclude that condition (2.24) holds with the constructed function $\tilde{W}_i$. Based on the results in this section, it can be seen that the explicit form of Lyapunov function can be constructed. Nevertheless, note that there may exist other form of Lyapunov function, i.e., the Lyapunov function design is not unique.

2.4.2 Stability Analysis Under Hybrid Systems Framework

Now, we are ready to state the main stability analysis results with respect to the hybrid system (with flow dynamics (2.13), jump dynamics (2.16), flow set (2.15), and jump set (2.17)). The stability will be shown by means of the designed Lyapunov function (2.19) in previous section, meanwhile, the upper bound of sampling period can be explicitly calculated. Next, based on the function $\phi_{l_i i}(\tau_i)$ in (2.19), the Assumptions 2.3–2.5, and the constructed Lyapunov function (2.19), we present the following main results.

Theorem 2.1 *Consider the hybrid system* (2.13) *and* (2.16) *satisfying Propositions* 2.1 *and* 2.2. *If there exist $h_i > 0$ and $0 < \bar{d}_i \leq h_i$, such that*

$$
\begin{aligned}
&\gamma_{0i}\phi_{0i}(h_i) \geq \lambda_i^2 \gamma_{1i}\phi_{1i}(0), \\
&\gamma_{1i}\phi_{1i}(\tau_i) \geq \gamma_{0i}\phi_{0i}(\tau_i), \text{ for } 0 \leq \tau_i \leq \bar{d}_i
\end{aligned}
\tag{2.36}
$$

hold for $\phi_{l_i i}(\tau_i)$ with initial conditions $\phi_{l_i i}(0) > 1$ and $\gamma_{1i}\phi_{1i}(0) \geq \gamma_{0i}\phi_{0i}(0)$, and λ_i is a constant as given in (2.20). *Then, the $(\delta, \bar{e}_a, \bar{e}_b)$ dynamics in the closed-loop state ξ is UGAS, which implies that leaderless/leader-following bipartite consensus is achieved.*

Proof Choose the Lyapunov function $U(\xi)$ as designed in (2.19). First, consider the evolution of $U(\xi)$ along flow dynamics of the hybrid system, that is $\langle \nabla U(\xi), \mathscr{F}(\xi)\rangle$ with $\mathscr{F}(\xi)$ given in (2.14). The arguments of $\tilde{W}_i$, H_{1i}, and H_{2i} are omitted for brevity. Based on Assumptions 2.3–2.5, for all $\tau \in \mathbb{R}^N$, $l \in \{0, 1\}^N$, and almost all $(\delta, \bar{e}_a, \bar{e}_b) \in \mathbb{R}^{nN} \times \mathbb{R}^{nN} \times \mathbb{R}^{nN}$, it obtains that

$$\begin{aligned}
&\langle \nabla U(\xi), \mathscr{F}(\xi)\rangle \\
&\le -\rho(\|\delta\|) - \sum_{i=1}^{N}\left(\frac{1}{\kappa_i}(H_{1i}^2(\delta, \bar{e}_a, \bar{e}_b) + H_{2i}^2(\delta, \bar{e}_a, \bar{e}_b)). - \frac{(1-\epsilon_i)\kappa_i\gamma_{l_i i}^2}{\max\{1, l_i/\lambda_i^2\}}(W_i^2(\bar{e}_{ai}) + W_i^2(\bar{e}_{bi}))\right) \\
&- \sum_{i=1}^{N}(2L_{l_i i}\gamma_{l_i i}\phi_{l_i i}(\tau_i)\tilde{W}_i^2 + \kappa_i\gamma_{l_i i}^2(\phi_{l_i i}^2 + 1)\tilde{W}_i^2) + \sum_{i=1}^{N} 2\gamma_{l_i i}\phi_{l_i i}(\tau_i)\tilde{W}_i(L_{l_i i}\tilde{W}_i + H_{1i}(\delta, \bar{e}_a, \bar{e}_b) + H_{2i}(\delta, \bar{e}_a, \bar{e}_b))
\end{aligned}$$

which further leads to

$$\begin{aligned}
&\langle \nabla U(\xi), \mathscr{F}(\xi)\rangle \\
&\le -\rho(\|\delta\|) - \sum_{i=1}^{N}(1-\epsilon_i)\kappa_i\gamma_{l_i i}^2\tilde{W}_i^2 - \sum_{i=1}^{N}\epsilon_i\kappa_i\gamma_{l_i i}^2\tilde{W}_i^2 + \sum_{i=1}^{N}\frac{(1-\epsilon_i)\kappa_i\gamma_{l_i i}^2}{\max\{1, l_i/\lambda_i^2\}}(W_i^2(\bar{e}_{ai}) + W_i^2(\bar{e}_{bi})) \\
&- \sum_{i=1}^{N}\left(\frac{H_{1i}(\delta, \bar{e}_a, \bar{e}_b)}{\sqrt{\kappa_i}} - \sqrt{\kappa_i}\gamma_{l_i i}\phi_{l_i i}(\tau_i)\tilde{W}_i\right)^2 - \sum_{i=1}^{N}\left(\frac{H_{2i}(\delta, \bar{e}_a, \bar{e}_b)}{\sqrt{\kappa_i}} - \sqrt{\kappa_i}\gamma_{l_i i}\phi_{l_i i}(\tau_i)\tilde{W}_i\right)^2 \\
&\overset{(2.24)}{\le} -\rho(\|\delta\|) - \sum_{i=1}^{N}\sigma_i(\tilde{W}_i).
\end{aligned} \tag{2.37}$$

where the function $\sigma_i(r) = \epsilon_i\kappa_i\gamma_{l_i i}^2 r^2$.

Then, consider the change of $U(\xi)$ when $\xi \in \mathbb{J}$, without loss of generality, for a certain agent i, there exist two types of jump dynamics. Specifically, according to the hybrid model of MAS, there exist two cases:

①: $\xi \in \mathbb{J}_i \wedge l_i = 0$

According to the hybrid system's jump dynamics (2.16), $l_i = 0$ indicates that $\mathscr{J}_i(\xi) = \mathscr{J}_{0i}(\xi)$, then, based on the expression of $\mathscr{J}_{0i}(\xi)$, we have

$$\begin{aligned}
&U(\xi^+) - U(\xi) \\
&= \gamma_{l_i^+ i}\phi_{l_i^+ i}(\tau_i^+)\tilde{W}_i^2(\bar{e}_{ai}^+, \bar{e}_{bi}^+, s_i^+, l_i^+) - \gamma_{l_i i}\phi_{l_i i}(\tau_i)\tilde{W}_i^2(\bar{e}_{ai}, \bar{e}_{bi}, s_i, l_i) \\
&= \gamma_{1i}\phi_{1i}(0)\tilde{W}_i^2(0, \bar{e}_{bi}, -\bar{e}_{bi}, 1) - \gamma_{0i}\phi_{0i}(h_i)\tilde{W}_i^2(\bar{e}_{ai}, \bar{e}_{bi}, s_i, 0) \\
&\overset{(2.20)}{\le} (\lambda_i^2\gamma_{1i}\phi_{1i}(0) - \gamma_{0i}\phi_{0i}(h_i))\tilde{W}_i^2(\bar{e}_{ai}, \bar{e}_{bi}, s_i, 0) \\
&\overset{(2.36)}{\le} 0.
\end{aligned} \tag{2.38}$$

②: $\xi \in \mathbb{J}_i \wedge l_i = 1$

According to the hybrid system's jump dynamics (2.16), $l_i = 1$ indicates that $\mathscr{J}_i(\xi) = \mathscr{J}_{1i}(\xi)$, then, based on the expression of $\mathscr{J}_{1i}(\xi)$, we have

$$
\begin{aligned}
&U(\xi^+) - U(\xi) \\
&= \gamma_{0i}\phi_{0i}(\tau_i)\tilde{W}_i^2(\bar{e}_{ai}, \bar{e}_{bi} + F(d_{k_i}^i)s_i, 0, 0) - \gamma_{l_i i}\phi_{l_i i}(\tau_i)\tilde{W}_i^2(\bar{e}_{ai}, \bar{e}_{bi}, s_i, 1) \\
&\overset{(2.20)}{\leq} (\gamma_{0i}\phi_{0i}(\tau_i) - \gamma_{1i}\phi_{1i}(\tau_i))\tilde{W}_i^2(\bar{e}_{ai}, \bar{e}_{bi}, s_i, 1) \\
&\overset{(2.36)}{\leq} 0.
\end{aligned}
\tag{2.39}
$$

From (2.37) to (2.39), it can be seen that the Lyapunov function $U(\xi)$ decreases along ξ's flow dynamics and jump dynamics. Then, following standard Lyapunov theory of hybrid systems, see e.g., [11], it is known that UGAS of the hybrid system is achieved. In other words, leaderless or leader-following bipartite consensus is achieved. This completes the proof. □

The proposed approach can be applied to MAS with asynchronous sampling mechanism and heterogeneous time delays, that is rarely considered simultaneously in the existing results. For instance, the consensus problem of MAS with synchronous samplings and homogeneous communication delays has been widely studied in [1, 3–6]. It's worth to be noted that, with the proposed hybrid system approach, if set the sampling periods h_i and let communication delays $d_{k_i}^i$ be the same for all agents, the obtained results of this chapter can straightforwardly recover the aforementioned existing results, i.e., synchronous samplings and homogeneous communication delays are special cases of asynchronous sampling mechanism and heterogeneous time delays. The asynchronous samplings associated with heterogeneous delays induce closed-loop system model with multiple state updates, that makes the existing results be inapplicable. The proposed hybrid system approach incorporates the sampling periods h_i and communication delays $d_{k_i}^i$ into the flow/jump sets, such that the heterogeneous delays induced modelling and stability analysis problems can be tackled under hybrid systems framework.

2.4.3 Parameter Design

According to the analysis and design method in this chapter, to achieve the objective of bipartite consensus under network environment with sampled and delay information transmission, there exist some parameters that should be selected appropriately. The main design parameters are the control gain K and the asynchronous sampling period h_i for each agent. It is obvious that $h_i > 0$, but it cannot be determined arbitrarily, i.e., there exists an upper bound $\bar{h}_i$ for h_i, how to determine $\bar{h}_i$ should be considered. Moreover, in order to determine the control gain K and $\bar{h}_i$, some parameters related to stability of closed-loop system are also needed to be designed, such as γ_{0i}, γ_{1i}, $\phi_{0i}(0)$, $\phi_{1i}(0)$, λ_i, and the maximum allowable time delay $\bar{d}_i$.

First, the control gain K in control laws (2.4) and (2.5) is the same for all agents. As shown in Proposition 2.2, it can be designed as $K = \mu B^T P$, where the positive constant μ and positive definite matrix P are obtained by solving Riccati equation

(2.34). Note that it is easy to find solutions μ and P for the Riccati equation, but one more condition, as given in (2.35), should be satisfied. That is, μ and P should be properly designed such that the matrix inequality condition (2.35) is satisfied.

Second, since asynchronous sampled-data based consensus control is considered, one important parameter that should be properly designed is the sampling period h_i. It is obvious that h_i should be upper bounded, thus, the key is to calculate its upper bound. The maximum allowable sampling period $\bar{h}_i$ (i.e., upper bound of h_i) can be calculated as follows

$$\bar{h}_i = \begin{cases} \dfrac{1}{L_{0i} p_{1i}} \arctan \dfrac{p_{1i}(1-\phi_{0i}^{-2}(0))}{1+p_{2i}}, & \text{If } \Delta_i > 0, \\ L_{0i}^{-1} \dfrac{1-\phi_{0i}^{-1}(0)}{1+\phi_{0i}^{-1}(0)}, & \text{If } \Delta_i = 0, \\ \dfrac{1}{L_{0i} p_{1i}} \operatorname{arctanh} \dfrac{p_{1i}(1-\phi_{0i}^{-2}(0))}{1+p_{2i}}, & \text{If } \Delta_i < 0, \end{cases} \tag{2.40}$$

where $p_{1i} = \sqrt{\|(\gamma_{0i}/L_{0i})^2 - 1\|}$, $p_{2i} = \phi_{0i}^{-2}(0) - 2\phi_{0i}^{-1}(0)L_{0i}^{-1}$, $\Delta_i = \kappa_i\gamma_{0i} - L_{0i}$. To calculate $\bar{h}_i$ in the above form, we can select $\phi_{0i}(0) > 1$ and assume that $\phi_{0i}(\bar{h}_i) = \phi_{0i}^{-1}(0) > 0$. According to the dynamics of $\phi_{0i}(\tau_i)$ in (2.19), it follows that $\bar{h}_i = -\int_{\phi_{0i}(0)}^{\phi_{0i}^{-1}(0)} \frac{d\phi_i}{2L_{0i}\phi_{0i}+\kappa_i\gamma_{0i}(1+\phi_{0i}^2)}$. Define $v_i = \phi_{0i} + \frac{L_{0i}}{\gamma_{0i}\kappa_i}$, we have $\bar{h}_i = -\int_{v_{1i}}^{v_{2i}} \frac{dv_i}{v_i^2 - \operatorname{sgn}(L_{0i}^2 - \kappa_i^2\gamma_{0i}^2)(L_{0i}p_{1i}\gamma_{0i}^{-1}\kappa_i^{-1})^2}$, where $v_{1i} = \phi_{0i}(0) + L_{0i}\gamma_{0i}^{-1}\kappa_i^{-1}$ and $v_{2i} = \phi_{0i}^{-1}(0) + L_{0i}\gamma_{0i}^{-1}\kappa_i^{-1}$, and $\operatorname{sgn}(\cdot)$ denotes the sign function. First, consider $\gamma_{0i}\kappa_i > L_{0i}$, by some routine integral operations, $\bar{h}_i = \frac{1}{L_{0i}p_{1i}} \arctan \frac{p_{1i}(1-\phi_{0i}^{-2}(0))}{1+p_{2i}}$. The other two cases with $\gamma_{0i}\kappa_i = L_{0i}$ and $\gamma_{0i}\kappa_i < L_{0i}$ can also be calculated similarly. Hence, if $\bar{h}_i$ is calculated using (2.40), it is ensured that $\phi_{0i}^{-1}(0) \le \phi_{0i}(\tau_i) \le \phi_{0i}(0)$, and an explicit upper bound of the sampling period is obtained. Then, the sampling period h_i can be selected arbitrarily within the range $(0, \bar{h}_i]$.

Third, the parameters γ_{0i} and γ_{1i} are usually selected to be positive and satisfy $\gamma_{0i} = \lambda_i\gamma_{1i}$. With this relationship, the conditions in (2.36) can be simplified as $\phi_{0i}(h_i) \ge \lambda_i\phi_{1i}(0)$ and $\phi_{1i}(\tau_i) \ge \lambda_i\phi_{0i}(\tau_i)$ for $0 \le \tau_i \le \bar{d}_i$, which indicates that $\phi_{0i}(h_i) \ge \lambda_i\phi_{1i}(0) \ge \lambda_i^2\phi_{0i}(0)$ should be satisfied. Thus, consider the initial conditions $\phi_{0i}(0)$ and $\phi_{1i}(0)$, first, select $\phi_{0i}(0)$ and λ_i such that $\phi_{0i}(h_i) \ge \lambda_i^2\phi_{0i}(0)$ holds, based on (2.40) and its proof, this can be achieved by selecting $\phi_{0i}(0) = \lambda_i^{-1}$ with $0 < \lambda_i < 1$, then the initial condition $\phi_{1i}(0)$ can be selected as $1 \le \phi_{1i}(0) \le \phi_{0i}(h_i)/\lambda_i$. Moreover, consider the condition $\phi_{1i}(\tau_i) \ge \lambda_i\phi_{0i}(\tau_i)$ for $0 \le \tau_i \le \bar{d}_i$, on one hand, it is possible that $\phi_{1i}(\tau_i) \ge \lambda_i\phi_{0i}(\tau_i)$ for $0 \le \tau_i \le h_i$, in this case, we have $\bar{d}_i = h_i$. On the other hand, if there exists a T_i such that $\phi_{1i}(\tau_i) \ge \lambda_i\phi_{0i}(\tau_i)$ for $0 \le \tau_i \le T_i < h_i$, it indicates that $\bar{d}_i = T_i$. Thus, there exists tradeoff between h_i and $\bar{d}_i$ for different parameters $L_{r_i i}$, λ_i, $\phi_{0i}(0)$ and $\phi_{1i}(0)$. If these parameters are given, the explicit MASP can be calculated based on (2.40), and the explicit MAD can be calculated based on (2.19). In the following section, tradeoff curves are shown to clarify this relationship.

The parameter design has been discussed in this section. In order to make the design be clear to readers, we present the following **design procedure**.

Step 1: Choose parameters $\mu > 0$, $\gamma_{0i} > 0$, $\gamma_{1i} > 0$, $0 < \lambda_i < 1$, $\kappa_i > 0$, $\varepsilon_i > 0$, $L_{0i} > 0$, $\sigma_{0i} > 0$, matrix $Q > 0$, initial conditions $\phi_{0i}(0) > 1$, $\phi_{1i}(0) > 0$.

Step 2: Solve the Riccati equation (2.34) to get P and let $K = \mu B^T P$. Verify the feasibility of the matrix inequality (2.35). If it is infeasible, goto *Step 1* and change the values of μ and P. Otherwise, go to next step.

Step 3: Calculate the upper bound $\bar{h}_i$ of the sampling period h_i according to (2.40), and determine the value of h_i within the range $(0, \bar{h}_i]$.

Step 4: Determine the maximum allowable delay $\bar{d}_i$ such that $0 < \bar{d}_i < \bar{h}_i$. Then, verify if the inequality conditions in (2.36) hold. If these conditions are satisfied, the design procedure is end. If they are not satisfied, change the values of h_i and $\bar{d}_i$ and goto *Step 3*.

After the design procedure, it is ensured that the hybrid system consisting of (2.13) and (2.16) is asymptotically stable. In other words, for any initial conditions of agents' states, the objective of bipartite consensus of MAS under sampled and delay information is achieved. Based on the obtained results and the proposed hybrid system approach in this chapter, the readers can learn about how to analyze and design the MAS under asynchronous sampling communication and transmission delays, and interested readers can try to generalize the proposed approach to more general MAS, e.g., the dynamics of agent include Lipschitz nonlinearities.

2.5 Simulation Study

First, consider a group of agents without leaders, the model of each agent has the form of (2.1), specifically, the system matrices are given by $A = [0\ 1; -1\ 0]$, $B = [1\ 1]^T$. The agents connect with each other via wireless communication network, the communication topology relationship is shown as in Fig. 2.2. The six agents are divided into two groups, that is $\mathcal{V}_1 = \{1, 2\}$, $\mathcal{V}_2 = \{3, 4, 5, 6\}$. From Fig. 2.2 and based on the concept of communication graph, it can be derived that the Laplacian matrix $\mathcal{L} = [3\ -1\ 2\ 0\ 0\ 0; -1\ 5\ 4\ 0\ 0\ 0; 2\ 4\ 8\ -2\ -1\ 0; 0\ 0\ -2\ 6\ 0\ 3; 0\ 0\ 0\ -1\ 1\ 0; 0\ 0\ 0\ -3\ 0\ 3]$, and the gauge transformation matrix is $M = \mathrm{diag}\{1, 1, -1, -1, -1, -1\}$.

Fig. 2.2 Communication topology

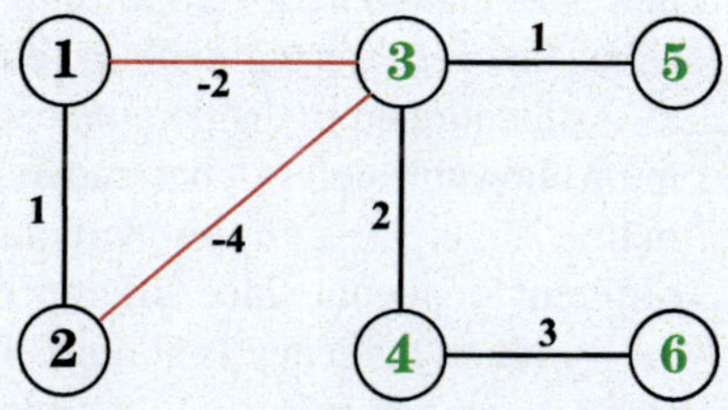

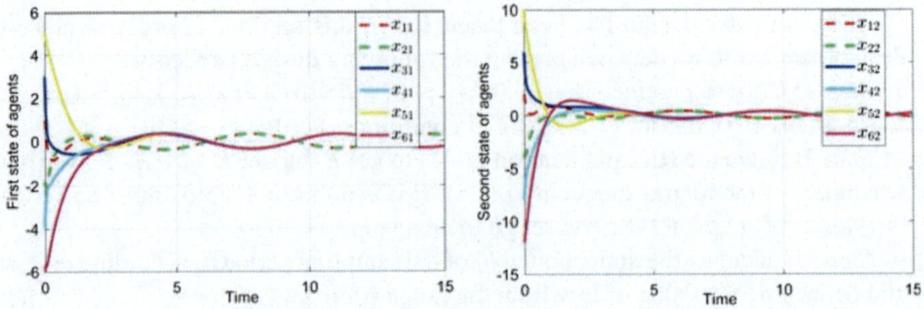

Fig. 2.3 Agents' states of leaderless MAS. States of agent 1–2 are denoted by *dash lines*. States of agent 3–6 are denoted by *dot dash lines*

According to the design procedure in Sect. 2.4.3, choose the $\mu = 0.15$ and matrix $Q = 10I$, and note that $\hat{\mathscr{L}} = \mathscr{L}_M = M\mathscr{L}M$ and $a(\hat{\mathscr{L}}) = 0.798$, by solving the Riccati equation (2.34), it can be computed that the control gain $K = [1.3041\ 0.4232]$. L_{0i} can be calculated as $L_{0i}|i \in \{1, 2, 3, 4, 5, 6\} = 4.5358, 7.6162, 12.3202, 9.1788, 3.6786, 4.5358$. For the internal variable ϕ_{0i}, choose initial conditions $\phi_{0i}(0)|i \in \{1, 2, 3, 4, 5, 6\} = 2, 4, 7, 7, 4, 3$, and let $\lambda_i = 1/\phi_{0i}(0)$. The rest parameters are selected as $\gamma_{01} = \gamma_{02} = 9, \gamma_{03} = \gamma_{04} = 8, \gamma_{05} = \gamma_{06} = 12, \gamma_{1i} = \gamma_{0i}/\lambda_i, \kappa_i = 0.8$, $L_{0i}|i \in \{1, 2, 3, 4, 5, 6\} = 4.5, 7.6, 12.3, 9.1, 3.6, 4.5$, $L_{1i} = L_{0i}/\lambda_i$. According to (2.40), the upper bound of sampling period can be computed as $\bar{h}_1 = 0.069$, $\bar{h}_2 = 0.075$, $\bar{h}_3 = 0.082$, $\bar{h}_4 = 0.064$, $\bar{h}_5 = 0.088$, $\bar{h}_6 = 0.071$. Based on the proposed method, it is known that as long as $h_i < \bar{h}_i$, the closed-loop stability can be ensured. Here, for simplicity, we choose the sampling periods 0.04 for agents 1–3 and 0.05 for agents 4–6. According to the design procedure in Sect. 2.4.3, choose $\phi_{1i}(0) = \phi_{0i}(h_i)/\lambda_i$ such that the conditions in (2.36) hold. To perform the simulation, the initial conditions of agents can be arbitrarily selected, here, set $x_i(0) = (-1)^{i-1}[1\ 2]^T \times i$, $i \in [1, 6]$. Then, with the designed parameters and selected initial conditions, we perform the simulation in MATLAB. The asynchronous sampling can be realized using 'Zero-Order Hold' module, the time-varying delay can be realized using 'Variable Time Delay' module. With the constructed closed-loop MAS model in MATLAB, the state trajectories of agents are shown in Fig. 2.3, which shows that the states of agents in the same group converge to the same value, i.e., these agents work cooperatively. In addition, the states of the rest agents that have antagonistic relationship converge to the same modulus but with opposite signs, that is consistent with the concept of bipartite consensus.

As mentioned in the parameter design section, the sampling period h_i and maximum allowable delay $\bar{d}_i$ has tradeoff relationship. To clearly shown such relationship, in Fig. 2.4, h_i versus $\bar{d}_i$ tradeoff curves are shown. From Fig. 2.4, tradeoff curves are different for agents, since different design parameters are chosen for agents. However, the variation tendency is similar. That is, when the sampling period h_i is relatively small, neighbor agents can get more information of a certain agent, the maximum allowable delay $\bar{d}_i$ can be the same as h_i. As the sampling period h_i becomes larger,

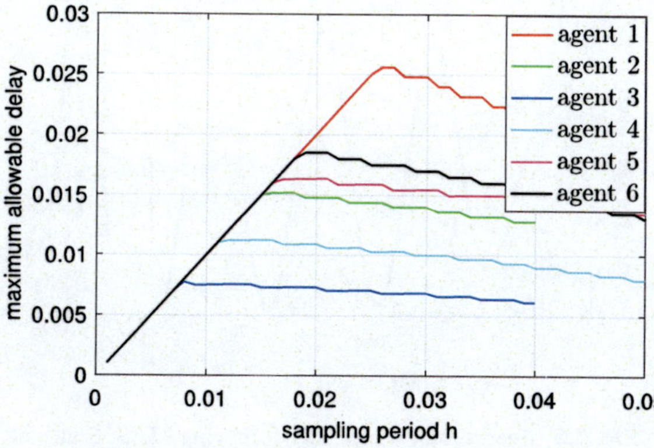

Fig. 2.4 Tradeoff curves of h_i versus $\bar{d}_i$

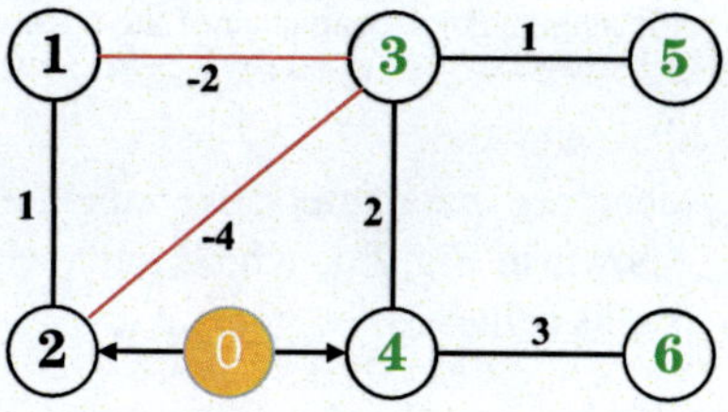

Fig. 2.5 Communication topology

neighbor agents get less and less information of a certain agent, in this case, in order to maintain system stability, the transmitted information is not allowed to be delayed too much, so, the maximum allowable delay $\bar{d}_i$ becomes smaller. This tradeoff curves can give us some illustrations for practical applications, that is, according to the operation environment and possibly appeared time delay, we can select appropriate sampling period on the basis of this tradeoff relationship.

Then, on the basis of the previous considered leaderless MAS, we further consider that there exists a leader in the MAS. In the leader-following MAS, the follower agents' dynamics is the same as before, and the leader's dynamics has the form of (2.2). With the leader agent, the communication relationship between agents is shown in Fig. 2.5, the leader agent 0 is only connected to agent 2 and agent 4. In this case, according to the communication graph, it can be calculated that $\hat{\mathscr{L}} = \mathscr{H}_M$ and $a(\hat{\mathscr{L}}) = 0.286$. Using the same parameters as in leaderless case, by solving the Riccati equation (2.34), we get the control gain matrix $K = [2.0835\ 0.9506]$. L_{0i} is the same as in leaderless case. For the internal variable ϕ_{0i}, the initial conditions are set to be $\phi_{0i}(0)|i \in \{1, 2, 3, 4, 5, 6\} = 6, 6, 9, 9, 12, 12$, and let $\lambda_i = 1/\phi_{0i}(0)$. The rest parameters are selected to be the same as in leaderless case. According to (2.40), it is calculated that $\bar{h}_1 = 0.0886$, $\bar{h}_2 = 0.0689$, $\bar{h}_3 = 0.0635$, $\bar{h}_4 = 0.0729$, $\bar{h}_5 = 0.097$, $\bar{h}_6 = 0.0898$, here, we choose the sampling periods 0.05 and 0.04 for agents 1–3 and agents 4–6 respectively. The initial states of agents are set to be $x_i(0) = (-1)^{i-1}[1\ 2]^T \times i,\ i \in [1, 6]$, $x_0(0) = [5\ \ -3]^T$. The simulation is performed in MATLAB, on the basis of the simulation of leaderless MAS, an extra leader agent model is constructed and added in the simulation module. With the

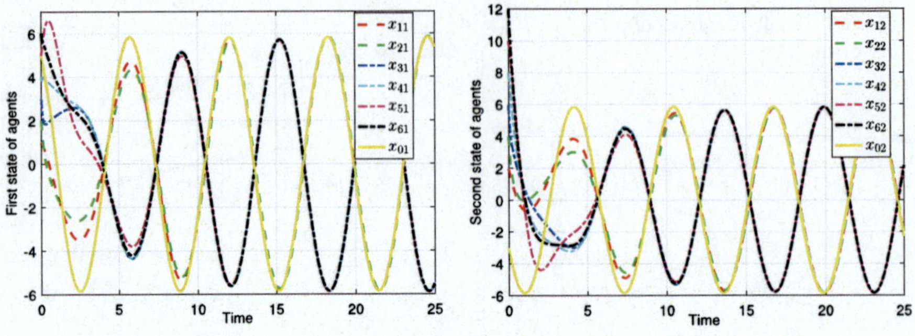

(a) (b)

Fig. 2.6 Agents' states of leader-following MAS. States of agent 1–2 and agent 3–6 are denoted by *dash lines* and *dot dash lines* respectively. Yellow lines denote leader's states. **(a)** first state of the agents; **(b)** second state of the agents

designed parameters, we run the simulation, the state trajectories of agents are shown in Fig. 2.6, which demonstrates that the leader-following bipartite consensus is achieved.

2.6 Conclusion

In this chapter, we developed a hybrid systems approach for bipartite consensus of linear MAS with asynchronous sampled and heterogeneous delayed data, both the leaderless and leader-following bipartite consensus problems are transformed into stability problem of a hybrid system. Lyapunov-based stability results and how to construct Lyapunov function are shown. The tradeoff curves between MASP and MAD can be obtained with the proposed approach. Simulation results are provided to demonstrate the obtained results.

References

1. Xing, M., Deng, F., Hu, Z.: Sampled-Data Consensus for Multiagent Systems With Time Delays and Packet Losses. IEEE Trans. Syst. Man Cybern. -Syst. **50**(1), 203–210 (2020)
2. Deng, C., M. J. Er., Yang, G., Wang, N.: Event-Triggered Consensus of Linear Multiagent Systems With Time-Varying Communication Delays. IEEE Trans. Cybern. **50**(7), 2916-2925 (2020)
3. Su, H., Zhang, J., Chen, X.: A Stochastic Sampling Mechanism for Time-Varying Formation of Multiagent Systems With Multiple Leaders and Communication Delays. IEEE Trans. Neural Netw. Learn. Syst. **30**(12), 3699-3707 (2019)
4. Ding, L. and Han, Q,-L., Guo, G.: Network-based leader-following consensus for distributed multi-agent systems. Automatica. **49**(7), 2281-2286 (2013)

5. Jiang, W., Chen, Y. and T. Charalambous.: Consensus of General Linear Multi-Agent Systems With Heterogeneous Input and Communication Delays. IEEE Control Syst. Lett. **5**(3), 851-856 (2020)
6. Qian, W., Gao, Y., Yang, Y.: Global Consensus of Multiagent Systems With Internal Delays and Communication Delays. IEEE Trans. Syst. Man Cybern. -Syst. **49**(10), 1961-1970 (2019)
7. Kaviarasan, B., Sakthivel, R., Wang, C., Alzahrani., F.: Resilient control design for consensus of nonlinear multi-agent systems with switching topology and randomly varying communication delays. Neurocomputing **311**(3), 155-163 (2018)
8. Heijmans, S.H.J., Postoyan, R., Nesic, D., Heemels., W.P.M.H.: Computing Minimal and Maximal Allowable Transmission Intervals for Networked Control Systems Using the Hybrid Systems Approach. IEEE Control Syst. Lett. **1**(1), 56-61(2017)
9. Heemels, W.P.M.H., Teel, A.R., Wouw, N. van de, Nešić., D.: Networked control systems with communication constraints: Tradeoffs between transmission intervals, delays, and performance. IEEE Trans. Autom. Control. **55**(8), 1781-1796 (2010)
10. Dolk, V. S., Borgers, D. P., Heemels., W. P. M. H.: Output-Based and Decentralized Dynamic Event-Triggered Control With Guaranteed $\mathcal{L}_p$-Gain Performance and Zeno-Freeness. IEEE Trans. Autom. Control. **62**(1), 34-49 (2017)
11. Goebel, R., Sanfelice, R. G., Teel., A. R.: Hybrid Dynamical Systems: Modeling, Stability, Robustness. Princeton Univ. Press. (2012)
12. Ning, B., Han, Q,-L., Zuo, Z.: Bipartite Consensus Tracking for Second-Order Multiagent Systems: A Time-Varying Function-Based Preset-Time Approach. IEEE Trans. Autom. Control. **66**(6), 2739-2745 (2020)
13. Altafini, C.: Consensus problems on networks with antagonistic interactions. IEEE Trans. Atuom. Control. **58**(4), 935-946 (2013)
14. Zhao, G., Hua, C.: Leaderless and Leader-Following Bipartite Consensus of Multi-Agent Systems with Sampled and Delayed Information. IEEE Trans. Neural. Netw. Learn. Syst. https://doi.org/10.1109/TNNLS.2021.3106015 (2021)

Chapter 3
Sampled-Data Consensus Control of MAS with Packet Losses

3.1 Introduction

In previous Chap. 2, the problem of time delay has been considered for the sampled-data control of MAS. It is formally proved that sampled-data control could have excellent tolerance and robustness to this typical network-induced issue. However, to only consider the problem of time delay is far from adequate, since as aforementioned in Chap. 1, the long-time delay may contribute to another type of network-induced issue, namely, the packet losses problem, in the sense that the information arrived after a long delayed time period is outdated and is not suitable for the current implementation of control, thus it should be discarded proactively in general, which implies that the packet losses problem arises. In addition, the packet disordering induced by time delay may also lead to packet losses. However, it should be noted that packet loss problem is essentially different from the pure time delay problem considered in Chap. 2, because with time delays, the transmitted information can always be received eventually despite outdated, while with packet losses, the transmitted information is missed regardless of proactively or reactively. And, we remark that the caused reasons for packet losses are various in general, such as, time delay, limited bandwidth, noise interference, cyber attack, etc., and the control algorithm designed for addressing packet losses sometimes is feasible to address the time delay problem to some degree (by dropping delayed data), but not vice versa. Thus, the packet losses problem belongs to another type of network-induced issue, which deserves a further investigation. For the sampled-data control subject to packet loss, various feasible solutions have been proposed in the currently available literature [1–5]. But most of them adopted such an assumption that the packet losses in all communication channels are *identical* (i.e., packet losses in all communication channels occur at the same time). This assumption is somewhat conservative in practical circumstance, because the information in a MAS network may be transmitted along different communication channels at the same time. Therefore, it is more realistic to consider the case that packet losses in each communication channel from agent i to agent j is independent of other channels (i.e., *nonidentical* packet losses), which can

G. Zhao et al., *Cooperative Control of Multi-agent Systems*,
https://doi.org/10.1007/978-981-97-0968-7_3

cover the identical packet losses as a special case. This poses many technical challenges to the model construction and stability analysis of sampled-data consensus control.

In this chapter, we address the sampled-data leaderless and leader-following consensus problems of MASs under nonidentical packet losses in a unified hybrid framework. The packet losses are considered to be non-identical, which means that packet losses in each communication channel can be independent of the other channels. This is essentially different from the existing results [1, 3–5], where identical packet losses were considered. Then, a hybrid systems approach is proposed to address the considered consensus problem, specifically, a unified hybrid model is constructed to describe the closed-loop MAS under nonidentical packet losses, such that the leaderless and leader-following consensus problems can be dealt with in a unified framework. Explicit relationship between sampling period and the number of maximum allowable successive packet losses (NMASPL) is given. Finally, simulation example is presented to verify the effectiveness of the obtained results.

3.2 Networked MAS with Packet Losses

This section formulates the leaderless and leader-following consensus problems for the sampled-data based MASs with non-identical packet losses. The non-identical assumption on packet losses is a less conservativeness condition, which actually allows different communication channels in a MAS network to have non-synchronized dropouts. The system dynamics of the considered MAS is depicted in the state space expression form

$$\dot{x}_i(t) = Ax_i(t) + Bu_i(t), \tag{3.1}$$

where $x_i(t) \in \mathbb{R}^n$, $u_i(t) \in \mathbb{R}^p$, denotes the system state and control input of the follower agents $i \in [1, N]$. A, B are matrices with appropriate dimensions and it is assumed that (A, B) is stabilizable. If there is a leader in the MAS, then, the leader can be labeled as 0, and its dynamics is described by

$$\dot{x}_0(t) = Ax_0(t), \tag{3.2}$$

where $x_0(t) \in \mathbb{R}^n$ is the leader's state.

The preliminaries about graph, which is used to describe the communication relationship between agents, has been introduced in Chap. 1. For proceeding this chapter, we make Assumption 3.1 to limit the communication connection among agents, and Assumption 3.2 to limit the number of maximum allowable successive packet losses (NMASPL).

Assumption 3.1 In the leaderless case, the communication graph $\mathcal{G}$ is connected and undirected. In leader-following case, at least one follower can receive information from the leader, i.e., there exists a spanning tree in the graph $\mathcal{G}$ with the leader agent 0 as the root.

Assumption 3.2 For the communication channel from agent j to agent i, the number of successive packet losses since the most recent successful transmission instant is upper bounded by $\rho_{\max}^{ij} \in \mathbb{N}_{>0}$, which is called the NMASPL.

In the presence of packet losses, Assumption 3.2 is needed to ensure desired stability property of closed-loop system, which is a reasonable assumption and also appears in [6–10]. Based on Assumption 3.1, we can obtain a useful technical lemma as follows:

Lemma 3.1 ([11]) *For connected undirected graph $\mathcal{G}$, if a vector $x \in \mathbb{R}^N$ satisfies $\mathbf{1}_N^T x = 0$, we have*

$$\min_{x \neq 0} \frac{x^T \hat{\mathcal{L}} x}{x^T x} = a(\hat{\mathcal{L}}),$$

where $\hat{\mathcal{L}} = \mathcal{L}$ in leaderless case and $\hat{\mathcal{L}} = \mathcal{H}$ in leader-following case, $a(\hat{\mathcal{L}})$ denotes the minimum positive eigenvalue of $\hat{\mathcal{L}}$.

As discussed in the Introduction, information packets being transmitted in the network may be randomly lost due to the unreliable communication networking environment. In a comprehensive MAS network, there are amount of communication channels allowing the simultaneous information transmissions from multiple nodes. Sometimes, these communication channels inevitably are damaged in part or in total, due to various uncertain factors including time delay, limited bandwidth, noise interference, cyber attack. It is more attractive to consider a more realistic case that the packet loss for each communication channel from agent i ($i \in [1, N]$) to agent j, $j \in \mathcal{N}_i$ can be different from each other (i.e., nonidentical packet losses). In what follows, Fig. 3.1 is presented to illustrate the case of nonidentical packet losses. Specially, in Fig. 3.1, when agent i broadcasts information to its neighbors at t_k^i, it is possible that the information transmitted in channel ① is lost, but is not lost in channels ② and ③. Then, at next sampling instant t_{k+1}^i, it is possible that the information transmitted in channel ① and ② is not lost, but is lost in channel ③. Such situation is more general in practical applications than the existing results [1, 3, 4].

Given the introduced nonidentical packet losses problem, our immediate concern is to appropriately describe the case with proper mathematical expressions, and then design corresponding consensus protocols for the leaderless MAS and leader-following MAS respectively, to achieve leaderless consensus in the sense that $\lim_{t\to\infty} \|x_i(t) - x_j(t)\| = 0$, $i, j \in [1, N]$, and to achieve leader-following consensus in the sense that $\lim_{t\to\infty} \|x_i(t) - x_0(t)\| = 0$, $i \in [1, N]$.

To this end, we first introduce an auxiliary binary variable m_{ij}, and relates its value to whether there exist packet losses in the communication channel from agent j to agent i. That is, by letting $m_{ij} = 1$, we denote the no packet loss case that the information transmission from agent j to agent i is successful, and by letting

Fig. 3.1 Diagrammatic sketch of packet losses in MAS

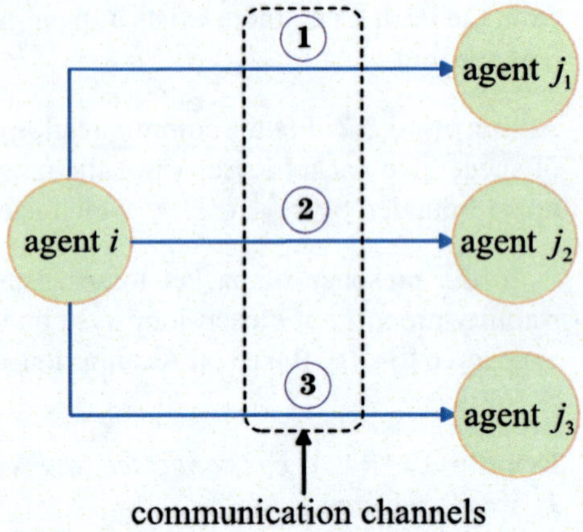

$m_{ij} = 0$, we denote the packet loss case that information transmission fails. Denote $\hat{x}_{ij}(t)$ as the estimation of agent j's state based on sampled-data, its dynamics is given by

$$\begin{aligned}\dot{\hat{x}}_{ij}(t) &= A\hat{x}_{ij}(t), t \in (t_k^j, t_{k+1}^j],\\ \hat{x}_{ij}(t^+) &= \begin{cases} x_j(t), t = t_k^j \wedge m_{ij} = 1 \\ \hat{x}_{ij}(t), t = t_k^j \wedge m_{ij} = 0 \end{cases}\end{aligned} \tag{3.3}$$

in which the estimation $\hat{x}_{ij}(t)$ is updated to be the newly received $x_j(t)$ when $m_{ij} = 1$, otherwise, $\hat{x}_{ij}(t)$ is not updated and remains unchanged. t_k^j denotes the kth sampling instant of agent j, $\wedge$ denotes logical operator 'and'. Moreover, it should be noted that m_{ij} is just a flag to denote whether the packet loss occurs or not, it is not needed in practice when implementing the estimator (3.3).

Then, based on the well-defined estimator (3.3), according to whether there exists leader agent or not, we design the control protocols as

$$u_i(t) = K \sum_{j \in \mathcal{N}_i} a_{ij}(\hat{x}_{ij}(t) - x_i(t)), \tag{3.4}$$

and

$$u_i(t) = K \left(\sum_{j \in \mathcal{N}_i} a_{ij}(\hat{x}_{ij}(t) - x_i(t)) + b_i(x_0(t) - x_i(t)) \right), \tag{3.5}$$

for leaderless MAS and leader-following MAS respectively, where K is the gain matrix to be designed.

3.3 Hybrid Model of MAS

In last section, we have analyzed the sampled-data based packet losses problem in detail, modeled the packet losses in mathematical form correctly, and proposed the consensus protocols for leaderless and leader-following MASs. But, how to constitute the closed-loop state, and reformulate a suitable hybrid system for stability analysis is still uncompleted, thus in this section, the MAS dynamics in leaderless or leader-following case is first derived. Then, by defining appropriate consensus error variables and exploiting property of error dynamics between sampled state and real state, the closed-loop system is reformulated as a hybrid system, which has a stronger description ability in the sense that flow dynamics and jump dynamics can be contained in modelling.

3.3.1 Dynamics of Leaderless MAS

For the leaderless MAS, define the combinational consensus error variable and the measurement error between estimated state and real state as $\delta_i = \sum_{j=1}^{N} a_{ij}(x_j - x_i)$ and $e_{ij} = \hat{x}_{ij} - x_j$. Then, from (3.1), the system dynamics can be represented as

$$\begin{aligned}
\dot{x}_i &= Ax_i + BK(\delta_i + (\mathscr{A}_r^i \otimes I_n)e_i) \\
\dot{\delta}_i &= A\delta_i - BK(\mathscr{L}_r^i \otimes I_n)(\delta + \mathscr{A}_d e) \\
\dot{e}_{ij} &= Ae_{ij} - BK(\delta_j + (\mathscr{A}_r^j \otimes I_n)e_j)
\end{aligned}$$

and

$$\begin{aligned}
\dot{\delta} &= (I_N \otimes A - \mathscr{L} \otimes BK)\delta - (\mathscr{L} \otimes BK)\mathscr{A}_d e \\
\dot{e} &= (I_{N^2} \otimes A)e - \mathbf{1}_N \otimes ((I_N \otimes BK)(\delta + \mathscr{A}_d e))
\end{aligned} \tag{3.6}$$

where

$$\begin{aligned}
\delta &= \left[\delta_1^T \ \delta_2^T \ \cdots \ \delta_N^T\right]^T, \ e = \left[e_1^T \ e_2^T \ \cdots \ e_N^T\right]^T, \\
e_i &= \left[e_{i1}^T \ e_{i2}^T \ \cdots \ e_{iN}^T\right]^T, \ \mathscr{A}_d = \mathrm{diag}_{i \in [1,N]}\{\mathscr{A}_r^i \otimes I_n\}.
\end{aligned} \tag{3.7}$$

From the definition of (3.3), it can be known that $\hat{x}_{ij}$ has jump dynamics at the successful transmission instants, that is

$$e_{ij}^+(t) = \begin{cases} 0, & t = t_k^j \ \wedge \ m_{ij} = 1 \\ e_{ij}(t), & t = t_k^j \ \wedge \ m_{ij} = 0 \end{cases} \tag{3.8}$$

Thus, the system dynamics of MAS with sampled-data and packet losses in leaderless case can be described by (3.6) and (3.8). The mathematical model can describe the case that all communication channels have independent packet losses, which is realistic under network environment, and it does not need acknowledgement scheme to confirm whether the sending information is successfully received or not, each agent only needs to broadcast its information to neighbors at sampling instants.

3.3.2 Dynamics of Leader-Following MAS

With the participation of leader agent, the group behavior of MAS is no more dependent on the initial values of the system states, instead it will be dominated by the leader agent, which implies that all the followers' states will converge to the leader's state. Thus, the following consensus error variable is sufficient to tackle the consensus tracking problem

$$\delta_i = x_i - x_0. \tag{3.9}$$

The error variable between estimated state and real state is defined in the same form as in leaderless case. Then, based on (3.1), (3.2), (3.5) and (3.9), the system dynamics can be represented as

$$\begin{aligned}
\dot{x}_i &= Ax_i - BK((\mathscr{L}_r^i \otimes I_n)\delta - (\mathscr{A}_r^i \otimes I_n)e_i + b_i\delta_i) \\
\dot{\delta}_i &= A\delta_i - BK((\mathscr{L}_r^i \otimes I_n)\delta - (\mathscr{A}_r^i \otimes I_n)e_i + b_i\delta_i) \\
\dot{e}_{ij} &= Ae_{ij} + BK((\mathscr{L}_r^j \otimes I_n)\delta - (\mathscr{A}_r^j \otimes I_n)e_j + b_j\delta_j)
\end{aligned}$$

and

$$\begin{aligned}
\dot{\delta} &= (I_N \otimes A - \mathscr{H} \otimes BK)\delta + (I_N \otimes BK)\mathscr{A}_d e \\
\dot{e} &= (I_{N^2} \otimes A)e + \mathbf{1}_N \otimes ((\mathscr{H} \otimes BK)\delta)\mathbf{1}_N \otimes ((I_N \otimes BK)\mathscr{A}_d e)
\end{aligned} \tag{3.10}$$

where $\delta, e, e_i, \mathscr{A}_d$ are defined in (3.7). From the definition of (3.3), the jump dynamics of e_{ij} can be described by (3.8).

Similar as leaderless case, the system dynamics of MAS with sampled-data and packet losses in leader-following case can be described by (3.10) and (3.8). According to (3.6), (3.10), and (3.8), we can see that the system dynamics in leaderless and leader-following cases are represented as the same form, only the system matrices are different. Thus, the sampled-data leaderless and leader-following consensus problems of MAS under packet losses can be reformulated as one unified hybrid system.

3.3.3 Reformulation as Hybrid Model

Observe that the afore-constituted system dynamics in the form of (3.6) and (3.8) for leaderless case, and in the form of (3.10) and (3.8) for leader-following case, is not yet appropriate for consensus analysis of the sampled-data MAS with non-identical packet losses, as the closed-loop state fails to contain some necessary discrete variables, such as the timer variable τ_{ij} and counting variable κ_{ij}. In this subsection,with a suitable closed-loop state, the closed-loop system is reformulated as a hybrid system, by which the MAS dynamics is completely described via flow dynamics and jump dynamics.

First, we introduce an internal variable $\kappa_{ij} \in \mathbb{N}_{\geq 0}$ to record the number of successive packet losses when agent j sending information to agent i, κ_{ij} evolves according to $\dot{\kappa}_{ij} = 0$ at non-sampling instant, $\kappa_{ij}^{+} = 0$ ($\kappa_{ij}^{+} = \kappa_{ij} + 1$) if the information transmission is successful (unsuccessful) at sampling instant. Another internal variable $\tau_{ij} \in \mathbb{R}_{\geq 0}$ is introduced to keep track of the time elapsed since the most recent successful transmission instant, the dynamics of τ_{ij} can be described by $\dot{\tau}_{ij} = 1$ at non-sampling instant, $\tau_{ij}^{+} = 0$ ($\tau_{ij}^{+} = \tau_{ij}$) if the information transmission from agent j to agent i is successful (unsuccessful) at sampling instant.

Based on the definition of m_{ij}, the jump dynamics of e_{ij}, τ_{ij}, and κ_{ij} can be formally represented as

$$\begin{aligned} &e_{ij}^{+} = 0, \tau_{ij}^{+} = 0, \kappa_{ij}^{+} = 0, \text{ if } \tau_{ij} = h_j(\kappa_{ij} + 1) \wedge m_{ij} = 1, \\ &e_{ij}^{+} = e_{ij}, \tau_{ij}^{+} = \tau_{ij}, \kappa_{ij}^{+} = \kappa_{ij} + 1, \text{ if } \tau_{ij} = h_j(\kappa_{ij} + 1) \wedge m_{ij} = 0. \end{aligned} \tag{3.11}$$

The above equations and conditions describe how e_{ij}, τ_{ij}, and κ_{ij} change when there exist packet losses or not at sampling instants of agent j.

Based on the variable definitions in previous Sects. 3.3.1–3.3.2, and (3.11), for the sampled-data MAS with non-identical packet losses, the closed-loop state ξ can be constituted as

$$\begin{aligned} &\tau_i = \left[\tau_{i1}\ \tau_{i2}\ \cdots\ \tau_{iN}\right]^T, \ \tau = \left[\tau_1^T\ \tau_2^T\ \cdots\ \tau_N^T\right]^T, \ \kappa_i = \left[\kappa_{i1}\ \kappa_{i2}\ \cdots\ \kappa_{iN}\right]^T, \\ &\kappa = \left[\kappa_1^T\ \kappa_2^T\ \cdots\ \kappa_N^T\right]^T, \ \xi = \left[\delta^T\ e^T\ \tau^T\ \kappa^T\right]^T, \end{aligned}$$

Reviewing the basic hybrid model (1.1) given in Chap. 1, then according to (3.6), (3.10), (3.8) and (3.11), the closed-loop MAS (in both leaderless and leader-following cases) can be represented as

$$\begin{aligned} &\dot{\xi} = \mathscr{F}(\xi), \text{ If } \xi \in \mathbb{F}, \\ &\xi^{+} \in \bigcup_{i=1}^{N} \bigcup_{j=1}^{N} \mathscr{J}_{ij}(\xi), \text{ If } \xi \in \mathbb{J}, \end{aligned} \tag{3.12}$$

where the flow dynamics of the closed-loop state is given by

$$\mathscr{F}(\xi) = \left[f_\delta(\delta, e),\ f_e(\delta, e),\ \mathbf{1}_{N^2},\ \mathbf{1}_{N^2}\right], \tag{3.13}$$

where $f_\delta(\delta, e) = (I_N \otimes A - \mathscr{L} \otimes BK)\delta - (\mathscr{L} \otimes Bk)\mathscr{A}_d e$ and $f_e(\delta, e) = (I_{N^2} \otimes A)e - \mathbf{1}_N \otimes ((I_N \otimes BK)(\delta + \mathscr{A}_d e))$ in leaderless case, and $f_\delta(\delta, e) = (I_N \otimes A - \mathscr{H} \otimes BK)\delta + (I_N \otimes Bk)\mathscr{A}_d e$ and $f_e(\delta, e) = (I_{N^2} \otimes A)e + \mathbf{1}_N \otimes ((\mathscr{H} \otimes BK)\delta)$ $\mathbf{1}_N \otimes ((I_N \otimes BK)\mathscr{A}_d e)$ in leader-following case.

The jump dynamics is given by

$$\mathscr{J}_{ij}(\xi) = \left[\delta,\ M_{ij}^e e,\ M_{ij}^{\tau\kappa}\tau,\ M_{ij}^{\tau\kappa}\kappa + (R_i^T \otimes I_N)(1 - m_{ij})\Lambda_j \mathbf{1}_N\right], \tag{3.14}$$

where $M_{ij}^e = (I_N - \Lambda_i)I_{nN} + (R_i^T \otimes I_{nN})(I_{nN} - m_{ij}(\lambda_{ij} \otimes I_n))(R_i \otimes I_{nN})$, $M_{ij}^{\tau\kappa} = (I_N - \Lambda_i)I_N + (R_i^T \otimes I_N)(I_N - m_{ij}\Lambda_j)(R_i \otimes I_N)$, $\Lambda_i \in \mathbb{R}^{N \times N}$ is a diagonal matrix with the iith entry being 1 and all the rest entries being zeros, $R_i \in \mathbb{R}^{1 \times N}$ is a row matrix with the ith entry being 1 and the rest entries being zeros.

The flow set $\mathbb{F}$ and jump set $\mathbb{J}$ are given by

$$\mathbb{F} = \bigcap_{i=1}^{N}\bigcap_{j=1}^{N} \mathbb{F}_{ij},\ \mathbb{J} = \bigcup_{i=1}^{N}\bigcup_{j=1}^{N}\left(\mathbb{J}_{ij}^0 \bigcup \mathbb{J}_{ij}^1\right), \tag{3.15}$$

where $\mathbb{F}_{ij}$, $\mathbb{J}_{ij}^0$, $\mathbb{J}_{ij}^1$ are designed as

$$\begin{aligned} &\mathbb{F}_{ij} := \{\xi \in \mathbb{X} | \tau_{ij} < h_j(\kappa_{ij} + 1)\}, \\ &\mathbb{J}_{ij}^0 := \{\xi \in \mathbb{X} | \tau_{ij} = h_j(\kappa_{ij} + 1) \wedge m_{ij} = 0\},\ \mathbb{J}_{ij}^1 := \{\xi \in \mathbb{X} | \tau_{ij} = h_j(\kappa_{ij} + 1) \wedge m_{ij} = 1\}, \end{aligned} \tag{3.16}$$

where $\mathbb{X} = \{(\delta, e, \tau, \kappa) \in \mathbb{R}^{nN} \times \mathbb{R}^{nN^2} \times \mathbb{R}^{N^2} \times \mathbb{R}^{N^2}\}$.

Based on the constructed hybrid model, it is clear that the leaderless and leader-following consensus problems of MAS can be transformed into stability problem of hybrid system (3.12). Thus, the stability definition for hybrid systems (3.12) is formally given as follows.

Definition 3.1 Consider the hybrid system (3.12), the set $\{\xi \in \mathbb{X} | \delta = 0_{nN}, e = 0_{nN^2}\}$ is said to be UGAS if there exists a function $\beta \in \mathscr{KL}$, such that for initial condition $\xi(0) \in \mathbb{X}$, the solution ξ satisfies

$$\|(\delta(t), e(t))\| \leq \beta(\|(\delta(0), e(0))\|, t). \tag{3.17}$$

3.4 Stability Analysis and Parameter Design

This section will focus on showing the stability conditions for the hybrid system (3.12), and the criterions for selecting sampling periods h_i and calculating the NMASPL $\rho_{\max}^{ij}$. The basic idea for stability analysis is to construct a positive semi-definite function S such that $\dot{S} \leq -\alpha_1(\delta) - \alpha_2(e)$ holds during flow and $S^+ \leq S$ holds during jumps, where $\alpha_1(\cdot), \alpha_2(\cdot)$ are class $\mathcal{K}_\infty$ functions.

To impel the stability analysis, we introduce an internal function $\phi_{ij}(\tau_{ij}) : \mathbb{R}_{\geq 0} \rightarrow \mathbb{R}$ as

$$\dot{\phi}_{ij}(\tau_{ij}) = -\gamma_{ij}\left(\phi_{ij}^2(\tau_{ij}) + 1\right), \tag{3.18}$$

where $\gamma_{ij} > 0$ is a constant parameter that can be selected. It is obvious that $\phi_{ij}(\tau_{ij})$ is a strictly decreasing function, and there always exists an upper bound $\mathfrak{T}_{ij}(\gamma_{ij}, \phi_{ij}(0))$ for the independent variable τ_{ij} in the sense that if $\phi_{ij}(\tau_{ij})$ is used to construct Lyapunov function, $\phi_{ij}(\tau_{ij}) > 0$ is required to be satisfied. For any pre-specified $\phi_{ij}(0) > 0$, by integrating both side of (3.18), the value range of $\mathfrak{T}_{ij}(\gamma_{ij}, \phi_{ij}(0))$ can be calculated as $0 \leq \tau_{ij} \leq \mathfrak{T}_{ij}(\gamma_{ij}, \phi_{ij}(0)) < \frac{\arctan(\phi_{ij}(0))}{\gamma_{ij}}$.

Theorem 3.1 *Consider the hybrid system (3.12), with initial state set $\mathbb{X}_0 = \{\xi \in \mathbb{X} | \kappa_{ij} = 0\}$. Let $K = \mu B^T P$, where the constant $\mu > 0$, and the symmetric matrix $P > 0$ is a solution to the following Riccati equation*

$$PA + A^T P - 2\mu a(\hat{\mathscr{L}}) PBB^T P + Q = 0, \tag{3.19}$$

where Q is a positive definite matrix. If the number of maximum allowable successive packet losses $\rho_{\max}^{ij}$ satisfies

$$\rho_{\max}^{ij} < \mathfrak{T}_{ij}(\gamma_{ij}, \phi_{ij}(0))/h_j - 1, \tag{3.20}$$

and there exist constants $0 < \sigma < 1$, $\epsilon > 0$, $\gamma_{ij} > 0$, such that the following matrix inequality condition holds

$$\begin{bmatrix} \varXi_{11} & \varXi_{12} \\ * & \varXi_{22} \end{bmatrix} \leq 0 \tag{3.21}$$

where

$$\begin{aligned} \varXi_{11} &= -(1-\sigma)(I_N \otimes Q) + \sum_{i=1}^{N}\sum_{j=1}^{N} \bar{R}_{ij}^{\delta T} \bar{R}_{ij}^{\delta}, \\ \varXi_{12} &= (I_N \otimes P)\hat{A}_{12} + \sum_{i=1}^{N}\sum_{j=1}^{N} \bar{R}_{ij}^{\delta T} \bar{R}_{ij}^{e}, \\ \varXi_{22} &= -\varGamma_\epsilon + \sum_{i=1}^{N}\sum_{j=1}^{N} \bar{R}_{ij}^{eT} \bar{R}_{ij}^{e}, \end{aligned} \tag{3.22}$$

and $\varGamma_\epsilon = diag_{i\in[1,N]}\{\varGamma_{\epsilon i}\}$, $\varGamma_{\epsilon i} = diag_{j\in[1,N]}\{(\gamma_{ij}^2 - \epsilon) \otimes I_n\}$, $\bar{R}_{ij}^{\delta} = (R_j \otimes I_n)\hat{E}_\delta$, $\bar{R}_{ij}^{e} = (R_j \otimes A)(R_i \otimes I_{nN}) + (R_j \otimes I_n)\hat{E}_e$, and the matrices $\hat{E}_\delta = -I_N \otimes BK$,

$\hat{E}_e = -(I_N \otimes BK)\mathscr{A}_d$, $\hat{A}_{12} = -(\mathscr{L} \otimes BK)\mathscr{A}_d$ in leaderless case, and $\hat{E}_\delta = \mathscr{H} \otimes BK$, $\hat{E}_e = -(I_N \otimes BK)\mathscr{A}_d$, $\hat{A}_{12} = -(I_N \otimes BK)\mathscr{A}_d$ in leader-following case.

Then, the (δ, e) dynamics is UGAS, in other words, the sampled-data based leaderless and leader-following consensus of MAS under packet losses is achieved.

Proof Choose the candidate Lyapunov function

$$V(\xi) = \delta^T(I_N \otimes P)\delta + \sum_{i=1}^{N}\sum_{j=1}^{N} \gamma_{ij}\phi_{ij}(\tau_{ij})e_{ij}^T e_{ij}. \tag{3.23}$$

From $V(\xi)$, it is clear that the quadratic term $\delta^T(I_N \otimes P)\delta \geq 0$ holds all the time, and $\delta^T(I_N \otimes P)\delta = 0$ if and only if $\delta = 0$. Due to the NMASPL $\rho_{\max}^{ij}$ is supposed to be upper bounded by (3.20), based on the role of κ_{ij}, it is known that $\kappa_{ij} \leq \rho_{\max}^{ij}$. Then, according to (3.16), if there exist $\rho_{\max}^{ij}$ successive packet losses in the communication channel from agent j to agent i, the next transmission must be successful, and it implies that $\tau_{ij} = h_j(\kappa_{ij}+1) \leq h_j(\rho_{\max}^{ij}+1) = \mathfrak{T}_{ij}(\gamma_{ij}, \phi_{ij}(0))$, so $\phi_{ij}(\tau_{ij}) > 0$ is guaranteed. Due to $\gamma_{ij} > 0$ and $e_{ij}^T e_{ij} \geq 0$, we can conclude that $V(\xi)$ is a radially unbounded function with respect to (δ, e), and $V(\xi)$ is a suitable Lyapunov function for stability analysis.

Consider the candidate Lyapunov function $V(\xi)$ given in (3.23), when jump dynamics occurs, there exist two possible cases: $\xi \in \mathbb{J}_{ij}^0$ or $\xi \in \mathbb{J}_{ij}^1$, $i, j \in [1, N]$, the first case corresponds that packet transmitted from agent j to agent i loses, and the second case means no packet loss and the transmission is successful. According to the jump dynamics in (3.14), for $\xi \in \mathbb{J}_{ij}^0$, we have $V(\xi^+) - V(\xi) = 0$, since $\delta^+ = \delta$, $e_{ij}^+ = e_{ij}$, $\tau_{ij}^+ = \tau_{ij}$. For $\xi \in \mathbb{J}_{ij}^1$, we have $V(\xi^+) - V(\xi) \leq 0$, since $\delta^+ = \delta$, $e_{ij}^+ = 0$, $\tau_{ij}^+ = 0$. Thus, it implies that $V(\xi^+) - V(\xi) \leq 0$ for all $\xi \in \mathbb{J}$.

Consider the evolution of $V(\xi)$ along the flow dynamics of ξ, due to ξ has jump dynamics, it is not differentiable everywhere, so, we use the generalized directional derivative $\langle \nabla V(\xi), \mathscr{F}(\xi)\rangle$ instead of $\dot{V}(\xi)$. In addition, note that although $V(\xi)$ is not differentiable with respect to κ_{ij}, this does not have effect on the following analysis, since $\dot{\kappa}_{ij} = 0$ during flow. For τ and almost all (δ, e), we have

$$\begin{aligned}
&\langle \nabla V(\xi), \mathscr{F}(\xi)\rangle \\
&= \delta^T((I_N \otimes P)(I_N \otimes A - \hat{\mathscr{L}} \otimes BK) + (I_N \otimes A - \hat{\mathscr{L}} \otimes BK)^T(I_N \otimes P))\delta + 2\delta^T(I_N \otimes P)\hat{A}_{12}e \\
&\quad - \sum_{i=1}^{N}\sum_{j=1}^{N} \gamma_{ij}^2(\phi_{ij}^2(\tau_{ij}) + 1)e_{ij}^T e_{ij} + \sum_{i=1}^{N}\sum_{j=1}^{N} 2\gamma_{ij}\phi_{ij}(\tau_{ij})e_{ij}^T \dot{e}_{ij}.
\end{aligned} \tag{3.24}$$

Based on Lemma 3.1, Riccati equation (3.19), and $K = \mu B^T P$, the following conditions are obtained

$$\begin{aligned}
&\delta^T((I_N \otimes P)(I_N \otimes A - \hat{\mathscr{L}} \otimes BK) + (I_N \otimes A - \hat{\mathscr{L}} \otimes BK)^T(I_N \otimes P))\delta \\
&= \delta^T(I_N \otimes (PA + A^T P) - 2\mu\hat{\mathscr{L}} \otimes PBB^T P)\delta \\
&\leq -\delta^T(I_N \otimes Q)\delta.
\end{aligned} \tag{3.25}$$

Besides, note that $e_{ij} = (R_j \otimes I_n)e_i$ and $e_i = (R_i \otimes I_{nN})e$, it follows from the flow dynamics $\mathscr{F}(\xi)$ and the definitions of $\bar{R}^{\delta}_{ij}$, $\bar{R}^{e}_{ij}$, $\hat{E}_{\delta}$, and $\hat{E}_e$ that

$$\begin{aligned} \dot{e}_{ij} &= (R_j \otimes I_n)(R_i \otimes I_{nN})\dot{e} \\ &= (R_j \otimes I_n)((I_N \otimes A)e_i + \hat{E}_{\delta}\delta + \hat{E}_e e) \\ &= (R_j \otimes A)(R_i \otimes I_{nN})e + (R_j \otimes I_n)\hat{E}_{\delta}\delta + (R_j \otimes I_n)\hat{E}_e e \\ &= \bar{R}^{\delta}_{ij}\delta + \bar{R}^{e}_{ij}e \end{aligned} \tag{3.26}$$

Due to $\gamma_{ij} > 0$ and $\phi_{ij}(\tau_{ij}) > 0$ as discussed before, combine (3.24) with (3.25) and (3.26), we have

$$\begin{aligned} &\langle \nabla V(\xi), \mathscr{F}(\xi)\rangle \\ \leq & -\delta^T(I_N \otimes Q)\delta + 2\delta^T(I_N \otimes P)\hat{A}_{12}e - \sum_{i=1}^{N}\sum_{j=1}^{N}\gamma_{ij}^2(\phi_{ij}^2(\tau_{ij})+1)\left\|e_{ij}\right\|^2 \\ &+ \sum_{i=1}^{N}\sum_{j=1}^{N} 2\gamma_{ij}\phi_{ij}(\tau_{ij})\left\|e_{ij}\right\|\left\|\bar{R}^{\delta}_{ij}\delta + \bar{R}^{e}_{ij}e\right\|. \end{aligned} \tag{3.27}$$

For the matrix inequality condition given in (3.21), calculate $\begin{bmatrix}\delta \\ e\end{bmatrix}^T \begin{bmatrix}\Xi_{11} & \Xi_{12} \\ * & \Xi_{22}\end{bmatrix}\begin{bmatrix}\delta \\ e\end{bmatrix}$, it can be derived that

$$\begin{aligned} &-\delta^T(I_N \otimes Q)\delta + 2\delta^T(I_N \otimes P)\hat{A}_{12}e \\ \leq & -\delta^T(I_N \otimes \sigma Q)\delta + e^T\Gamma_{\epsilon}e + \sum_{i=1}^{N}\sum_{j=1}^{N}\left(\delta^T\bar{R}^{\delta T}_{ij}\bar{R}^{\delta}_{ij}\delta + 2\delta^T\bar{R}^{\delta T}_{ij}\bar{R}^{e}_{ij}e + e^T\bar{R}^{eT}_{ij}\bar{R}^{e}_{ij}e\right) \\ = & -\delta^T(I_N \otimes \sigma Q)\delta + \sum_{i=1}^{N}\sum_{j=1}^{N}(\gamma_{ij}^2 - \epsilon)\left\|e_{ij}\right\|^2 - \sum_{i=1}^{N}\sum_{j=1}^{N}\left\|\bar{R}^{\delta}_{ij}\delta + \bar{R}^{e}_{ij}e\right\|^2. \end{aligned} \tag{3.28}$$

From (3.27) and (3.28), we obtain

$$\begin{aligned} &\langle \nabla V(\xi), \mathscr{F}(\xi)\rangle \\ \leq & -\delta^T(I_N \otimes \sigma Q)\delta - \sum_{i=1}^{N}\sum_{j=1}^{N}\epsilon\left\|e_{ij}\right\|^2 - \sum_{i=1}^{N}\sum_{j=1}^{N}\left(\gamma_{ij}\phi_{ij}(\tau_{ij})e_{ij} - \left\|\bar{R}^{\delta}_{ij}\delta + \bar{R}^{e}_{ij}e\right\|\right)^2 \\ \leq & -\lambda_{\min}(\sigma Q)\left\|\delta\right\|^2 - \epsilon\left\|e\right\|^2 \end{aligned} \tag{3.29}$$

Thus, according to $V(\xi^+) - V(\xi) \leq 0$ for $\xi \in \mathbb{J}$ and $\langle \nabla V(\xi), \mathscr{F}(\xi)\rangle \leq -\lambda_{\min}(\sigma Q)$ $\|\delta\|^2 - \epsilon\|e\|^2$ for $\xi \in \mathbb{F}$, and based on the standard Lyapunov stability theory of hybrid systems [12, 13], we obtain that the (δ, e) dynamics is UGAS, which means that the leaderless and leader-following consensus of MAS under packet losses is achieved. □

Remark 3.1 When implementing the controllers (3.4) and (3.5), the control variables (i.e., estimation states in (3.3)) are updated passively as long as it receives some new information, which implies that it does not need to feedback an acknowledgement to confirm the current reception. In other words, each agent broadcasts its information to neighbors at sampling instants, but does not need to know if the transmitted data is successfully received or not. In addition, the occurrence of packet losses in different communication channels (from agent i to agent j) can be random and independent of each other, that is more closed to practical circumstance and more general than the existing results [1–3].

Remark 3.2 The relationships among the NMASPL $\rho^{ij}_{\max}$, the sampling period h_j, and other design parameters have been exhibited explicitly by the condition (3.20). From this condition, the NMASPL $\rho^{ij}_{\max}$ is constrained, and it requires that the maximally allowable successive packet losses are at most $\mathfrak{T}_{ij}(\gamma_{ij}, \phi_{ij}(0))/h_j - 1$. It can be seen that that $\rho^{ij}_{\max}$ is related to the values of γ_{ij}, $\phi_{ij}(0)$, h_j, and once the parameters γ_{ij} and $\phi_{ij}(0)$ are determined, the NMASPL $\rho^{ij}_{\max}$ is inversely proportional to the sampling period h_j. So, there is tradeoff between the NMASPL and sampling period.

Remark 3.3 Obviously, the validity of Theorem 3.1 is tied to matrix inequality (3.21), and its solution is the key to ensure stability of hybrid system (3.12). Provided that the pair (P, Q, μ) is the solution of Riccati equation (3.19), then, select $P_1 = sP$, $Q_1 = sQ$, $\mu_1 = \mu/s$, with $s > 0$, using (P_1, Q_1, μ_1) instead of (P, Q, μ) in Riccati equation (3.19), we can obtain

$$\begin{aligned} &P_1A + A^T P_1 - 2\mu_1 a(\hat{\mathscr{L}})P_1BB^T P_1 + Q_1 \\ &= sPA + sA^T P - 2\frac{\mu}{s}a(\hat{\mathscr{L}})s^2 PBB^T P + sQ \\ &= s(PA + A^T P - 2\mu a(\hat{\mathscr{L}})PBB^T P + Q). \end{aligned} \tag{3.30}$$

This implies that (P_1, Q_1, μ_1) is also a group of solution to Riccati equation (3.19). In this regard, if the selection of (P, Q, μ) cannot guarantee $\Xi_{11} < 0$, we can select $Q_1 = sQ$, while the gain matrix $K_1 = \mu_1 B^T P_1 = \mu/sB^T sP = K$ remains unchanged. Then, there always exists sufficiently large $s > 0$ such that $\Xi_{11} < 0$ holds with Q_1 instead of Q. In addition, note that the parameter γ_{ij} is independent of the rest parameters in matrix inequality (3.21) and γ_{ij} can be arbitrarily selected, thus, we can always select γ_{ij} such that $\Xi_{22} < 0$ holds. According to matrix theory and the above analysis, there always exist sufficiently large $s > 0$ and $\gamma_{ij} > 0$, such that the matrix inequality (3.21) has solutions.

3.5 Further Results

In the former of this chapter, for ease of exposition, we have supposed that the communication channels between the leader agent and its immediate followers are free of packet losses, namely, the leader's state is directly available (see (3.5)). In this section, we withdraw this assumption, and extend the proposed hybrid system approach to the case with packet losses among leader and followers.

Given the case the communication channels between the leader and its immediate follower also is unreliable, i.e., packet losses may occur, the consensus control protocol is modified as

$$u_i(t) = K\left(\sum_{j\in\mathcal{N}_i} a_{ij}(\hat{x}_{ij}(t) - x_i(t)) + b_i(\hat{x}_{i0}(t) - x_i(t))\right) \tag{3.31}$$

Contrary to (3.5), $\hat{x}_{i0}(t)$ is used here due to packet losses. $\hat{x}_{i0}(t)$ denotes the estimation of leader's state as seen by agent i, its dynamics can be described by (3.3) by letting $j = 0$.

Define $e_{i0} = \hat{x}_{i0} - x_0$ and redefine $e_i = [e_{i0}^T\ e_{i1}^T\ \cdots\ e_{iN}^T]$, $\mathscr{A}_d = \mathrm{diag}_{i\in[1,N]}\{[b_i I_n\ \mathscr{A}_r^i \otimes I_n]\}$, and note that the jump dynamics of e_{i0} can be described by (3.8) with $j = 0$. Then, based on (3.1), (3.2), (3.31), and (3.9), we have

$$\begin{aligned}\dot{\delta} &= (I_N \otimes A - \mathscr{H} \otimes BK)\delta + (I_N \otimes BK)\mathscr{A}_d e\\ \dot{e} &= (I_N \otimes (I_{N+1} \otimes A)e + \mathbf{1}_N \otimes (E_L[0_n^T\ \delta^T]^T) - \mathbf{1}_N \otimes ((I_N \otimes BK)\mathscr{A}_d e)\end{aligned} \tag{3.32}$$

where $E_L = [0_{n\times n}\ 0_{n\times nN};\ 0_{nN\times n}\ \mathscr{H} \otimes BK]$.

Compare (3.32) with (3.10), one can find that the dynamics of δ and the dynamics of e are similar, only some system matrices are slightly different. According to (3.32), we can also construct a hybrid system model for the leader-following MAS in the same form as (3.12) and (3.15), only with some slightly modifications, such as $j \in [0,\ N]$ instead of $j \in [1,\ N]$. Then, the main results in Theorem 3.1 still hold. Therefore, we can conclude that, even if there exist packet losses among leader and followers, the proposed approach is still applicable.

In addition, in view of the considered connected and undirected graph in Assumption 3.2, we would like to clarify that the proposed method can also be extended to the case with directed graph provided that the communication graph $\mathscr{G}$ is directed and strongly connected. Specifically, based on the results of Lemmas 1 and 2 in [11], suppose that the directed graph $\mathscr{G}$ is strongly connected, there exists a vector $\zeta = [\zeta_1\ \zeta_2\ \cdots\ \zeta_N]^T$ with all the elements being positive such that $\zeta^T\mathscr{L} = 0$ and $\sum_{i=1}^N \zeta_i = 1$. Define $Z = \mathrm{diag}\{\zeta_1\ \zeta_2\ \cdots\ \zeta_N\}$, then, the matrix $(Z\mathscr{L} + \mathscr{L}^T Z)$ is symmetric and we have the condition $\min_{x^T\zeta=0, x\neq 0} \frac{x^T\hat{\mathscr{L}}x}{x^T Z x} \geq a(\mathscr{L}) > 0$ where $\hat{\mathscr{L}} = \frac{1}{2}(Z\mathscr{L} + \mathscr{L}^T Z)$ in leaderless case and $\hat{\mathscr{L}} = \frac{1}{2}(Z\mathscr{L} + \mathscr{L}^T Z) + \mathscr{B}$ in leader-following case. Then, substitute $\delta^T(I_N \otimes P)\delta$ in Lyapunov function $V(\xi)$ with

$\delta^T(Z \otimes P)\delta$, and let the Riccati equation be $PA + A^TP - 2\mu a(\mathscr{L})PBB^TP + Q = 0$, then, the main results in Theorem 3.1 can still be obtained.

3.6 Simulation Study

In this section, simulation results are given to show the effectiveness of the proposed approach for MAS under packet losses, both the cases with and without leader are considered.

First, consider leaderless MAS with 6 agents. The communication topology among agents is shown in Fig. 3.2. The agents' dynamics is described in the form of (3.1) with system matrices

$$A = \begin{bmatrix} 0 & 1 \\ -1 & 0 \end{bmatrix}, B = \begin{bmatrix} 1 \\ 1 \end{bmatrix}. \tag{3.33}$$

From Fig. 3.2, it is clear that there exist 8 bidirectional communication channels (i.e., 16 communication channels from some agent i to agent j). As discussed before, the packet loss occurs in each channel may be different from others, for convenience, we define $\mathbf{cha}_{ij}$ to denote the communication channel from agent j to agent i. Based on Lemma 3.1, $a(\mathscr{L})$ can be obtained as $a(\mathscr{L}) = 1.382$. Select $\mu = 0.1$, $Q = 15I$, based on Riccati equation (3.19), the gain matrix K can be computed as $K = \mu B^T P = [1.0072 \ 0.2662]$. The asynchronous sampling periods of agents are selected as $h_1 = h_2 = 0.01$s, $h_3 = h_4 = 0.02$s, $h_5 = h_6 = 0.03$s. The communication link is assumed to be unreliable and there exist packet losses, according to (3.20), the maximum allowable successive packet losses for each communication channel is related to parameters γ_{ij}, $\phi_{ij}(0)$ and sampling period h_i. To show that the packet losses can be nonidentical, select these parameters as $\gamma_{ij} = 10$ for i, j such that $\mathscr{E}_{ij} \in \mathscr{E}$, $\phi_{ij}(0) = 10 + i$ for $i \in [1, 6]$ and $j \in \mathscr{N}_i$. Then, the NMASPL $\rho_{\max}^{ij}$ for each communication channel can be calculated based on (3.20), the results are shown in Table 3.1 (e.g., the NMASPL for $\mathbf{cha}_{12}$ is 13).

To implement the simulation, the initial conditions $x_i(0)$ are selected as $x_i(0) = (-1)^i \times [1 \ 2]^T$, and the initial conditions of rest parameters (such as $\hat{x}_{ij}, \kappa_{ij}, \tau_{ij}$) are

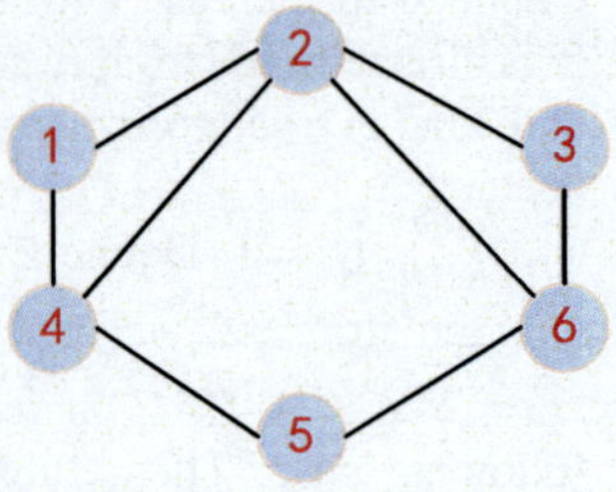

Fig. 3.2 Topology of leaderless MAS

Table 3.1 Packet losses in communication channels. The black numbers denote the NMASPL, the blue numbers denote the actual number of maximum successive packet losses in simulation

cha$_{12}$	**cha**$_{14}$	**cha**$_{21}$	**cha**$_{23}$	**cha**$_{24}$	**cha**$_{25}$	**cha**$_{32}$	**cha**$_{36}$
13	6	13	6	6	3	13	3
9	6	6	2	5	2	5	3
cha$_{41}$	**cha**$_{42}$	**cha**$_{45}$	**cha**$_{54}$	**cha**$_{56}$	**cha**$_{62}$	**cha**$_{63}$	**cha**$_{65}$
13	13	3	6	4	14	6	4
9	5	3	3	4	5	3	4

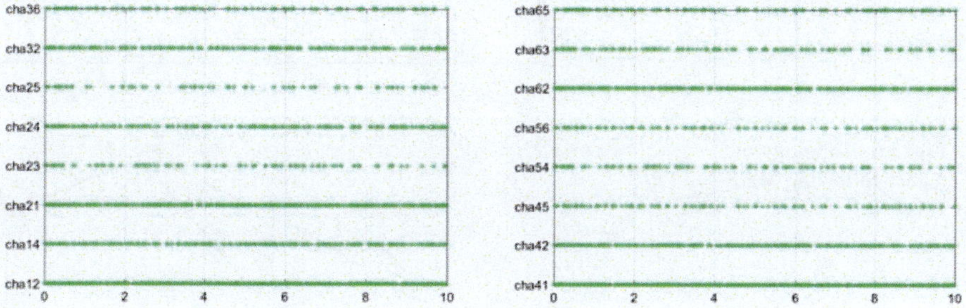

Fig. 3.3 Packet losses in communication channels. cha$_{ij}$ denotes the channel from j to i

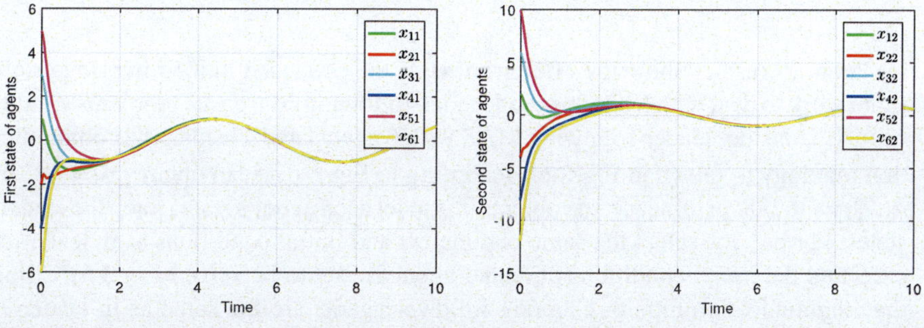

Fig. 3.4 The states of leaderless MAS

set to be zeros. In addition, select $\sigma = 0.01, \epsilon = 0.01$, based on the given parameters, it can be verified that the stability conditions given in Theorem 3.1 are satisfied. The simulation results are shown in Figs. 3.3, 3.4 and 3.5. Figure 3.3 presents the packet losses in each channel **cha**$_{ij}$, the packet losses are randomly generated in simulation, and the NMASPL is not larger than $\rho^{ij}_{\max}$. Specifically, the NMASPL and actual number of maximum successive packet losses in communication channels are shown in Table 3.1. Moreover, the state trajectories of agents and control inputs u_i are shown in Figs. 3.4 and 3.5 respectively. According to the simulation results, it is clear shown that the consensus of leaderless MAS under nonidentical packet losses is achieved.

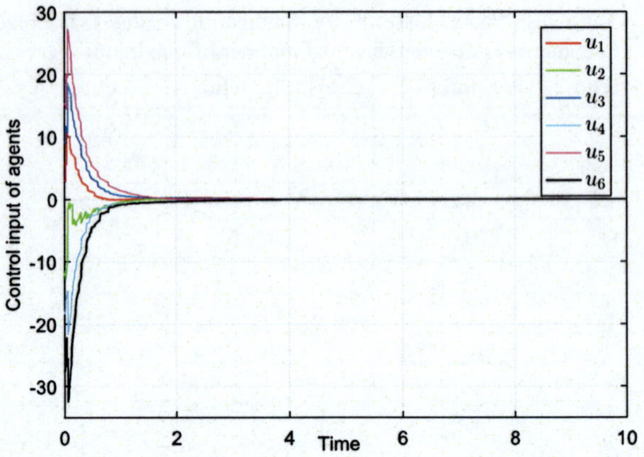

Fig. 3.5 Control input u_i in leaderless case

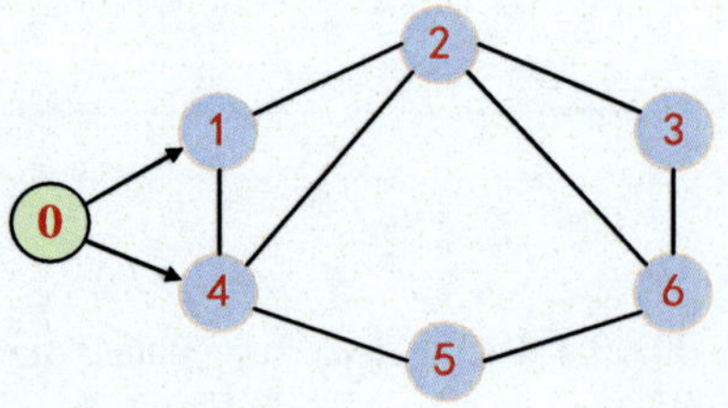

Fig. 3.6 Topology of leader-following MAS

Then, to further show the effectiveness of the proposed unified hybrid systems framework to deal with leaderless and leader-following consensus problems simultaneously, consider leader-following MAS with 6 agents and a leader, the communication topology is shown in Fig. 3.6. According to Fig. 3.6, it can obtain $\hat{\mathscr{L}} = \mathscr{H}$ and $a(\mathscr{L}) = 0.228$. In this case, the objective is to let each agent's states track the leader's states. For brevity, select the same parameters and initial conditions as in leaderless case, and the initial conditions of leader agent are set to be $x_0(0) = [-3\ 5]^T$. Since the communication topology among follower agents are the same as in leaderless case, the NMASPL in each communication channel is the same as given in Table 3.1. Under packet losses, the simulation results are shown in Figs. 3.7 and 3.8, we can see that the leader-following consensus is achieved. The simulation results show the effectiveness of the proposed approach.

3.7 Conclusion

This chapter investigates sampled-data consensus problem of MAS under packet losses, asynchronous sampling mechanism is used and nonidentical packet losses is considered. The nonidentical packet losses can cover the case of identical packet losses, that is more general in practical applications. A unified hybrid model is constructed for both leaderless and leader-following MAS to describe the closed-loop

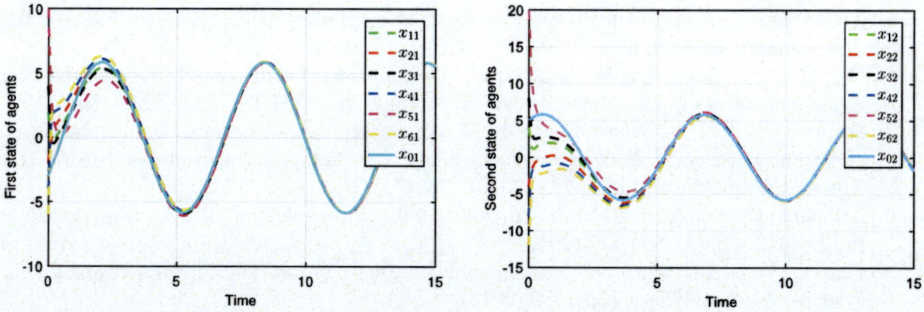

Fig. 3.7 The states of leader-following MAS

Fig. 3.8 Control input u_i in leader-following case

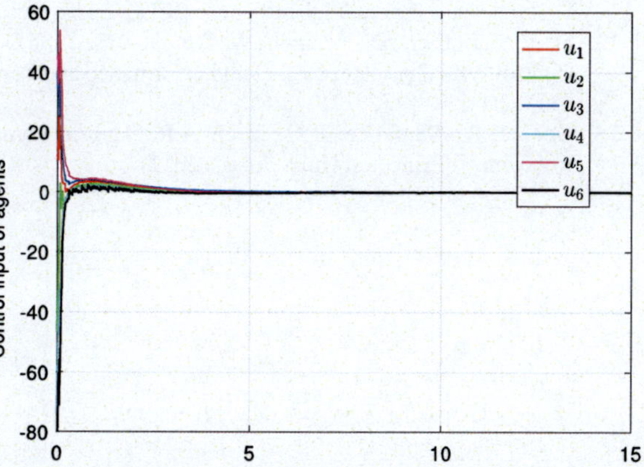

dynamics, such that leaderless and leader-following consensus of MAS under packet losses can be tackled in a unified framework. Lyapunov-based stability conditions are obtained. Moreover, explicit relationship between successive packet losses and sampling period is also presented.

References

1. Jiang, X., Xia, G., Feng, Z., Jiang, Z.: Consensus Tracking of Data-Sampled Nonlinear Multi-Agent Systems With Packet Loss and Communication Delay. IEEE Trans. Netw. Sci. Eng., **8**(1), 126-137 (2021)
2. Xing, M., Deng, F., Hu, Z.: Sampled-Data Consensus for Multiagent Systems With Time Delays and Packet Losses. IEEE Trans. Syst., Man, Cybern: Syst., **50**(1), 203-210 (2020)
3. Wang, F., Wen, G., Peng, Z., Huang, T., Yu, Y.: Event-Triggered Consensus of General Linear Multiagent Systems With Data Sampling and Random Packet Losses. IEEE Trans. Syst., Man, Cybern: Syst., **51**(2), 1313-1321 (2021)
4. Zhang, W., Tang, Y., Huang, T., Kurths, J.: Sampled-Data Consensus of Linear Multi-agent Systems With Packet Losses. IEEE Trans. Neural Netw. Learn. Syst., **28**(11), 2516-2527 (2017)

5. Xu, L., Mo, Y., Xie, L.: Distributed Consensus Over Markovian Packet Loss Channels. IEEE Trans. Autom. Control. **65**(1), 279-286 (2020)
6. Guinalda, M., Lehmann, D., Sanchez, J., Dormido, S., Johansson, K.: Distributed event-triggered control for non-reliable networks. J. Frankl. Inst. **351**(12), 5250-5273 (2014)
7. Heemels, W. Teel, A., Wouw, N., Nesic, D.: Networked control systems with communication constraints: Tradeoffs between transmission intervals, delays and performance. IEEE Trans. Autom. Control. **55**(8), 1781-1796 (2010)
8. Lehmann, D., Lunze, J.: Event-based control with communication delays and packet losses. Int. J. Control **85**(5), 563-577 (2012)
9. Peng, C., Yang, Y.: Event-triggered communication and control co-design for networked control systems. Autom., **49**(5), 1326-1332 (2013)
10. Dolk, V., Borgers, D., Heemels, W.: Event-triggered control systems under packet losses. Automatica **80**, 143-155 (2017)
11. Zhao, G., Hua, C.: A Hybrid Dynamic Event-Triggered Approach to Consensus of Multiagent Systems With External Disturbances. IEEE Trans. Autom. Control, **66**(7), 3213-3220 (2021)
12. Dolk, V., Borgers, D., Heemels, W.: Output-Based and Decentralized Dynamic Event-Triggered Control With Guaranteed $\mathcal{L}_p$-Gain Performance and Zeno-Freeness. IEEE Trans. Autom. Control, **62**(1), 34-49 (2017)
13. Goebel, R., Sanfelice, R.G., Teel, A.R.: Hybrid Dynamical Systems: Modeling, Stability, Robustness. Princeton Univ. Press. (2012)

Chapter 4
Sampled-Data Consensus Control of MAS Under Switching Network

4.1 Introduction

In Chaps. 2 and 3, for MASs, we have investigated the problem of how to regulate the information transmissions over a connected network, and presented the sufficient conditions to make the proposed sampled-data based control protocol be feasible to address some network induced issues including time delays and packet losses. It is remarkable that these network induced issues can be viewed as in-process problems, that is, the communication connections has been preconstructed in the sense that what information is available for a certain agent is specified via a fixed communication graph, and the occurrence of these issues is only during the communicating process. The fixed graph implies that communication connections among agents do not change again, while if we want to take into account the communication link failure and link creation, reconfigurable networks and nearest neighbor coupling, the communication topology will be no more fixed but time-dependent. Such changes of connections among agents lead to network switching phenomenon, which is different from in-process problems, and can interrupt the communicating process at any time and change the current connection structure. In the context of sampled-data control and switching network, the sampled-data consensus problem was studied based on sampled position data in [1] and based on impulsive control laws in [2]. The robust consensus protocol for multiagent networks with asynchronous sampling was developed in [3]. Note that the aforementioned results only considered the leaderless MAS. Regarding the sampled-data leader-following consensus problem, [4] proposed weighted consensus protocols for double-integrator MAS, and [5] developed delay-dependent stability conditions for general linear MAS. Reference [6] studied consensus of nonlinear MAS with stochastic sampling. Furthermore, taken the network switching issue into consideration, [7] studied the sampled-data consensus problem of first-order MAS under static network case, and then it was extended to switching network case [8]. Under the assumption of switched and jointly connected communication graph, [9] utilized asynchronous sampling based control laws to achieve consensus of single/double-integrator MAS respectively. For general lin-

G. Zhao et al., *Cooperative Control of Multi-agent Systems*,
https://doi.org/10.1007/978-981-97-0968-7_4

ear MAS, the sampled-data consensus of MAS with time-varying topologies was studied in [10]. More recently, an asynchronous sampled-data consensus control law was proposed in [11] for MAS over static/switching directed networks, and explicit upper bound of sampling periods (UBSP) of all agents can be calculated based on a technical lemma. However, the obtained UBSP is a little conservative due to all agents share a common UBSP, and it should be noted that the UBSP is a constraint condition imposed on the selection of sampling periods, which means that the maximal sampling period is no more than UBSP. Thus, to reduce the number of sampling, how to develop novel approach to improve the UBSP is worth of further investigation.

In addition, we should remark that in practice the transient performance has the same importance as the asymptotic consensus behavior, also deserves our particular concerns, since it is an important index reflecting the relationship between response speed and reliability of the automatic system. Besides, note that sampled-data control may result in degradation of transient performance compared with continuous control in general. Thus, how to design control protocol to improve transient performance is an interesting topic. Recently, reset control approach, which was originally proposed in [12] and subsequently used to improve transient performance of linear systems [13–15], has been used to improve transient performance of MAS [16, 17]. In this chapter, a hybrid system approach is adopted to address the sampled-data leader-following consensus problem of MAS under static and switching directed networks. We propose an asynchronous sampled-data based consensus protocol consisting of both linear and reset control parts, where the reset control part always contributes to improve transient performance (e.g., settling time). The proposed hybrid control approach can not only be applied to MAS over static/switching directed networks, but also guarantees a larger UBSP compared with [11], which can be calculated independent of other agents.

4.2 Networked MAS Under Static/Switching Network

This section formulates the sampled-data based leader-following consensus problem of MAS under static and switching networks. Consider a group of MAS consisting of one leader agent and N follower agents, the leader-following consensus of MAS can be achieved in the sense that all the followers' state trajectories asymptotically approach to and remain the same trajectory as the leader i.e., $\lim_{t\to\infty} \|x_i(t) - x_0(t)\| = 0$ for any initial conditions $x_i(0)$, $i \in [1, N]$, as long as the control protocol regarding managing how the agents interact is developed. Based on the analysis in Introduction, it is clear that the communication connections among MAS are typically described by graphs, which can be either fixed or time-varying in practical scenarios. It should be emphasized that the control protocols designed for addressing the aforementioned in-process network issues cannot be directly used to cope with the network induced graph switching issue, despite they are all called network issues. The dynamics of the leader and that of the follower agents are described by

$$\dot{x}_0(t) = Ax_0(t), \tag{4.1}$$

$$\dot{x}_i(t) = Ax_i(t) + Bu_i(t), \tag{4.2}$$

respectively, where $x_0(t) \in \mathbb{R}^n$ and $x_i(t) \in \mathbb{R}^n, i \in [1, N]$, $u_i(t) \in \mathbb{R}$ denotes the control input. A, B are matrices with appropriate dimensions and it is assumed that (A, B) is stabilizable.

The preliminaries about static graph and switching graph have been introduced in Chap. 1. Note that in the switching network case, the communication graph $\bar{\mathscr{G}}_{\sigma(t)}$ is time-varying as $\sigma(t)$ changes, and if $\sigma(t)$ is a constant signal, the communication topology is static, in this case, we use $\bar{\mathscr{G}}$ and $\bar{\mathscr{A}}$ to denote $\bar{\mathscr{G}}_{\sigma(t)}$ and $\bar{\mathscr{A}}_{\sigma(t)}$, respectively. For proceeding this chapter, we make the following Assumptions on the communication graphs.

Assumption 4.1 ([18]) The graph $\bar{\mathscr{G}}$ is directed and contains a spanning tree with the root node being the leader node 0.

Assumption 4.2 ([11]) For any $p \in \mathscr{P}$, the graph $\bar{\mathscr{G}}_p$ is directed and contains a spanning tree with the root node being the leader node 0.

Based on Assumptions 4.1–4.2 and Lemma 4 in [19], it is known that $\mathscr{H}$ ($\mathscr{H}_p$) is a $\mathscr{M}$-matrix, i.e., the non-diagonal entries of $\mathscr{H}$ ($\mathscr{H}_p$) are non-positive and all its eigenvalues have positive real part. Then, it has the following technical lemma.

Lemma 4.1 ([18]) *There exists a positive definite diagonal matrix* $\mathscr{Q} = diag\{q_1, q_2, \ldots, q_N\}$ *such that* $\mathscr{Q}\mathscr{H} + \mathscr{H}^T\mathscr{Q}$ *(or* $\mathscr{Q}\mathscr{H}_p + \mathscr{H}_p^T\mathscr{Q}$*) is positive definite.*

Given the above discussions, for the static network case, we define the combination error variable

$$z_i(t) = \sum_{j \in \mathscr{N}_i^{in}} (\hat{x}_j(t) - \hat{x}_i(t)), \tag{4.3}$$

and for the switching network case, we define

$$z_i(t) = \sum_{j \in \mathscr{N}_i^{in}(t)} (\hat{x}_j(t) - \hat{x}_i(t)) \tag{4.4}$$

Then, the sampled-data based consensus control protocol can be designed as follows

$$u_i(t) = Kz_i(t) + K_r x_{ri}(t), \tag{4.5}$$

where $Kz_i(t)$ is the proportional control part, and $K_r x_{ri}(t)$ is the reset control part. $\hat{x}_i(t)$, $\hat{x}_j(t)$ denote the open-loop estimate of agent i and j. The dynamics of $\hat{x}_i(t)$, $i \in [0, N]$, is given by

$$\begin{cases} \dot{\hat{x}}_i(t) = A\hat{x}_i(t), & t \in (t_k^i, t_{k+1}^i] \\ \hat{x}_i^+(t) = x_i(t), & t = t_k^i \end{cases} \tag{4.6}$$

where t_k^i denotes agent i's sampling instant, and the sampling period $h_i = t_{k+1}^i - t_k^i$. Based on the consensus control protocol (4.5), each agent i transmits information to its neighbors $j, j \in \mathcal{N}_i^{out}(t)$, at t_k^i, and passively receives information from its neighbors (i.e., agent $j, j \in \mathcal{N}_i^{in}(t)$) at t_k^j. Thus, this control scheme can be implemented under network environment.

Define $e_i(t) = \hat{x}_i(t) - x_i(t)$, then, to incorporate the dynamics of $e_i(t)$ into the closed-loop system and to get the upper bound of h_i, we introduce a timer variable $\tau_{hi}(t) \in \mathbb{R}$

$$\left.\begin{aligned} \dot{\tau}_{hi} &= 1 \\ \dot{e}_i &= \dot{\hat{x}}_i - \dot{x}_i \end{aligned}\right\} \text{ If } \tau_{hi} < h_i, \qquad \left.\begin{aligned} \tau_{hi}^+ &= 0 \\ e_i^+ &= 0 \end{aligned}\right\} \text{ If } \tau_{hi} = h_i. \tag{4.7}$$

From (4.6) and (4.7), it can be known that $e_i(t)$ and τ_{hi} have jump dynamics at sampling instant t_k^i. For this reason, the concept of generalized directional derivative (i.e., Clarke derivative [20]) is introduced for these discontinuous variable. $x_{ri} \in \mathbb{R}^n$ is the reset state, which is utilized to improve transient performance, the dynamics of x_{ri} is given by

$$\left.\begin{aligned} \dot{\tau}_{ri} &= 1_n \\ \dot{x}_{ri} &= a_r x_{ri} + b_r z_i \end{aligned}\right\} \text{ If } (x_{ri}, z_i, \tau_{ri}) \in \bigcap_{m=1}^{n} \mathbb{F}_{rim}, \qquad \left.\begin{aligned} \tau_{ri}^+ &= R_{\tau i}\tau_{ri} \\ x_{ri}^+ &= R_{ri}x_{ri} \end{aligned}\right\} \text{ If } (x_{ri}, z_i, \tau_{ri}) \in \bigcup_{m=1}^{n} \mathbb{J}_{rim}, \tag{4.8}$$

where a_r, b_r are constant parameters to be determined, $\tau_{ri} = \text{col}(\tau_{ri1}, \ldots, \tau_{rim}, \ldots, \tau_{rin}) \in \mathbb{R}^n$, with $\tau_{rim}, m \in [1, n]$, denoting the internal timer variable, that is used to record the time elapsed since the last reset instant of state x_{rim}. The flow set $\mathbb{F}_{rim}$ and jump set $\mathbb{J}_{rim}$ (for $i \in [1, N], m \in [1, n]$) are defined as follows

$$\begin{aligned} \mathbb{F}_{rim} &= \{(x_{ri}, z_i, \tau_{ri}) | (x_{rim}z_{im} \geq 0 \wedge \tau_{hi} = h_i \wedge \tau_{rim} < \Delta_{im}h_i) \vee \tau_{hi} < h_i\} \\ \mathbb{J}_{rim} &= \{(x_{ri}, z_i, \tau_{ri}) | (x_{rim}z_{im} \leq 0 \wedge \tau_{hi} = h_i) \vee \tau_{rim} \geq \Delta_{im}h_i\}. \end{aligned} \tag{4.9}$$

$R_{\tau i}, R_{ri} \in \mathbb{R}^{n \times n}$ denote reset map matrices, with $R_{\tau i} = \text{diag}\{r_{\tau i1}, \ldots, r_{\tau im}, \ldots, r_{\tau in}\}$, $R_{ri} = \text{diag}\{r_{ri1}, \ldots, r_{rim}, \ldots, r_{rin}\}$, where $r_{\tau im}$ and $r_{rim}, m \in [1, n]$, are determined by

$$\begin{cases} r_{\tau im} = 1, & \text{If } (x_{rim}, z_{im}, \tau_{rim}) \in \mathbb{F}_{rim} \\ r_{\tau im} = 0, & \text{If } (x_{rim}, z_{im}, \tau_{rim}) \in \mathbb{J}_{rim} \end{cases}$$

$$\begin{cases} r_{rim} = 1, & \text{If } (x_{rim}, z_{im}, \tau_{rim}) \in \mathbb{F}_{rim} \\ r_{rim} = 0, & \text{If } (x_{rim}, z_{im}, \tau_{rim}) \in \mathbb{J}_{rim} \wedge x_{rim} z_{im} \leq 0 \\ r_{rim} = p_{im}, & \text{If } (x_{rim}, z_{im}, \tau_{rim}) \in \mathbb{J}_{rim} \wedge \tau_{rim} \geq \Delta_{im} h_i \end{cases}$$

where $0 < p_{im} < 1$, $(x_{rim}, z_{im}, \tau_{rim}) \in \mathbb{F}_{rim}/\mathbb{J}_{rim}$ means that the judging conditions in (4.9) hold for a certain $m \in [1, n]$. How to determine $\Delta_{im} \in \mathbb{N}_{>0}$ will be specified later. In addition, the gain matrices K, K_r and the upper bound of h_i will also be specified latter.

4.3 Consensus Control Under Static Network

In this section, we first consider the static network case. Based on the suitable combination error variable (4.3), we design an asynchronous sampled-data hybrid consensus control protocol (4.5) and make a persistently effective interaction rule among the fixed graph. Subsequently, the asymptotical leader-following consensus of MAS and better transient consensus performance are guaranteed. To facilitate the stability analysis, several internal variables are introduced properly, then a hybrid model is constructed to completely describe the flow and jump dynamics of closed-loop system. Based on this model and internal decreasing functions, Lyapunov-based stability conditions are derived, and larger upper bound of sampling period can be obtained.

The leader-following consensus of MAS requires that all the followers can track the leader's state trajectory. It means that if we can prove the state difference between any follower and the leader converge to 0, then the group behavior of MAS would attain synchronization. Therefore, the state difference between leader's state and followers' state can be viewed as the consensus error and used for stability analysis, that is,

$$\delta_i(t) = x_i(t) - x_0(t), i \in [1, N]. \tag{4.10}$$

According to (4.1), (4.2), and (4.5), we have

$$\begin{aligned} \dot{\delta}_i &= A\delta_i + BKz_i + BK_r x_{ri}, \\ \dot{e}_i &= Ae_i - BKz_i - BK_r x_{ri}. \end{aligned} \tag{4.11}$$

Note that the error e_i has jump dynamics at agent i's sampling instants, the internal reset element state x_{ri} has jump dynamics when reset conditions are met, and the timer variables τ_{hi}, τ_{ri} also have jump dynamics. Thus, only with the conventional continuous-time system modelling method cannot completely describe the dynamics of closed-loop state of MAS containing continuous-discrete variables e_i, x_{ri}, τ_{hi},

τ_{ri}. Thus, in view of the hybrid system model (1.1) and the previously introduced modelling conveniences with it, we will model the closed-loop MAS toward a hybrid dynamic system.

Let

$$\begin{aligned}&\delta = \mathrm{col}(\delta_1, \ldots, \delta_N),\ e = \mathrm{col}(e_1, \ldots, e_N),\ x_r = \mathrm{col}(x_{r1}, \ldots, x_{rN}),\\&\tau_h = \mathrm{col}(\tau_{h1}, \ldots, \tau_{hN}),\ \tau_r = \mathrm{col}(\tau_{r1}, \ldots, \tau_{rN}),\ \xi = \mathrm{col}(\delta, x_r, e, \tau_h, \tau_r),\end{aligned} \tag{4.12}$$

where col$(\cdots)$ denotes a column vector. Then, according to (4.1)–(4.5), (4.7)–(4.8), (4.11), and note that $z_i = -(\mathscr{H}^i_{row} \otimes I_n)(\delta + e)$, where $\mathscr{H}^i_{row}$ denotes the ith row of matrix $\mathscr{H}$, the closed-loop MAS dynamics can be described by the following hybrid systems form

$$\begin{aligned}&\dot{\xi} = \mathscr{F}(\xi),\ \text{If}\ \xi \in \mathbb{F},\\&\xi^+ \in \bigcup_{i=1}^{N}\bigcup_{m=1}^{n} \mathscr{J}_{im}(\xi),\ \ \text{If}\ \xi \in \mathbb{J},\end{aligned} \tag{4.13}$$

where the flow set $\mathbb{F} = \bigcap_{i=1}^N \mathbb{F}_i$, and $\mathbb{F}_i = \{\xi \in \mathbb{X} | \tau_{hi} < h_i\}$, where $\mathbb{X} := \{(\delta, x_r, e, \tau_h, \tau_r) \in \mathbb{R}^{nN} \times \mathbb{R}^{nN} \times \mathbb{R}^{nN} \times \mathbb{R}^N_{\geq 0} \times \mathbb{R}^{nN}_{\geq 0}\}$. The flow dynamics $\mathscr{F}(\xi)$ has the following form

$$\mathscr{F}(\xi) = \begin{bmatrix} (I_N \otimes A - \mathscr{H} \otimes BK)\delta - (\mathscr{H} \otimes BK)e + (I_N \otimes BK_r)x_r \\ a_r x_r - b_r(\mathscr{H} \otimes I_n)(\delta + e) \\ (I_N \otimes A + \mathscr{H} \otimes BK)e + (\mathscr{H} \otimes BK)\delta - (I_N \otimes BK_r)x_r \\ \mathbf{1}_N \\ \mathbf{1}_{nN} \end{bmatrix}.$$

The jump set $\mathbb{J} := \bigcup_{i=1}^N \mathbb{J}_i$, and $\mathbb{J}_i = \{\xi \in \mathbb{X} | \tau_{hi} = h_i\}$. The jump dynamics $\mathscr{J}_{im}(\xi)$ has the following form

$$\mathscr{J}_{im}(\xi) = \begin{cases} \mathscr{J}^1_{im}(\xi),\ \text{If}\ \xi \in \mathbb{J}_i \wedge (x_{rim}, z_{im}, \tau_{rim}) \in \mathbb{J}_{rim} \wedge x_{rim} z_{im} \geq 0 \wedge \tau_{rim} < \Delta_{im} h_i \\ \mathscr{J}^2_{im}(\xi),\ \text{If}\ \xi \in \mathbb{J}_i \wedge (x_{rim}, z_{im}, \tau_{rim}) \in \mathbb{J}_{rim} \wedge x_{rim} z_{im} \leq 0 \wedge \tau_{rim} < \Delta_{im} h_i \\ \mathscr{J}^3_{im}(\xi),\ \text{If}\ \xi \in \mathbb{J}_i \wedge \tau_{rim} \geq \Delta_{im} h_i \end{cases} \tag{4.14}$$

where

$$\begin{aligned}&\mathscr{J}^1_{im}(\xi) = \mathrm{col}(\delta, x_r, (\Lambda_i \otimes I_n)e, \Lambda_i \tau_h, \tau_r),\\&\mathscr{J}^2_{im}(\xi) = \mathrm{col}(\delta, ((I_N - \Lambda_i) \otimes \Lambda_m + \Lambda_i \otimes I_n)x_r, (\Lambda_i \otimes I_n)e, \Lambda_i \tau_h, ((I_N - \Lambda_i) \otimes \Lambda_m + \Lambda_i \otimes I_n)\tau_r),\\&\mathscr{J}^3_{im}(\xi) = \mathrm{col}(\delta, (I_{nN} - (1 - p_{im})(I_N - \Lambda_i) \otimes (I_n - \Lambda_m))x_r,\\&\qquad (\Lambda_i \otimes I_n)e, \Lambda_i \tau_h, ((I_N - \Lambda_i) \otimes \Lambda_m + \Lambda_i \otimes I_n)\tau_r),\end{aligned}$$

where $\Lambda_i \in \mathbb{R}^{N \times N}$ ($\Lambda_m \in \mathbb{R}^{n \times n}$) is diagonal matrix, the ith (mth) diagonal element is zero and the rest diagonal elements are ones.

Before proceeding further, we introduce two internal functions inspired by [21], which are useful for stability analysis. With respect to the timer variable τ_{hi}, the internal function $\phi_{hi}(\tau_{hi}) : \mathbb{R}_{\geq 0} \to \mathbb{R}_{\geq 0}$, $i \in [1, N]$ is introduced as

$$\dot{\phi}_{hi}(\tau_{hi}) = -\theta_{hi}(\phi_{hi}^2(\tau_{hi}) + 1), \tag{4.15}$$

where θ_{hi} and the initial $\phi_{hi}(0)$ are chosen to be positive. It is obvious that the function $\phi_{hi}(\tau_{hi})$ is strictly decreasing, thus for any given $\phi_{hi}(0)$, it can always find an explicit upper bound $\bar{h}_i$ such that $\phi_{hi}(\bar{h}_i) = 0$ and $\phi_{hi}(\tau_{hi})$ becomes negative as long as τ_{hi} exceeds this value. Note that the timer variable τ_{hi} plays a significant role to divide the flow set $\mathbb{F}_i$ and jump set $\mathbb{J}_i$, therefore, it is needed to incorporate the function $\phi_{hi}(\tau_{hi})$ to construct Lyapunov function. From (4.13), it is known that the jump dynamics only occurs at the sampling instants period $\tau_{hi} = h_i$, and within time period $0 \leq \tau_{hi} \leq h_i \leq \bar{h}_i$, the function $\phi_{hi}(\tau_{hi})$ is always nonnegative, thus, the function $\phi_{hi}(\tau_{hi})$ is suitable for the analysis of sampled-data control. The sampling period h_i can be arbitrarily selected from the value range $(0, \bar{h}_i]$, where the upper bound $\bar{h}_i = \arctan(\phi_{hi}(0))/\theta_{hi}$ is obtainable by integrating both side of (4.15).

Next, consider the timer variable τ_{rim}, we introduce another internal function $\phi_{rim}(\tau_{rim})$, which is similar as (4.15). The dynamics of $\phi_{rim}(\tau_{rim})$ is given by

$$\dot{\phi}_{rim}(\tau_{rim}) = -b_r\theta_{rim}(\phi_{rim}^2(\tau_{rim}) + 1), \tag{4.16}$$

for $i \in [1, N]$, $m \in [1, n]$, where θ_{rim} and the initial condition $\phi_{rim}(0)$ are chosen to be positive. The function $\phi_{rim}(\tau_{rim})$ will also be used to construct Lyapunov function, according to the reset conditions in (4.8), τ_{rim} is always within the range $0 \leq \tau_{rim} \leq \Delta_{im}h_i$. Then, given a specified h_i, the integer parameter Δ_{im} can be determined by the condition

$$0 < \Delta_{im} < \frac{\arctan(\phi_{rim}(0))}{b_r\theta_{rim}h_i}. \tag{4.17}$$

Observe that (A, B) is stabilizable and (I_n, A) is observable, it can be known from [11, Eq. (16)] that there always exists a positive symmetric matrix $P > 0$ such that the following Riccati equation holds

$$PA + A^T P - \mu_1 PBB^T P + \mu_2 I_n = 0 \tag{4.18}$$

where $\mu_1 > 0$, $\mu_2 > 0$ are constants. Similar to [11], define $q_M = \lambda_{\max}(\mathcal{Q})$, $\lambda_1 = \lambda_{\min}(\mathcal{Q}\mathcal{H} + \mathcal{H}^T\mathcal{Q})$, $\alpha = \mu_1 q_M/\lambda_1$.

Theorem 4.1 *Consider the MAS* (4.1) *and* (4.2) *in the static network case satisfying Assumption 4.1. For the consensus protocol* (4.5) *with* $K = \alpha B^T P$, *let* $\theta_{hi} > 0$, $\theta_{rim} > 0$, $\phi_{hi}(0) > 0$, $\phi_{rim}(0) > 0$, *if the sampling period* h_i *satisfies* $0 \leq h_i \leq \bar{h}_i = \arctan(\phi_{hi}(0))/\theta_{hi}$, *the integer parameter* Δ_{im} *satisfies condition* (4.17), *and there* *exist matrix K_r, constants $0 < \rho < 1$, $\epsilon > 0$, $0 < p_{im} < 1$ $a_r < 0$, $b_r > 0$, such that $p_{im}^2 \phi_{rim}(0) \le \phi_{rim}(\Delta_{im} h_i)$, and the following inequality conditions are satisfied*

$$\begin{bmatrix} \Omega_{11} & \Omega_{12} & \mathcal{Q} \otimes PBK_r + \sum_{i=1}^N E_{i1}^{eT} E_{i3}^e \\ * & \Omega_{22} & \sum_{i=1}^N E_{i2}^{eT} E_{i3}^e \\ * & * & -(1-\epsilon) b_r^2 \Theta_r^2 + \sum_{i=1}^N E_{i3}^{eT} E_{i3}^e + 2a_r \Phi_r(\tau_r) \end{bmatrix} \le 0 \tag{4.19}$$

with

$$\begin{aligned} \Omega_{11} &= -(1-\rho)(\mathcal{Q} \otimes \mu_2 I_n) + \sum_{i=1}^N E_{i1}^{eT} E_{i1}^e + \Omega_z, \\ \Omega_{12} &= -\alpha \mathcal{Q} \mathcal{H} \otimes PBB^T P + \sum_{i=1}^N E_{i1}^{eT} E_{i2}^e + \Omega_z, \\ \Omega_{22} &= -(1-\epsilon)(\Theta_h^2 \otimes I_n) + \sum_{i=1}^N E_{i2}^{eT} E_{i2}^e + \Omega_z, \\ \Omega_z &= \sum_{i=1}^N \sum_{m=1}^n E_i^{zT} E_m^{zT} E_m^z E_i^z \end{aligned}$$

where $E_{i1}^e = E_i(\mathcal{H} \otimes BK)$, $E_{i2}^e = E_i(I_N \otimes A + \mathcal{H} \otimes BK)$, $E_{i3}^e = -E_i(I_N \otimes BK_r)$, $E_i = [0, \ldots, I_n, \ldots, 0]_{n \times nN}$, $E_i^z = -\mathcal{H}_{row}^i \otimes I_n$, $E_m^z = [0, \ldots, 1, \ldots, 0]_{1 \times n}$, $\Theta_h = \text{diag}\{\theta_{h1}, \theta_{h2}, \ldots, \theta_{hN}\}$ *and* $\Theta_r = \text{diag}\{\theta_{r1}, \theta_{r2}, \ldots, \theta_{rN}\}$, $\theta_{ri} = \text{diag}\{\theta_{ri1}, \theta_{ri2}, \ldots, \theta_{rin}\}$. *Then, the (δ, x_r, e) dynamics is asymptotically stable, and the leader-following consensus is achieved.*

Proof Choose the candidate Lyapunov function

$$V(\xi) = \delta^T(\mathcal{Q} \otimes P)\delta + e^T \Phi_h(\tau_h) e + x_r^T \Phi_r(\tau_r) x_r, \tag{4.20}$$

where

$$\begin{aligned} \Phi_h(\tau_h) &= \text{diag}\{\theta_{h1}\phi_{h1}(\tau_{h1}), \ldots, \theta_{hN}\phi_{hN}(\tau_{hN})\} \otimes I_n, \\ \Phi_r(\tau_r) &= \text{diag}\{\Phi_{r1}(\tau_{r1}), \ldots, \Phi_{rN}(\tau_{rN})\}, \\ \Phi_{ri}(\tau_{ri}) &= \text{diag}\{\theta_{ri1}\phi_{ri1}(\tau_{ri1}), \ldots, \theta_{rin}\phi_{rin}(\tau_{rin})\}. \end{aligned} \tag{4.21}$$

Observe the form of (4.20), it is easy to find that the continuous-discrete error variable e_i and reset state x_{ri} have been incorporated into the construction of Lyapunov function design. Based on (4.15)–(4.17), $\theta_{hi}\phi_{hi}(\tau_{hi}) > 0$ and $\theta_{rim}\phi_{rim}(\tau_{rim}) > 0$ can be ensured. Hence, the combination of continuous-discrete error variables and the internal functions still guarantees the positive definiteness. Due to $\delta^T(\mathcal{Q} \otimes P)\delta$ is radially unbounded, $V(\xi)$ is also radially unbounded and $V(\xi) = 0$ if and only if $\delta = 0, e = 0, x_r = 0$. This implies that $V(\xi)$ is a suitable Lyapunov function for the sampled-data control based stability analysis.

Since there exist both flow dynamics and jump dynamics in the closed-loop state ξ, first, consider the flow dynamics, we have

$$
\begin{aligned}
\dot{V}(\xi) = & 2\delta^T((\mathscr{Q} \otimes P)(I_N \otimes A - \mathscr{H} \otimes BK))\delta - 2\delta^T(\mathscr{Q} \otimes P)((\mathscr{H} \otimes BK)e - (I_N \otimes BK_r)x_r) \\
& + \sum_{i=1}^{N} 2\theta_{hi}\phi_{hi}(\tau_{hi})e_i^T\dot{e}_i - \sum_{i=1}^{N} \theta_{hi}^2(\phi_{hi}^2(\tau_{hi}) + 1)e_i^T e_i \\
& + \sum_{i=1}^{N}\sum_{m=1}^{n} 2\theta_{rim}\phi_{rim}(\tau_{rim})x_{rim}(a_r x_{rim} + b_r z_{im}) \\
& - \sum_{i=1}^{N}\sum_{m=1}^{n} b_r^2\theta_{rim}^2(\phi_{rim}^2(\tau_{rim}) + 1)x_{rim}^2
\end{aligned}
\tag{4.22}
$$

In the above equation,

$$
\begin{aligned}
& 2\delta^T((\mathscr{Q} \otimes P)(I_N \otimes A - \mathscr{H} \otimes BK))\delta \\
= & \delta^T(\mathscr{Q} \otimes (PA + A^T P) - \alpha(\mathscr{Q}\mathscr{H} + \mathscr{H}^T\mathscr{Q}) \otimes PBB^T P)\delta \\
\leq & -\mu_2\delta^T(\mathscr{Q} \otimes I_n)\delta.
\end{aligned}
\tag{4.23}
$$

In addition, note that $\dot{e}_i = E_i\dot{e} = E_{i1}^e\delta + E_{i2}^e e + E_{i3}^e x_r$, and $z_{im} = E_m^z z_i = E_m^z E_i^z (\delta + e)$, it follows that

$$
\begin{aligned}
\dot{V}(\xi) \leq & -\mu_2\delta^T(\mathscr{Q} \otimes I_n)\delta - \sum_{i=1}^{N} \theta_{hi}^2(\phi_{hi}^2(\tau_{hi}) + 1)e_i^T e_i \\
& - 2\delta^T(\mathscr{Q} \otimes P)((\mathscr{H} \otimes BK)e - (I_N \otimes BK_r)x_r) \\
& + \sum_{i=1}^{N} 2\theta_{hi}\phi_{hi}(\tau_{hi})e^T E_i^T(E_{i1}^e\delta + E_{i2}^e e + E_{i3}^e x_r) \\
& + \sum\nolimits_{i=1}^{N}\sum\nolimits_{m=1}^{n} 2a_r\theta_{rim}\phi_{rim}(\tau_{rim})x_{rim}^2 \\
& + \sum_{i=1}^{N}\sum_{m=1}^{n} 2b_r\theta_{rim}\phi_{rim}(\tau_{rim})x_r^T E_i^{zT} E_m^{zT} E_m^z E_i^z(\delta + e) \\
& - \sum\nolimits_{i=1}^{N}\sum\nolimits_{m=1}^{n} b_r^2\theta_{rim}^2(\phi_{rim}^2(\tau_{rim}) + 1)x_{rim}^2.
\end{aligned}
\tag{4.24}
$$

Based on (4.19), let the matrix in (4.19) left multiplied by $[\delta^T \ e^T \ x_r^T]$ and right multiplied by $[\delta^T \ e^T \ x_r^T]^T$, we have

$$
\begin{aligned}
& -\mu_2\delta^T(\mathscr{Q} \otimes I_n)\delta - 2\delta^T(\mathscr{Q} \otimes P)(\mathscr{H} \otimes BK)e + 2\delta^T(\mathscr{Q} \otimes P)I_N \otimes BK_r)x_r \\
\leq & -\sum_{i=1}^{N} \| E_{i1}^e\delta + E_{i2}^e e + E_{i3}^e x_r \|^2 - \sum_{i=1}^{N}\sum_{m=1}^{n}(\delta + e)^T E_i^{zT} E_m^{zT} E_m^z E_i^z(\delta + e) \\
& + e^T(1 - \epsilon)\Theta_h^2 e + x_r^T\left((1 - \epsilon)b_r^2\Theta_r^2 + 2a_r\Phi_r(\tau_r)\right)x_r - \rho\delta^T(\mathscr{Q} \otimes \mu_2 I_n)\delta.
\end{aligned}
\tag{4.25}
$$

Note that $z_{im} = E_m^z z_i = E_m^z E_i^z(\delta + e)$, Θ_h, Θ_r and $\Phi_r(\tau_r)$ are diagonal matrices such that $e^T(\Theta_h^2 \otimes I_n)e = \sum_{i=1}^N e_i^T \theta_{hi}^2 e_i$, $x_r^T \Theta_r^2 x_r = \sum_{i=1}^N \sum_{m=1}^n x_{rim}^T \theta_{rim}^2 x_{rim}$, $x_r^T \Phi_r(\tau_r) x_r = \sum_{i=1}^N \sum_{m=1}^n x_{rim}^T \theta_{rim} \phi_{rim}(\tau_{rim}) x_{rim}$, thus, from (4.24) and (4.25), we obtain

$$
\begin{aligned}
\dot{V}(\xi) \leq & -\rho\delta^T(\mathcal{Q} \otimes \mu_2 I_n)\delta - \sum_{i=1}^N (\dot{e}_i - \theta_{hi}\phi_{hi}e_i)^T(\dot{e}_i - \theta_{hi}\phi_{hi}e_i) \\
& - \sum_{i=1}^N \sum_{m=1}^n (z_{im} - b_r\theta_{rim}\phi_{rim})^T (z_{im} - b_r\theta_{rim}\phi_{rim}) \\
& - \sum_{i=1}^N \theta_{hi}^2 \phi_{hi}^2(\tau_{hi}) e_i^T e_i - \sum_{i=1}^N \sum_{m=1}^n b_r^2 \theta_{rim}^2 \phi_{rim}^2(\tau_{rim}) x_{rim}^2 \\
& + e^T(1-\epsilon)\Theta_h^2 e + x_r^T \left((1-\epsilon)b_r^2\Theta_r^2 + 2a_r\Phi_r(\tau_r)\right) x_r \\
& + \sum_{i=1}^N \sum_{m=1}^n 2a_r \theta_{rim}\phi_{rim}(\tau_{rim}) x_{rim}^2 \\
\leq & -\rho\delta^T(\mathcal{Q} \otimes \mu_2 I_n)\delta - \epsilon e^T \Theta_h^2 e - \epsilon x_r^T \Theta_r^2 x_r
\end{aligned}
\tag{4.26}
$$

Then, consider the change of $V(\xi)$ when jump dynamics occurs, according to (4.13) and (4.14), the jump dynamics ξ^+ includes three cases. Let $\Gamma_i = \Lambda_i \otimes I_n$, $\Lambda_{im}^1 = (I_N - \Lambda_i) \otimes \Lambda_m + \Lambda_i \otimes I_n$, $\Lambda_{im}^2 = I_{nN} - (1 - p_{im})(I_N - \Lambda_i) \otimes (I_n - \Lambda_m)$. When $\mathcal{J}_{im}(\xi) = \mathcal{J}_{im}^1(\xi)$, from (4.14) and (4.20), we have

$$
\begin{aligned}
& V(\xi^+) - V(\xi) \\
= & \, \delta^T(\mathcal{Q} \otimes P)\delta + e^T \Gamma_i^T \Phi_h(\Lambda_i \tau_h)\Gamma_i e + x_r^T \Phi_r(\tau_r) x_r \\
& - \delta^T(\mathcal{Q} \otimes P)\delta - e^T \Phi_h(\tau_h) e - x_r^T \Phi_r(\tau_r) x_r \leq 0.
\end{aligned}
\tag{4.27}
$$

When $\mathcal{J}_{im}(\xi) = \mathcal{J}_{im}^2(\xi)$, we have

$$
\begin{aligned}
& V(\xi^+) - V(\xi) \\
= & \, \delta^T(\mathcal{Q} \otimes P)\delta - \delta^T(\mathcal{Q} \otimes P)\delta + e^T \Gamma_i^T \Phi_h(\Lambda_i \tau_h)\Gamma_i e \\
& - e^T \Phi_h(\tau_h) e + x_r^T \Lambda_{im}^{1T} \Phi_r(\Lambda_{im}^1 \tau_r) \Lambda_{im}^1 x_r - x_r^T \Phi_r(\tau_r) x_r \\
\leq & \, 0.
\end{aligned}
\tag{4.28}
$$

When $\mathcal{J}_{im}(\xi) = \mathcal{J}_{im}^3(\xi)$, we have

$$
\begin{aligned}
& V(\xi^+) - V(\xi) \\
= & \, \delta^T(\mathcal{Q} \otimes P)\delta - \delta^T(\mathcal{Q} \otimes P)\delta + e^T \Gamma_i^T \Phi_h(\Lambda_i \tau_h)\Gamma_i e \\
& - e^T \Phi_h(\tau_h) e + x_r^T \Lambda_{im}^{2T} \Phi_r(\Lambda_{im}^1 \tau_r) \Lambda_{im}^2 x_r - x_r^T \Phi_r(\tau_r) x_r \\
\leq & \, p_{im}^2 x_{rim}^2 \theta_{rim}\phi_{rim}(0) - x_{rim}^2 \theta_{rim}\phi_{rim}(\Delta_{im} h_i) \\
\leq & \, 0.
\end{aligned}
\tag{4.29}
$$

From (4.26) to (4.29), we can see that the Lyapunov function is decreasing along flow dynamics and is non-increasing along jump dynamics. Therefore, we can conclude that the MAS is asymptotically stable, and $\lim_{t\to\infty} V(\xi(t)) = 0$, which implies that $\lim_{t\to 0} \|\delta_i(t)\| = \lim_{t\to 0} \|x_i(t) - x_0(t)\| = 0$, namely the leader-following consensus is achieved. □

Remark 4.1 In view of the introduced internal function (4.15), the decreasing nature makes it be appropriate for stability analysis, as from (4.22) to (4.26) it can be seen that the combination of this internal function and error variable e_i guarantees the closed-loop performance. And, the decreasing time of $\phi_{hi}(\tau_{hi})$ from an given initial value $\phi_{hi}(0)$ to 0 leads to an explicitly allowable value range $(0, \bar{h}_i]$, within which the sampling period h_i ($\tau_{hi} \le h_i$) can be chosen independently for agent i. Note that the UBSP $\bar{h}_i = \arctan(\phi_{hi}(0))/\theta_{hi}$ can be calculated explicitly, which is essentially different from the calculation in [11], and generally larger than [11]. In [11], the sampling periods of all agents are set based on a unit sampling period $T_s > 0$. That is, each agent i's sampling period is an integral multiple of T_s. Such selection of sampling periods is necessary for deriving stability conditions in [11], however, selecting sampling period in this way is not necessary in practice. With the proposed method, it is only required that $0 < h_i < \bar{h}_i$ with $\bar{h}_i$ being the upper bound of h_i. Besides, since asynchronous sampling mechanism is used, the jump dynamics of $e_i(t)$, $i \in [1, N]$, is independent of each other.

Remark 4.2 In the designed consensus control protocol (4.5), an additional reset control part $K_r x_r(t)$ is newly added compared with the commonly used linear controllers, which is inspired by the reset control method [12, 14, 15]. It can be seen that if we let $K_r = 0$, then the control protocol (4.5) will be reduced to the general linear form. Thus, the designed consensus controller u_i inherits the merits of reset control in aspect of improving transient consensus performance. To be specific, (4.5) can be equivalently represented as $u_i = \sum_{m=1}^{n}(K_m z_{im} + K_{rm} x_{rim})$, where K_m, K_{rm} denote the mth entries of K, K_r. From (4.8), $\dot{x}_{rim} = a_r x_{rim} + b_r z_{im}$ when x_{rim} flows and $x_{rim}^+ = 0$ when x_{rim} resets. It implies that x_{rim} can be viewed as the integral of z_{im} since the last reset instant, so, if K_m and K_{rm} have the same sign, the integral action is beneficial for fast response time. In addition, the integral of z_{im} memorized via x_{rim} is cleared at reset instant, which is beneficial for reducing overshoot. Thus, we can select $K_r = \zeta K$, $\zeta > 0$, such that the introduction of reset control has potential to improve transient performance, this idea was also considered in [16, 17] for first-order MAS. But, as illustrated in [14, 22], how to theoretically show that reset control can achieve better performance than linear control is still an open problem.

Remark 4.3 According to jump dynamics of closed-loop state (4.14), when $\mathcal{J}_{im}(\xi) = \mathcal{J}_{im}^2(\xi)$, the reset state x_{rim} is reset to zero, then, for the difference equation (4.28), it naturally holds the property that the Lyapunov function is non-increasing. Different from the former case, when $\mathcal{J}_{im}(\xi) = \mathcal{J}_{im}^3(\xi)$, the reset state x_{rim} is reset to $p_{im} x_{rim}$ rather than 0. However in this case, similar property is not held for difference equation (4.29), thus, it required the constraint condition $p_{im}^2 \phi_{rim}(0) \le \phi_{rim}(\Delta_{im} h_i)$. Note that small p_{im} implies a relatively large Δ_{im} for a specified sampling period h_i, and vice versa, but small p_{im} may weaken the role of reset control. So, there is tradeoff between p_{im} and Δ_{im}. Since at each sampling instants e_i jumps to zero, thus at the reset instant $\tau_{rim} = \Delta_{im} h_i$, e_i and x_{rim} ($m \in [1, n]$) have the simultaneous jumps, where e_i will jump to zero. For this reason, by means of jump dynamics of e_i, we can make a slight modification to the aforementioned constraint condition. Specifically, the parameter p_{im} can be chosen such that the following constraint condition holds $p_{im}^2 \leq \frac{\theta_{hi}\phi_{hi}(\tau_{hi})e_i^T e_i}{n\theta_{rim}\phi_{rim}(0)x_{rim}^2} + \frac{\phi_{rim}(\Delta_{im}h_i)}{\phi_{rim}(0)}$, then, $V(\xi^+) - V(\xi) \leq 0$ still holds for $\mathscr{J}_{im}(\xi) = \mathscr{J}_{im}^3(\xi)$.

Remark 4.4 For the implementation of controller $u_i(t)$, some parameters (such as $K, K_r, a_r, b_r, \Delta_{im}, h_i, p_{im}, \theta_{hi}, \theta_{rim}$) are needed. The parameters K, K_r, a_r, b_r are needed to be computed and assigned to agents before the MAS starts to operate, and other parameters can be computed (based on (4.8)–(4.9), (4.15)–(4.17)) in a distributed way. But, to verify the stability condition (4.19), the parameters of all agents are needed. In addition, due to there exist nonlinear terms (such as $b_r \otimes \Theta_r^2$) related to decision variables, the optimal solution to (4.19) is difficult to compute. But, if some parameters (e.g., $a_r, b_r, K_r, \theta_{hi}, \theta_{ri}$) are pre-given, (4.19) is reduced to linear matrix inequality and a solution can be found. Specifically, if there exist P, μ_1, μ_2 such that Riccati equation (4.18) is satisfied, we can select $\bar{P} = \kappa P$, $\bar{\mu}_2 = \kappa\mu_2$, $\bar{\mu}_1 = \mu_1/\kappa$ (with $\kappa > 0$), and obtain

$$\bar{P}A + A^T\bar{P} - \bar{\mu}_1\bar{P}BB^T\bar{P} + \bar{\mu}_2 I_n = \kappa(PA + A^T P - \mu_1 PBB^T P + \mu_2 I_n).$$

The above equation means that $\bar{P}, \bar{\mu}_1, \bar{\mu}_2$ are also solutions of Riccati equation (4.18), but the gain matrix $\bar{K} = \frac{\bar{\mu}_1 d_M}{\lambda_1} B^T\bar{P} = \frac{\mu_1 d_M}{\lambda_1} B^T P = \alpha B^T P = K$ remains unchanged. Thus, using $\bar{\mu}_2$ instead of μ_2 and based on definition of Ω_{11}, the condition $\Omega_{11} < 0$ can be guaranteed with sufficiently large $\bar{\mu}_2$. In addition, note that θ_{hi}, θ_{ri} can be arbitrarily selected and are independent of other terms in the matrix in (4.19), which means that $\Omega_{22} < 0$ and $\Omega_{33} < 0$ can be guaranteed by selecting sufficiently large θ_{hi}, θ_{ri}. Therefore, if a solution of (4.19) cannot be found with pre-given parameters, we can increase μ_2 (μ_1 and P are changed accordingly), θ_{ri}, θ_{hi} until a solution is found.

4.4 Consensus Control Under Switching Network

In this section, we extend the results obtained in former section to the switching network case. First, in view of the preliminaries regarding switching graph introduced in Chap. 1, in this case, the communication graph $\mathscr{G}_{\sigma(t)}$ is time-varying and belongs to a finite graphs set. Then, by means of the suitable combination error variable (4.4) and consensus control protocol (4.5) for this case, and based on the similar modelling procedure, the closed-loop system under switching network is given by a hybrid system. Furthermore, it will be proved in this section that the system stability is guaranteed and the the results obtained in former section can be applied to the switching network case.

In the switching network case, suppose that Assumption 4.2 is satisfied, then, for the time-varying graph $\bar{\mathscr{G}}_p$, $z_i(t)$ in $u_i(t)$ can be represented as $z_i(t) = \sum_{j\in\mathscr{N}_i^{in}(t)} (\hat{x}_j(t) - \hat{x}_i(t)) = -(\mathscr{H}_{\sigma(t)}^{rowi} \otimes I_n)(\delta + e)$. Then, with the closed-loop state ξ defined in (4.12), we can construct a hybrid model for the MAS, which is represented as

$$\begin{aligned} \dot{\xi} &= \mathscr{F}_{\sigma(t)}(\xi), \text{ If } \xi \in \mathbb{F}, \\ \xi^+ &\in \bigcup_{i=1}^{N}\bigcup_{m=1}^{n} \mathscr{J}_{im}(\xi), \text{ If } \xi \in \mathbb{J}, \end{aligned} \tag{4.30}$$

where $\mathbb{F}$, $\mathbb{J}$, $\mathscr{J}_{im}(\xi)$ have the same form as in (4.13), and $\mathscr{F}_{\sigma(t)}(\xi)$ has the same form as $\mathscr{F}(\xi)$, but with $\mathscr{H}_{\sigma(t)}$ instead of $\mathscr{H}$.

Let $\bar{\lambda}_1 = \min_{p\in\mathscr{P}}\{\lambda_{min}(\mathscr{Q}\mathscr{H}_p + \mathscr{H}_p^T\mathscr{Q})\}$, $\bar{\alpha} = \mu_1\lambda_M/\bar{\lambda}_1$. Then, we have the following result.

Theorem 4.2 *Consider the MAS (4.1) and (4.2) in the switching case satisfying Assumption 4.2. Let $K = \bar{\alpha}B^T P$, $\theta_{hi} > 0$, $\theta_{rim} > 0$, $\phi_{hi}(0) > 0$, $\phi_{rim}(0) > 0$, if the sampling period h_i satisfies $0 \le h_i \le \bar{h}_i = \arctan(\phi_{hi}(0))/\theta_{hi}$, the integer parameter Δ_{im} satisfies condition (4.17), and there exist matrix K_r, constants $0 < \rho < 1$, $\epsilon > 0$, $0 < p_{im} < 1$ $a_r < 0$, $b_r > 0$, such that $p_{im}^2\phi_{rim}(0) \le \phi_{rim}(\Delta_{im}h_i)$, and the inequality condition (4.19) holds with $\mathscr{H}_{\sigma(t)}$ instead of $\mathscr{H}$, where $E_{i1}^e = E_i(\mathscr{H}_{\sigma(t)} \otimes BK)$, $E_{i2}^e = E_i(I_N \otimes A + \mathscr{H}_{\sigma(t)} \otimes BK)$, $E_i^z = -\mathscr{H}_{\sigma(t)}^{rowi} \otimes I_n$, the other parameters are the same as in Theorem 4.1. Then, the (δ, x_r, e) dynamics is asymptotically stable, and the leader-following consensus is achieved.*

Proof The proof is similar as the proof of Theorem 4.1. Choose the Lyapunov function $V(\xi) = \delta^T(\mathscr{Q} \otimes P)\delta + e^T\Phi_h(\tau_h)e + x_r^T\Phi_r(\tau_r)x_r$, we can compute the derivative of $\delta^T(\mathscr{Q} \otimes P)\delta$ along the flow dynamics of δ and get

$$\begin{aligned} \dot{V}(\xi) &= 2\delta^T((\mathscr{Q} \otimes P)(I_N \otimes A - \mathscr{H}_{\sigma(t)} \otimes BK))\delta \\ &\quad - 2\delta^T(\mathscr{Q} \otimes P)((\mathscr{H}_{\sigma(t)} \otimes BK)e - (I_N \otimes BK_r)x_r) \\ &\le \delta^T(\mathscr{Q} \otimes (PA + A^T P) - \mu_1\mathscr{Q} \otimes PBB^T P)\delta \\ &\quad - 2\delta^T(\mathscr{Q} \otimes P)((\mathscr{H}_{\sigma(t)} \otimes BK)e - (I_N \otimes BK_r)x_r) \\ &= -\mu_2\delta^T(\mathscr{Q} \otimes I_n)\delta - 2\delta^T(\mathscr{Q} \otimes P)(\mathscr{H}_{\sigma(t)} \otimes BK)e \\ &\quad + 2\delta^T(\mathscr{Q} \otimes P)(I_N \otimes BK_r)x_r \end{aligned} \tag{4.31}$$

Then, the rest proof follows the proof of Theorem 4.1. □

Remark 4.5 For switching network, the switching of communication graph $\bar{\mathscr{G}}_{\sigma(t)}$ depends on the switching signal $\sigma(t)$, which takes values from a finite indices set $\mathscr{P}$. It is worth mentioning that at the next switching instant, $\sigma(t)$ can be arbitrary element of the set $\mathscr{P}$, and it does not need to change orderly. This means that there is no additional restrictions on how $\sigma(t)$ changes, and it only requires that $\sigma(t) \in \mathscr{P}$ and Assumption 4.2 (as also used in [11]) holds. In addition, at the sampling instants of a certain agent i, $e_i(t)$ and $\tau_{hi}(t)$ jump to zeros, $x_{ri}(t)$ and $\tau_{ri}(t)$ jump according to (4.8), but the jump dynamics is independent of $\sigma(t)$. Thus, the switching of the graph and the sampling operate separately and are not affected each other.

Remark 4.6 By utilizing the hybrid systems approach to address the sampled-data leader-following consensus problem, it shows some distinct advantages: First, the error $e_i(t)$ can be dealt with as closed-loop state, thus the technical lemma 3.1 in [11] is not needed. Second, the static network case and switching case can be modeled in a unified hybrid system, thus the stability analysis is simple, besides, it has potential to improve transient performance by adding a reset element. Third, in both Theorems 4.1 and 4.2, an explicit UBSP is presented to indicate the value range for sampling periods, despite [11] also can do, the obtained UBSP in this chapter can be larger. This will be shown via simulation.

4.5 Simulation Study

In this section, to show the effectiveness of the proposed approach, an example borrowed from recent work [11] is investigated. Consider MAS with one leader and four followers, the dynamics has the form of (4.1) and (4.2), with system matrices

$$A = \begin{bmatrix} -0.38 & 0.72 \\ -0.68 & 0.42 \end{bmatrix}, \ B = \begin{bmatrix} 0.26 \\ 0.31 \end{bmatrix}.$$

Static Network Case: In this case, the static communication graph $\bar{\mathscr{G}}$ is shown in Fig. 4.1, where node 0 denotes the leader. It can be seen that Assumption 4.1 is satisfied and $\mathscr{H} = [1\,0\,0\,0;\ 0\,2\,-1\,0;\ -1\,0\,2\,-1;\ 0\,-1\,0\,1]$. Select $\mu_1 = 1$, $\mu_2 = 1$, $\mathscr{Q} = 15I$, I is the identity matrix. The matrix P can be solved from Eq. (4.18) with $P = [7.2138\ -3.6897;\ -3.6897\ 6.3388]$, and $K = \alpha B^T P = [0.8874\ 1.2195]$. The other parameters are selected as $K_r = 2K$, $a_r = -1$, $b_r = 3$, $\rho = 10^{-3}$, $\epsilon = 10^{-6}$, $\theta_{hi} = 32$, $\theta_{rim} = 5$, $\phi_{hi}(0) = 25$, $\phi_{rim}(0) = 3$ for $i \in [1, 4]$, $m \in [1, 2]$. $p_{im} = 0.2$, $\Delta_{im} = 5$ for $i \in [1, 2]$. $p_{im} = 0.3$, $\Delta_{im} = 8$ for $i \in [3, 4]$. With these parameters, it can be verified that the conditions $p_{im}^2 \phi_{rim}(0) \le \phi_{rim}(\Delta_{im} h_i)$ and (4.19) in

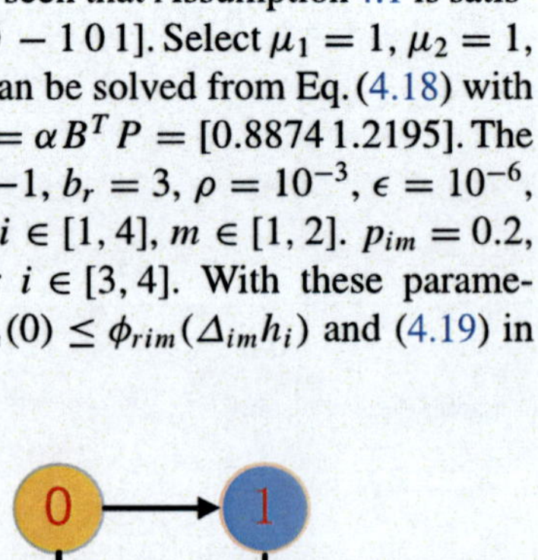

Fig. 4.1 Communication graph

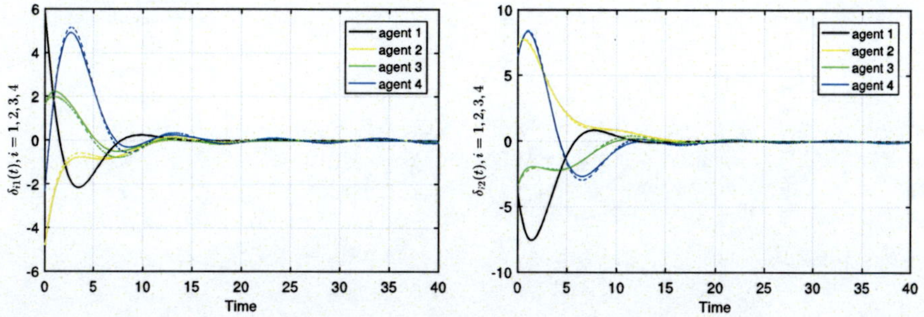

Fig. 4.2 Consensus errors under static network. Left: [11]; Right: our results

Theorem 4.1 are satisfied. Then, it can be calculated that $\bar{h}_i = 0.0478$, i.e., the maximum allowable sampling period is 0.0478 for each agent i. By contrast, with the method in [11], $\bar{h}_i = 0.0186$ is obtained, which is more conservative than our method.

The initial conditions are chosen as $x_0(0) = [1.2\ -0.8]^T$, $x_1(0) = [2.4\ -1.6]^T$, $x_2(0) = [-1.4\ 2.6]^T$, $x_3(0) = [1.8\ -2.5]^T$, $x_4(0) = [-0.2\ 1.3]^T$, $x_{ri}(0) = [0\ 0]^T$, $\tau_{hi}(0) = 0$, $\tau_{rim}(0) = [0\ \ 0]^T$. The sampling periods of follower agents are $h_1 = 0.015$, $h_2 = 0.016$, $h_3 = 0.017$, $h_4 = 0.018$. In order to show the superiority of the proposed method, the comparison scheme includes two aspects: (1). To illustrate the effectiveness of reset control part, consensus errors are given to show that transient performance (e.g., settling time, overshoot) can be improved compared with related work [11]; (2). To show that improved UBSP can be achieved with the proposed hybrid systems approach, comparison results (in contrast with [11]), with different parameters μ_1, μ_2, $\mathcal{Q}$, will be provided. For comparison, the initial conditions $x_i(0)$ and sampling periods h_i are the same as in [11]. The consensus errors $\delta_i(t)$ are shown in Fig. 4.2[1] and the state trajectories are shown in Fig. 4.3, which indicate that the leader-following consensus is achieved and improved transient performance compared with existing work is achieved. The control inputs $u_i(t)$ are shown in Fig. 4.4, from which, we can see that though the control input efforts are similar, the control signal $u_i(t)$ has larger vibrations with the proposed approach. The relatively large vibrations implies that the physical implementation of the proposed control scheme is more difficult than the control scheme in [11], which can be viewed as a drawback of the proposed approach. In addition, the comparison of settling time and overshoot[2] are given in Table 4.1, which clearly shows the effectiveness of reset control scheme.

Next, we show the superiority of the proposed method in improving UBSP $\bar{h}_i$ compared with [11]. Note that $\bar{h}_i$ in [11] mainly depends on parameters μ_1, μ_2, but $\bar{h}_i$ in our result depends on $\phi_{hi}(0)$ and θ_{hi} rather than μ_1, μ_2, $\phi_{hi}(0)$ and θ_{hi} can be

[1] x-coordinate denotes time and the unit is second, y-coordinate denotes the amplitude of corresponding signals.

[2] The settling time is defined as the time required for the consensus errors δ_i of the follower agents to reach and stay within a neighborhood of the origin and 0.02 is used here. Due to the consensus errors converge to the origin, so, the overshoot is given by absolute values rather than percent overshot.

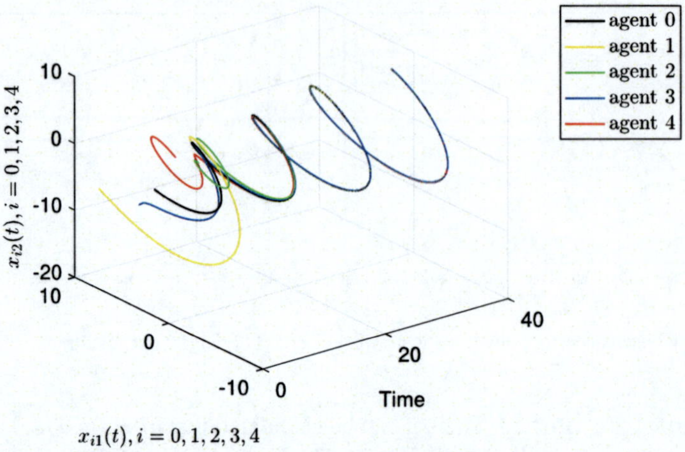

Fig. 4.3 State trajectories of MAS

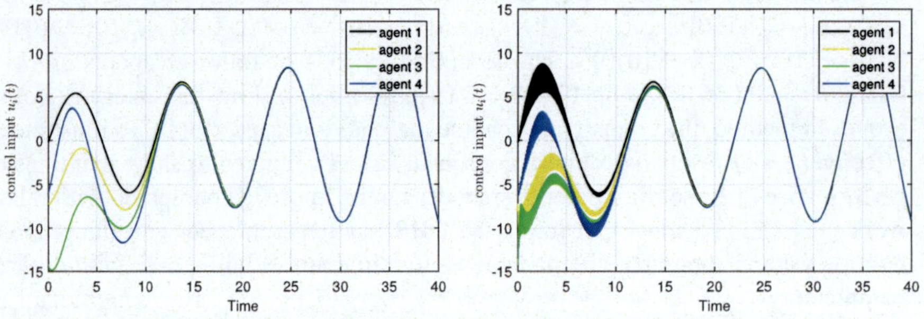

Fig. 4.4 Control input $u_i(t)$. Left: [11]; Right: our results

Table 4.1 Performance comparison (static network case). ST and OS denote settling time and overshoot respectively

		Agent 1	Agent 2	Agent 3	Agent 4
ST	$\delta_{i1}(t)$:[11]/our	12.53/12.28	26.78/20.78	31.10/22.27	32.21/21.89
	$\delta_{i2}(t)$:[11]/our	15.92/11.40	29.88/23.81	30.56/25.41	35.16/24.49
OS	$\delta_{i1}(t)$:[11]/our	0.60/0.44	0.14/0.04	0.61/0.29	2.70/2.37
	$\delta_{i2}(t)$:[11]/our	0.29/0.17	0.11/0.005	0.41/0.13	0.46/0.16

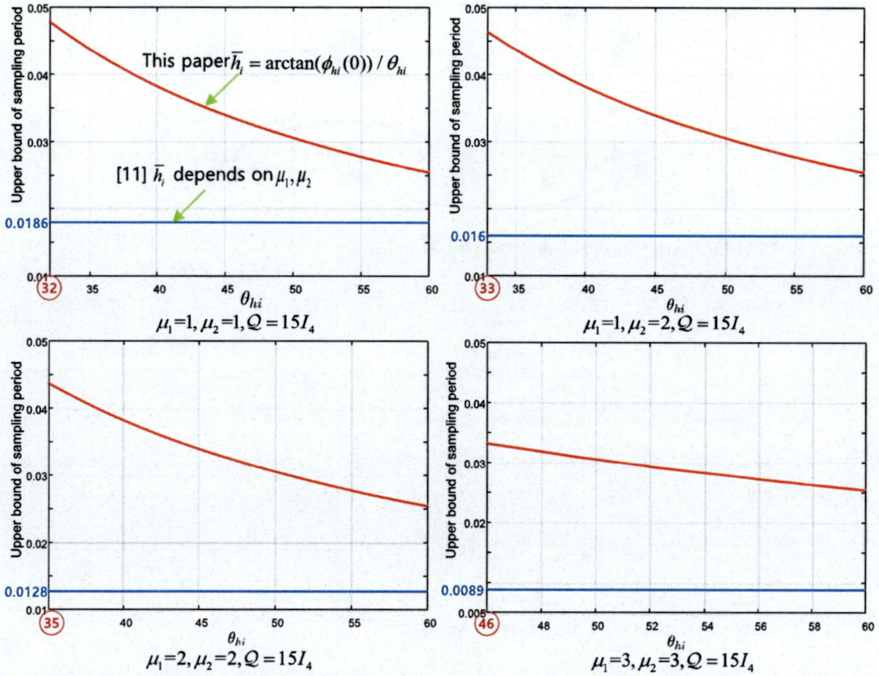

Fig. 4.5 UBSP with different methods

selected such that the stability conditions in Theorem 4.1 are satisfied. Then, Fig. 4.5 shows the maximum allowable UBSP with several group of μ_1, μ_2. The blue line denotes the UBSP with the method in [11], the red line denotes the UBSP with the proposed method. From the abscissa axis, the numbers in the red circles denotes minimum θ_{hi} that makes the stability conditions in Theorem 4.1 hold. From Fig. 4.5, it clearly shows that the UBSP with the proposed hybrid systems approach is much larger than with the method in [11].

Switching Network Case: Consider time-varying communication graph $\bar{\mathscr{G}}_{\sigma(t)}$, where

$$\sigma(t) = \begin{cases} 1, \text{ if } lT_0 \le t \le (l + \frac{2}{3})T_0 \\ 2, \text{ if } (l + \frac{2}{3})T_0 \le t \le (l + 1)T_0 \end{cases} \tag{4.32}$$

where $l = 0, 1, 2, \ldots$ and $T_0 = 1$, the two graphs are shown in Fig. 4.6. The matrices $\mathscr{H}_1$ and $\mathscr{H}_2$ can be easily obtained based on Fig. 4.6. Choose $\mu_1 = 1$, $\mu_2 = 1$, $\mathscr{Q} = 15I$, it can be verified that $\mathscr{Q}\mathscr{H}_1 + \mathscr{H}_1^T\mathscr{Q}$ and $\mathscr{Q}\mathscr{H}_2 + \mathscr{H}_2^T\mathscr{Q}$ are both positive definite, so, Assumption 4.2 holds. K is calculated as $K = [0.9483\ 1.3033]$. Select $\theta_{hi} = 35$ and $\phi_{hi}(0) = 25$, the stability conditions in Theorem 4.2 are satisfied and the

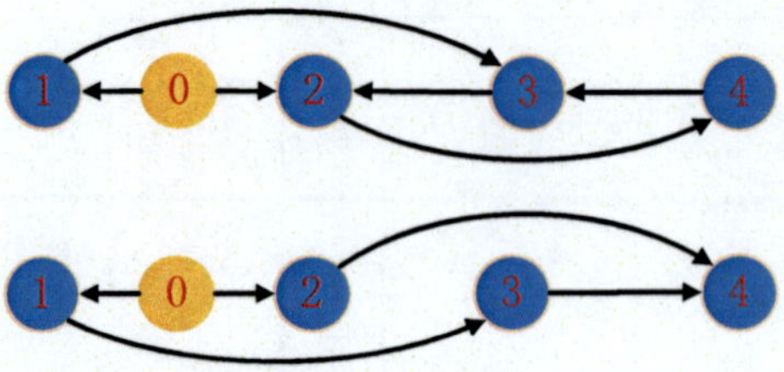

Fig. 4.6 Two communication graphs. Top: $\bar{\mathscr{G}}_1$. Bottom: $\bar{\mathscr{G}}_2$

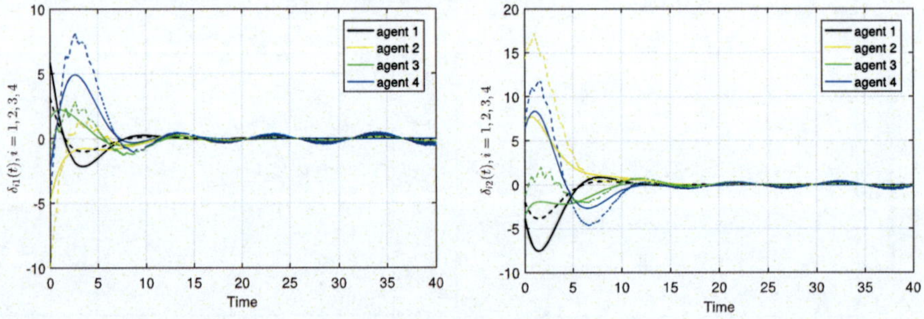

Fig. 4.7 Consensus errors under switching network

UBSP is $\bar{h}_i = 0.0437$ (note that the UBSP is $\bar{h}_i = 0.0167$ with the method in [11]). Let $h_1 = 0.026, h_2 = 0.028, h_3 = 0.030, h_4 = 0.032$, $K_r = K$, the other parameters and initial conditions are selected the same as in static network case.

The simulation results are shown in Figs. 4.7, 4.8 and 4.9. As it is expected, the leader-following consensus is achieved and the consensus errors δ_i converge to the origin asymptotically. Moreover, the comparison of settling time and overshoot is shown in Table 4.2. The control input comparison is similar as in static network case, so, it is omitted here due to space limitation. From simulation results, we can see that though at the cost of relatively large vibrations of control signals, the proposed approach provides an effective way to improve transient performance and UBSP.

4.6 Conclusion

In this chapter, we investigate the leader-following consensus problem of MAS via asynchronous sampled-data control. A hybrid systems framework is proposed for consensus analysis. With the hybrid systems approach, improved UBSP compared with the existing work is obtained, better transient performance of the consensus errors is achieved, and the proposed approach can be applied to both static and switching network cases.

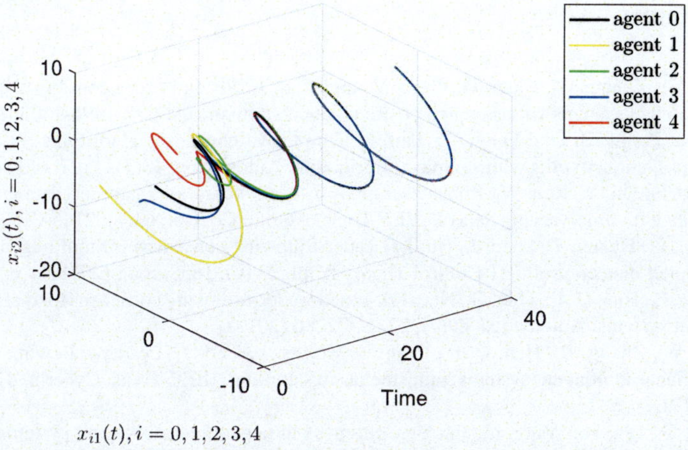

Fig. 4.8 State trajectories of MAS under switching network

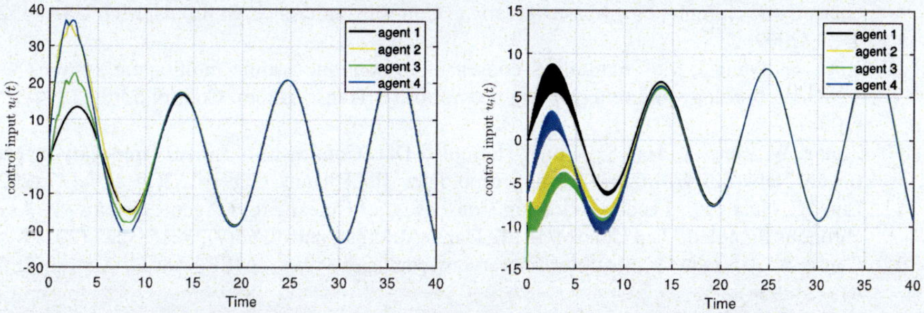

Fig. 4.9 Control input $u_i(t)$. Left: [11]; Right: our results

Table 4.2 Performance comparison (switching network case). ST and OS denote settling time and overshoot respectively

		Agent 1	Agent 2	Agent 3	Agent 4
ST	$\delta_{i1}(t)$:[11]/our	12.51/10.67	15.84/9.97	12.19/11.83	16.36/14.09
	$\delta_{i2}(t)$:[11]/our	15.67/11.07	14.52/14.42	16.17/14.57	19.85/12.89
OS	$\delta_{i1}(t)$:[11]/our	0.54/0.33	0.24/0.03	0.24/0.12	2.00/1.51
	$\delta_{i2}(t)$:[11]/our	0.24/0.09	0.019/0.003	0.12/0.04	3.31/3.02

References

1. Yu, W., Zheng, W., Chen, G., Ren, W., and Cao, J.: Second-order consensus in multi-agent dynamical systems with sampled position data. Automatica, **47**(7), 1496-1503 (2011).
2. Guan, Z., Li, Z. amd Feng, G., Jian, M.: Impulsive consensus algorithms for second-order multi-agent networks with sampled information. Automatica, **48**(7), 1397-1404 (2012).
3. Xiao, F., Shi, Y., Ren, W.: Robustness analysis of asynchronous sampled-data multiagent networks with time-varying delays. IEEE Trans. Autom. Control, **63**(7), 2145-2152 (2018).
4. Tang, Z., Huang, T., Shao, J., Hu, J.: Leader-following consensus for multi-agent systems via sampled-data control. IET Control Theory Appl., **5**(14), 1658-1665 (2011).
5. Ding, L., Han, Q.-L., Guo, G.: Network-based leader-following consensus for distributed multi-agent systems. Automatica, **49**(7), 2281–C2286 (2013).
6. He, W., Zhang, B., Han, Q.-L., Qian, F., Kurths, J., Cao, J.: Leader-following consensus of nonlinear multiagent systems with stochastic sampling, IEEE Trans. Cybern., **47**(2), 327-338 (2017).
7. Xie, G., Liu, H., Wang, L., Jia, Y.: Consensus in networked multiagent systems via sampled control: fixed topology case. American Control Conference, St. Louis, MO, USA, 3902-3907 (2009)
8. Xie, G., Liu, H., Wang, L., Jia, Y.: Consensus in networked multi-agent systems via sampled control: switching topology case. American Control Conference, St. Louis, MO, USA, 4525-4530 (2009)
9. Xiao, F., Wang L.: Asynchronous consensus in continuous-time multi-agent systems with switching topology and time-varying delays. IEEE Trans. Autom. Control **53**(8), 1804-1816 (2008)
10. Zhang, W., Tang, Y., Han, Q., Liu, Y.: Sampled-Data Consensus of Linear Time-Varying Multiagent Networks With Time-Varying Topologies. IEEE Trans. Cybern. **52**(1), 128-137 (2022)
11. Liu, W., Huang, J.: Leader-Following Consensus for Linear Multi-Agent Systems via Asynchronous Sampled-Data Control. IEEE Trans. Autom. Control, **65**(7), 3215-3222 (2020)
12. Clegg, J.: A nonlinear integrator for servomechanisms. Trans. AIEE, Part II, Appl. Ind., **77**, 41-42 (1958)
13. Beker, O., Hollot, C., Chait, Y.: Fundamental properties of reset control systems. Automatica **40**(6), 905-915 (2004)
14. Zhao, G., Nesic, D., Tan, Y., Hua, C.: Overcoming overshoot performance limitations of linear systems with reset control. Automatica **101**, 27-35 (2019)
15. Nesic, D., Teel, A., Zaccarian, L.: Stability and performance of SISO control systems with first-order reset elements. IEEE Trans. Autom. Control **56**(11), 2567-2582 (2011)
16. Zhao, G., Hua, C., Guan, X.: Reset Control for Consensus of Multi-Agent Systems with Asynchronous Sampling. IEEE Trans. Syst., Man, Cybern: Syst., **51**(8), 4911-4919 (2021)
17. Meng, X., Xie, L., Soh, Y.: Reset control for synchronization of multi-agent systems. Automatica **104**, 3189-195 (2019)
18. You, X., Hua, C., Guan, X.: Adaptive Leader-Following Consensus for Second-Order Time-Varying Nonlinear Multiagent Systems. IEEE Trans. Cybern., **47**(6), 1532-1539 (2017)
19. Hu, J., Hong,Y.: Leader-following coordination of multi-agent systems with coupling time delays. Physica A: Statistical Mechanics and its Applications, **374**(2), 853-863 (2007)
20. Clarke, H.: Optimization and nonsmooth analysis, SIAM, Philadelphia, PA (1990)
21. Carnevale, D., Teel, A., Nesic, D.: A Lyapunov Proof of an Improved Maximum Allowable Transfer Interval for Networked Control Systems. IEEE Trans. Autom. Control, **52**(5), 892-897 (2007)
22. Zhao, G., Nesic, D., Tan, Y., Wang J.: Open problems in reset control. IEEE Conf. Decision and Control, 3326-3331, Florence, Italy, (2020)

Chapter 5
Output Feedback Sampled-Data Control of MAS

5.1 Introduction

In previous chapters, particular attentions and concerns of us have concentrated on addressing the various communication issues, that resulted from the imperfect network environments, i.e., the time delays and packet losses occurred during communication process and the network switchings associated with communication structure changes, and concentrated on revealing the tradeoff relationship between the stability and performance of closed-loop system and the tolerability to network imperfection. Based on these efforts, we can guarantee the hybrid sampled-data control method to operate more safely and more smoothly. For this moment, it should be noted that the networked control of MAS poses the challenges mainly from two aspects: one is the above considered network imperfections, and the other is the scarce communication resource. In practice, the networked MASs usually have limited communication resources, such as bandwidth and node energy. It implies that traditional continuous communication, which requires real-time information of neighbors, may cause excessive resources consumption and is not easy to be implemented under network environment. Along this line, in this chapter, we will consider how to reduce the information transmission, here we should remark that the sampled-data control itself is a resource-efficient control method in essence, as it can greatly reduce the communication by sampling and sharing states according to a fixed time period. However, to further reduce the resource usage, we can alternatively consider to reduce the volume of information by transmitting part of the states or only the output information.

As is well known, on the premise of full states available, considerable number of literature have devoted to the sampled-data consensus control, e.g., [1–6]. Nevertheless, full states are available means that all states are required to be measurable. It is problematic in implementation for some practical systems in the sense that there may exist some unmeasurable states, or the cost of measurement is high. Thus, observer-based output feedback control was proposed with the attempt to reduce state measurements. For example, a distributed observer-based adaptive control law was designed in [7] for MAS under switching topology. Dynamic event-triggered

G. Zhao et al., *Cooperative Control of Multi-agent Systems*,
https://doi.org/10.1007/978-981-97-0968-7_5

controller with varying thresholds was proposed in [8], where observers were utilized to estimate agents' states. The observer-based consensus control law design was also considered in [9–11] for MAS under denial-of-service attack or with disturbances. However, it is remarkable that the volume of information transmissions with observer is essentially the same as that with full state feedback, not to mention the additional requirements for implementing the observer (the input and output information of original system and specialized implementation hardware are needed). These limitations can be overcome by using dynamic output feedback (DOF) control, which is more superior than observer-based control from the viewpoint of volume of transmitted information through network [12–15]. In addition, there are two types of information exchange fashions between agents, the first one is related to absolute information, i.e., neighbors' absolute state/output information needs to be measured, e.g., [8, 16]. In the second type, only the relative information between neighboring agents is needed, see [17, 18] for example. As discussed in [16], the absolute information cannot always be measured, e.g., in deep-space exploration [19], but the relative information between neighboring agents can. Thus, in this chapter, we assume that the sampled relative output information is available for control design, and consider the sampled-data dynamic output feedback consensus control problem of MAS, the objective is to achieve state consensus, while full state measurement or observer design is not needed. First, a sampled-data DOF control law is designed in order to achieve state consensus, such that full state measurement [3–6] and state observer design [7, 8, 10] are not needed, and continuous information exchange [12–14] is also avoided. Second, the proposed control law can be implemented based on sampled relative output information between neighboring agents, it is applicable for the case where relative information is available but absolute information not. Third, a hybrid system consisting of flow/jump dynamics is constructed to completely describe the MAS dynamics, rigorous Lyapunov stability conditions are derived based on this model and the maximum allowable sampling period (MASP) of each agent are given.

5.2 Networked MAS and Problem Formulation

This section formulates the sampled-data DOF consensus control problem of the MAS. The DOF control means that, for one certain agent, the only available information to constitute its consensus controller is the outputs of its immediate neighbors, at the same time, the agent merely needs to send out its output information. Compared with full state feedback control, the DOF is more challenging to be implemented, as the less information available, the more difficult it is to control. However, there is no denying that DOF still is attractive in MAS design, as the volume of information can be sharply cut down, and sometimes in practical applications the full state information of agent is not measurable, or the measurement cost is high. By contrast, the system output is usually related to part of the states, whose measurement is more easier and corresponds to lower cost. In this chapter, inspired by [8, 12, 13], we would like to develop a sampled-data DOF consensus control protocol for the MAS given by

$$\begin{aligned} \dot{x}_i(t) &= Ax_i(t) + Bu_i(t), \\ y_i(t) &= Cx_i(t), \end{aligned} \tag{5.1}$$

where $i \in [1, N]$ denotes the MAS consists of N agents. $x_i(t) \in \mathbb{R}^n$ denotes the state variable, $u_i(t) \in \mathbb{R}^m$ denotes the agent's input, $y \in \mathbb{R}^p$ is the output. (A, B) is stabilizable.

The preliminaries about graph have been introduced in Chap. 1. For proceeding this chapter, we make the following Assumption on the communication graph.

Assumption 5.1 There is no leader agent, and the communication graph $\mathcal{G}$ is connected and undirected.

Given the MAS (5.1), suppose that the relative output information among agents can be measured by digital sensors and each agent has its own sampling period (i.e., agent's absolute full state information $x_i(t)$ is not available). Then, the sampled-data based DOF consensus controller is designed as the following form. For $t \in (t_k^i, t_{k+1}^i]$,

$$\begin{aligned} u_i(t) &= Kv_i(t), \\ \dot{v}_i(t) &= (A + BK)v_i(t) + \mu G \sum_{j \in \mathcal{N}_i} (y_{vi}(t_k^i) - y_{vj}(t_k^i)) - \mu G \sum_{j \in \mathcal{N}_i} (y_i(t_k^i) - y_j(t_k^i)), \\ y_{vi}(t) &= Cv_i(t), \end{aligned} \tag{5.2}$$

where $v_i(t)$ is controller state, K and G are gain matrices to be designed, $\mu > 0$ is a constant parameter. t_k^i is agent i's sampling instant, and $h_i = t_{k+1}^i - t_k^i$ denotes the sampling period of agent i, which indicates that asynchronous sampling is used. $\sum_{j \in \mathcal{N}_i} (y_{vi}(t_k^i) - y_{vj}(t_k^i))$ and $\sum_{j \in \mathcal{N}_i} (y_i(t_k^i) - y_j(t_k^i))$ denote the sampled information of $q_{y_v i}(t)$ and $q_{yi}(t)$ respectively, with $q_{y_v i}(t) = \sum_{j \in \mathcal{N}_i} (y_{vi}(t) - y_{vj}(t))$ and $q_{yi}(t) = \sum_{j \in \mathcal{N}_i} (y_i(t) - y_j(t))$.

Based on Assumption 5.1, it can obtain that the resulting Laplacian matrix $\mathcal{L}$ is semi-positive and symmetric. Then, according to the graph theory and Lyapunov stability theory, the leaderless consensus of MAS with respect to such $\mathcal{L}$ can ultimately be achieved by means of the proposed sampled-data based DOF consensus controller (5.2) in the sense that $\lim_{t \to \infty} \|x_i(t) - x_j(t)\| = 0$, $i, j \in [1, N]$. In what follows, the main task is to specify the system gain matrices K and G, and to determine the maximum allowable sampling period (MASP) for the agents.

5.3 Sampled-Data DOF Control

In this section, we first analyze the dynamics of the closed-loop MAS. Then, a hybrid system model is constructed to describe the closed-loop system state with both flow dynamics and jump dynamics. Based on this hybrid model, Lyapunov stability conditions and how to choose appropriate sampling periods are developed.

Note from (5.1) and (5.2) that $y_i(t) = Cx_i(t)$ and $y_{vi}(t) = Cv_i(t)$, in order to describe the closed-loop dynamics appropriately, define

$$\begin{aligned} q_{xi}(t) &= \sum_{j\in\mathcal{N}_i}(x_i(t) - x_j(t)), q_{vi}(t) = \sum_{j\in\mathcal{N}_i}(v_i(t) - v_j(t)), \\ \bar{q}_{xi}(t) &= q_{xi}(t_k^i), e_{xi}(t) = \bar{q}_{xi}(t) - q_{xi}(t), t \in (t_k^i, t_{k+1}^i], \\ \bar{q}_{vi}(t) &= q_{vi}(t_k^i), e_{vi}(t) = \bar{q}_{vi}(t) - q_{vi}(t), t \in (t_k^i, t_{k+1}^i]. \end{aligned} \tag{5.3}$$

The above definition implies that $e_{xi}(t) = \bar{q}_{xi}(t) - q_{xi}(t)$ (or $e_{vi}(t) = \bar{q}_{vi}(t) - q_{vi}(t)$) at non-sampling instant, and $e_{xi}(t)$ (or $e_{vi}(t)$) jumps to zero at sampling instant, and $q_{y_v i}(t) = Cq_{vi}(t)$ and $q_{yi}(t) = Cq_{xi}(t)$. Then, based on (5.1), (5.2), and (5.3), we have

$$\begin{aligned} \dot{x}_i &= Ax_i + BKv_i, \\ \dot{v}_i &= (A + BK)v_i + \mu GC\bar{q}_{vi} - \mu GC\bar{q}_{xi} \\ &= (A + BK)v_i + \mu GC(q_{vi} + e_{vi}) - \mu GC(q_{xi} + e_{xi}), \end{aligned} \tag{5.4}$$

that is

$$\begin{bmatrix} \dot{x}_i \\ \dot{v}_i \end{bmatrix} = \begin{bmatrix} A & BK \\ 0 & A + BK \end{bmatrix} \begin{bmatrix} x_i \\ v_i \end{bmatrix} + \mu \begin{bmatrix} 0 & 0 \\ -GC & GC \end{bmatrix} \begin{bmatrix} q_{xi} \\ q_{vi} \end{bmatrix} + \mu \begin{bmatrix} 0 & 0 \\ -GC & GC \end{bmatrix} \begin{bmatrix} e_{xi} \\ e_{vi} \end{bmatrix}. \tag{5.5}$$

Define

$$\begin{aligned} z_i &= \begin{bmatrix} x_i^T & v_i^T \end{bmatrix}^T, z_{ei} = \begin{bmatrix} e_{xi}^T & e_{vi}^T \end{bmatrix}^T, \\ \bar{A} &= \begin{bmatrix} A & BK \\ 0 & A + BK \end{bmatrix}, H = \begin{bmatrix} 0 & 0 \\ -GC & GC \end{bmatrix}, \end{aligned} \tag{5.6}$$

and note that

$$\begin{bmatrix} q_{xi} \\ q_{vi} \end{bmatrix} = \begin{bmatrix} (\mathcal{L}_{row}^i \otimes I_n)x \\ (\mathcal{L}_{row}^i \otimes I_n)v \end{bmatrix} = (\mathcal{L}_{row}^i \otimes I_{2n})z, \tag{5.7}$$

where $\mathcal{L}_{row}^i$ denotes the ith row of matrix $\mathcal{L}$, and

$$x = \begin{bmatrix} x_1^T & x_2^T & \cdots & x_N^T \end{bmatrix}^T, \ v = \begin{bmatrix} v_1^T & v_2^T & \cdots & v_N^T \end{bmatrix}^T, \ z = \begin{bmatrix} z_1^T & z_2^T & \cdots & z_N^T \end{bmatrix}^T. \tag{5.8}$$

Then, based on (5.5)–(5.8), it is easy to derive that

$$
\begin{aligned}
\dot{z}_i &= \bar{A}z_i + \mu H(\mathscr{L}_{row}^i \otimes I_{2n})z + \mu H z_{ei}, \\
\dot{z} &= (I_N \otimes \bar{A} + \mu\mathscr{L} \otimes H)z + (I_N \otimes \mu H)z_e.
\end{aligned}
\tag{5.9}
$$

As is well known, the consensus of MAS meas that all agents' states tend to the same value. Furthermore, with respect to undirected and symmetric $\mathscr{L}$, the consensus of MAS just converge to the average value of the states coincidentally. Thus, the consensus error variable δ_i is defined as follows

$$
\delta_i = z_i - \frac{1}{N}\sum_{i=1}^{N} z_i = z_i - \frac{1}{N}(1_N^T \otimes I_{2n})z, \tag{5.10}
$$

and

$$
\delta = \left(\left(I_N - \frac{1}{N}1_N 1_N^T\right) \otimes I_{2n}\right) z = (M \otimes I_{2n})z, \tag{5.11}
$$

where $M = I_N - \frac{1}{N}1_N 1_N^T$, and $\delta = \begin{bmatrix}\delta_1^T & \cdots & \delta_N^T\end{bmatrix}^T$. Based on (5.9)–(5.11), the dynamics of δ can be represented as

$$
\dot{\delta} = (I_N \otimes \bar{A} + \mu\mathscr{L} \otimes H)\delta + (M \otimes \mu H)z_e, \tag{5.12}
$$

and from the definition of z_{ei} and $z_e = [z_{e1}^T \; z_{e2}^T \; \cdots \; z_{eN}^T]^T$, we have

$$
\begin{aligned}
\dot{z}_{ei} &= \begin{bmatrix}\dot{e}_{xi} \\ \dot{e}_{vi}\end{bmatrix} = -\begin{bmatrix}(\mathscr{L}_{row}^i \otimes I_n)\dot{x} \\ (\mathscr{L}_{row}^i \otimes I_n)\dot{v}\end{bmatrix} = -(\mathscr{L}_{row}^i \otimes I_{2n})\dot{z}, \\
\dot{z}_e &= -(\mathscr{L} \otimes \bar{A} + \mu\mathscr{L}^2 \otimes H)\delta - (\mathscr{L} \otimes \mu H)z_e.
\end{aligned}
\tag{5.13}
$$

It should be pointed out that e_{xi} and e_{vi} jump to zeros at t_k^i. So, z_e actually is not differentiable at t_k^i, and the derivative $\dot{z}_{ei}$ is a generalized directional derivative (i.e., Clarke derivative [20]). Note that (5.12) and (5.13) cannot completely describe the system dynamics, in the following, a system model with hybrid dynamics will be constructed in order to describe the MAS dynamics.

5.3.1 Hybrid Model

In previous section, the flow dynamics of consensus error δ and error variable z_e are explicitly given in (5.12) and (5.13). However, note that there also exist jump dynamics, such as z_e has jump dynamics at sampling instants. Moreover, in order to analyze MASP for the agents, it is expected that the sampling mechanism is also integrated into the closed-loop model. Thus, the system dynamics in previous section is not appropriate for stability analysis. Next, we construct a system model with hybrid dynamics for MAS with sampling mechanism, which includes flow dynamics, jump dynamics, flow set and jump set. The hybrid model is able to completely describe the MAS dynamics and incorporate the sampling mechanism into the flow/jump set, and it is useful for stability analysis and MASP analysis.

Recall the definition of hybrid system (1.1) presented in Chap. 1, and for convenience, the general form of hybrid system is introduced again

$$\mathscr{H}: \begin{cases} \dot{\xi} = \mathscr{F}(\xi), & \text{If } \xi \in \mathbb{F} \\ \xi^{+} \in \mathscr{J}(\xi), & \text{If } \xi \in \mathbb{J} \end{cases} \tag{5.14}$$

where ξ denotes the closed-loop state, $\mathscr{F}(\xi)/\mathscr{J}(\xi)$ describes the flow/jump dynamics, $\mathbb{F}/\mathbb{J}$ denotes the flow/jump set. To constitute a suitable closed-loop state ξ for the subsequent analyzing, we first introduce an internal timer variable $\tau_i \in \mathbb{R}_{\geq 0}$, which is utilized to record the time elapsed since the last sampling instant of agent i, this is useful for describing the flow set and jump set with sampling mechanism. The dynamics of τ_i is given by

$$\begin{cases} \dot{\tau}_i = 1, & \text{if } \tau_i < h_i, \\ \tau_i^{+} = 0, & \text{if } \tau_i = h_i, \end{cases} \tag{5.15}$$

where h_i denotes the sampling period of agent i, that is to be determined later.

Combination of the role of τ_i, a suitable closed-loop state can be designed as $\xi = [\delta^T \; z_e^T \; \tau^T]^T$, where $\tau = [\tau_1^T \; \tau_2^T \; \cdots \; \tau_N^T]^T$. The flow dynamics and flow set of ξ are

$$\mathscr{F}(\xi) = \begin{bmatrix} f_\delta(\delta, z_e) \\ f_e(\delta, z_e) \\ 1_N \end{bmatrix}, \quad \mathbb{F} = \bigcap_{i=1}^{N} \mathbb{F}_i, \text{ with } \mathbb{F}_i = \{\xi \in \mathbb{X} | \tau_i < h_i\}, \tag{5.16}$$

and, the jump dynamics and jump set are

$$\mathscr{J}(\xi) = \bigcup_{i \in [1,N]} \mathscr{J}_i(\xi), \quad \mathbb{J} = \bigcup_{i=1}^{N} \mathbb{J}_i, \text{ with } \mathbb{J}_i = \{\xi \in \mathbb{X} | \tau_i = h_i\}, \tag{5.17}$$

where $f_\delta(\delta, z_e) = (I_N \otimes \bar{A} + \mu\mathscr{L} \otimes H)\delta + (M \otimes \mu H)z_e$, $f_e(\delta, z_e) = -(\mathscr{L} \otimes \bar{A} + \mu\mathscr{L}^2 \otimes H)\delta - (\mathscr{L} \otimes \mu H)z_e$, and

$$\mathscr{J}_i(\xi) = \begin{bmatrix} \delta \\ (I_{2nN} - \Lambda_i \otimes I_{2n})z_e \\ (I_N - \Lambda_i)\tau \end{bmatrix}. \tag{5.18}$$

$\mathbb{X} := \{(\delta, z_e, \tau) \in \mathbb{R}^{2nN} \times \mathbb{R}^{2nN} \times \mathbb{R}^N_{\geq 0}\}$, $\Lambda_i \in \mathbb{R}^{N\times N}$ is a diagonal matrix, whose ith diagonal element is 1 and all the rest elements are zeros. Note that the jump dynamics $\mathscr{J}(\xi)$ is defined as union of $\mathscr{J}_i(\xi)$, this is because the agents have different sampling periods h_i, so that the jump dynamics of z_{ei}, τ_i respectively z_{ej}, τ_j typically happens at different instants. Thus, it is necessary to describe $\mathscr{J}(\xi)$ as union of $\mathscr{J}_i(\xi)$. Moreover, we remark that the adopted sampling manner is asynchronous, which is more general and can cover the synchronous sampling mechanism as a special case.

In the hybrid model, δ denotes the consensus error variable, as long as δ converges to the origin, it means that consensus is achieved. Thus, based on the hybrid model, the consensus of MAS can be obtained via analyzing the stability of a hybrid system.

Definition 5.1 Consider the MAS with sampled-data based DOF control, whose dynamics is described by (5.14). The set $\{\xi \in \mathbb{X} | \delta = 0_{2nN}, z_e = 0_{2nN}\}$ is said to be uniformly globally asymptotically stable (UGAS) if there exists a function $\beta \in \mathscr{KL}$, such that for any initial condition $\xi(0) \in \mathbb{X}$, the solution ξ satisfies

$$\|(\delta(t), z_e(t))\| \leq \beta(\|(\delta(0), z_e(0))\|, t). \tag{5.19}$$

5.3.2 Stability Analysis

To facilitate the stability analysis, the following auxiliary function $\phi_i(\tau_i)$ is introduced

$$\dot{\phi}_i(\tau_i) = -\gamma_i \left(\frac{\phi_i^2(\tau_i)}{\sigma_i} + 1 \right), \tag{5.20}$$

where $\gamma_i > 0, 0 < \sigma_i < 1$. It is obvious that $\phi_i(\tau_i)$ is monotonically decreasing. Note that the timer variable τ_i is included in the closed-loop state ξ, the function $\phi_i(\tau_i)$ will be used to construct Lyapunov function, thus, it is required that $\phi_i(\tau_i)$ is always positive. According to the dynamics of $\phi_i(\tau_i)$, we can derive an upper bound of τ_i, e.g., $\tau^i_{\max}$, such that $\phi_i(\tau_i) > 0$ for any $\tau_i < \tau^i_{\max}$. Specifically, based on (5.20), after some routine calculation, we can obtain that $\tau^i_{\max} = \frac{\sqrt{\sigma_i}}{\gamma_i} \arctan \frac{\phi_i(0)}{\sqrt{\sigma_i}}$.

Next, before giving the main results, the following lemma is proposed, which is useful for formally proving the main results.

Lemma 5.1 *Suppose that (A, B) is stabilizable and (C, A) is detectable, then, there exists a matrix K such that $A + BK$ is Hurwitz, and there exist constants $\alpha > 0$, $\kappa > 0$, and matrices $P > 0$, $Q > 0$, such that the following matrix inequality holds*

$$\Omega_1 + \alpha \Omega_2 < 0, \tag{5.21}$$

where

$$\Omega_1 = \begin{bmatrix} P(A+BK)+(A+BK)^T P & PBK \\ * & QA+A^T Q - \kappa C^T C \end{bmatrix}, \; \Omega_2 = \begin{bmatrix} P & 0 \\ * & Q \end{bmatrix}. \quad (5.22)$$

Proof According to Schur complement lemma, the matrix inequality (5.21) holds is equivalent to

$$\begin{aligned} \Omega_{11} &= P\left(A + \frac{\alpha}{2} I + BK\right) + \left(A + \frac{\alpha}{2} I + BK\right)^T P < 0, \\ \text{and } \Omega_{22} &= Q\left(A + \frac{\alpha}{2} I\right) + \left(A + \frac{\alpha}{2} I\right)^T Q - \kappa C^T C - (PBK)^T \Omega_{11}^{-1} PBK < 0. \end{aligned} \quad (5.23)$$

Since $A + BK$ is Hurwitz, the parameter α can be chosen such that $\frac{\alpha}{2} I + \lambda_{\max}(A + BK) < 0$, this is able to guarantee that $\Omega_{11} < 0$ holds.

Due to (C, A) is detectable, it is known that there exists $Q > 0$ and G_0 such that $A + G_0 C$ is Hurwitz, and

$$\begin{aligned} &(A + G_0 C)^T Q + Q(A + G_0 C) < 0 \\ \Leftrightarrow &A^T Q + QA + C^T Y + YC < 0, \end{aligned} \quad (5.24)$$

where $Y = QG_0$. According to Finsler's lemma [21], the above inequality is equivalent to

$$A^T Q + QA - \kappa_0 C^T C < 0, \quad (5.25)$$

that is $G_0 = -\frac{1}{2}\kappa_0 Q^{-1} C^T$. Based on the above formula, for any $\kappa > \kappa_0$, $A^T Q + QA - \kappa C^T C < 0$ is satisfied. Thus, there exists sufficiently small α such that $Q(A + \frac{\alpha}{2} I) + (A + \frac{\alpha}{2} I)^T Q - \kappa C^T C < 0$ holds. Next, take note of the term $(PBK)^T \Omega_{11}^{-1} PBK$ in Ω_{22}, if $\Omega_{22} < 0$ is not satisfied, we can choose $P_0 = sP$, $s > 0$, with P_0 instead of P, such that $(P_0 BK)^T \Omega_{11}^{-1} P_0 BK = s(PBK)^T \Omega_{11}^{-1} PBK$. Then, there always exists a $s > 0$ such that $\Omega_{22} < 0$ holds. Therefore, we can obtain $\Omega_1 + \alpha \Omega_2 < 0$. □

Theorem 5.1 *Consider the hybrid system (5.14), where $\mathscr{F}(\xi)$, $\mathscr{J}(\xi)$, $\mathbb{F}$, $\mathbb{J}$ are defined in (5.16) and (5.17). Let $G = -\kappa Q^{-1} C^T$ and define $\bar{P} = [P + Q \; -Q; -Q \; Q]$, where $\kappa > 0$ and the matrices $P > 0$, $Q > 0$ are obtained from Lemma 5.1. The gain matrix K is chosen such that $A + BK$ is Hurwitz, and the sampling period h_i is chosen to be $h_i < \tau_{\max}^i$. Moreover, there exist parameters $\gamma_i > 0$, $0 < \varepsilon < 1$, $\alpha > 0$, $\sigma_i > 0$, such that the following matrix inequality condition holds*

$$\begin{bmatrix} -(1-\varepsilon)\alpha\lambda_2(I_N \otimes \bar{P}) + \sum_{i=1}^N \sigma_i S_{\delta_i}^T S_{\delta_i} & \mathscr{L} \otimes \mu \bar{P} H + \sum_{i=1}^N \sigma_i S_{\delta_i}^T S_{e_i} \\ * & -(1-\varepsilon)\Gamma_i^2 \otimes I_{2n} + \sum_{i=1}^N \sigma_i S_{e_i}^T S_{e_i} \end{bmatrix} < 0, \quad (5.26)$$

where $\Gamma_i = \mathrm{diag}\{\gamma_1, \gamma_2, \ldots, \gamma_N\}$. $S_{\delta_i} = -(R_i \otimes I_{2n})((\mathscr{L} \otimes \bar{A} + \mu \mathscr{L}^2 \otimes H))$, $S_{ei} = -(R_i \otimes I_{2n})(\mathscr{L} \otimes \mu H)$, where R_i denotes a row matrix with the ith element being 1 and all the rest elements are zeros. Then, the (δ, z_e) dynamics is UGAS, which implies that consensus of MAS is achieved.

Proof Choose the following candidate Lyapunov function

$$V(\xi) = \delta^T(\mathscr{L} \otimes \bar{P})\delta + \sum_{i=1}^{N} \gamma_i \phi_i(\tau_i) z_{ei}^T z_{ei}. \tag{5.27}$$

First, we discuss the property of (5.27) to guarantee it a suitable function. It can be seen that auxiliary function $\phi_i(\tau_i)$ has been incorporated, which is actually a decreasing function with respect to the timer variable τ_i, that is, as τ_i grows, the value of $\phi_i(\tau_i)$ will become smaller and smaller, and there do exist a threshold $\tau_{\max}^i > 0$ for τ_i, at which $\phi_i(\tau_{\max}^i) = 0$ and if $\tau_i > \tau_{\max}^i$ exceeds this threshold, $\phi_i(\tau_i)$ becomes negative. According to (5.15), it is known that the sampling periods is selected as h_i for each agent i, and $\tau_i \leq h_i$ is always satisfied. Thus, to guarantee the positiveness of auxiliary function $\phi_i(\tau_i)$, we just need to specify the condition $h_i < \tau_{\max}^i$ for the sampling periods h_i. In addition, from the radial unboundedness of $\delta^T(\mathscr{L} \otimes \bar{P})\delta$ and $z_{ei}^T z_{ei}$, we can see that $V(\xi)$ is also radially unbounded. So, $V(\xi)$ is well defined. Then, to guarantee the stability of the closed-loop MAS, it is required that the Lyapunov function $V(\xi)$ is decreasing along both flow dynamics and jump dynamics.

Consider the evolution of $V(\xi)$ along the flow dynamics of ξ

$$\begin{aligned}
&\langle \nabla V(\xi), \mathscr{F}(\xi) \rangle \\
=&2\delta^T(\mathscr{L} \otimes \bar{P})((I_N \otimes \bar{A} + \mu \mathscr{L} \otimes H)\delta + (M \otimes \mu H)z_e) \\
&- \sum_{i=1}^{N} \gamma_i^2 \left(\frac{\phi_i^2(\tau_i)}{\sigma_i} + 1 \right) z_{ei}^T z_{ei} + \sum_{i=1}^{N} 2\gamma_i \phi_i(\tau_i) z_{ei}^T \dot{z}_{ei}.
\end{aligned} \tag{5.28}$$

Define $\Delta = \text{diag}\{\lambda_1, \lambda_2, \ldots, \lambda_N\}$, where λ_i, $i \in [1, N]$, denote the eigenvalues of $\mathscr{L}$. By means of Jordan decomposition, there exists an orthogonal matrix W, such that $\Delta = W^T \mathscr{L} W$. Define $\delta_w = (W \otimes I_{2n})\delta$, we can derive that

$$\begin{aligned}
&2\delta^T(\mathscr{L} \otimes \bar{P})(I_N \otimes \bar{A} + \mu \mathscr{L} \otimes H)\delta \\
=&2\delta_w^T(\Delta \otimes \bar{P}\bar{A} + \mu \Delta^2 \otimes \bar{P}H)\delta_w \\
=&2\sum_{i=1}^{N} \delta_{wi}^T(\lambda_i \bar{P}\bar{A} + \mu \lambda_i^2 \bar{P}H)\delta_{wi} \\
=&\sum_{i=1}^{N} \lambda_i \delta_{wi}^T(\bar{P}\bar{A} + \bar{A}^T\bar{P} + 2\mu\lambda_i \bar{P}H)\delta_{wi} \\
\leq&\sum_{i=1}^{N} \lambda_i \delta_{wi}^T(\bar{P}\bar{A} + \bar{A}^T\bar{P} + \bar{P}H)\delta_{wi},
\end{aligned} \tag{5.29}$$

where the condition $\mu\lambda_i \leq 1/2$ is used. Define $T = [I\ 0;\ I\ I]$, some calculation yields $T^T(\bar{P}\bar{A} + \bar{A}^T\bar{P} + \bar{P}H)T = \Omega_1$, where Ω_1 is defined in (5.22). It follows from (5.29) and Lemma 5.1 that

$$
\begin{aligned}
&2\delta^T(\mathscr{L} \otimes \bar{P})(I_N \otimes \bar{A} + \mu\mathscr{L} \otimes H)\delta \\
\leq &\sum_{i=1}^{N} \lambda_i \delta_{wi}^T T^{-T} \Omega_1 T^{-1} \delta_{wi} \\
\leq &\sum_{i=1}^{N} \lambda_i \delta_{wi}^T T^{-T} \alpha \Omega_2 T^{-1} \delta_{wi} \\
= &- \alpha \delta_w^T (\Delta \otimes \bar{P}) \delta_w \\
= &- \alpha \delta^T (\mathscr{L} \otimes \bar{P}) \delta \\
\leq &- \alpha \lambda_2 \delta^T (I_N \otimes \bar{P}) \delta.
\end{aligned}
\tag{5.30}
$$

Based on the matrix inequality condition (5.26), let the matrix be right and left multiplied by $[\delta^T\ z_e^T]^T$ and its transpose respectively, after some tedious calculation, and note that $\mathscr{L}M = M$, it can be derived that

$$
\begin{aligned}
&2\delta^T(\mathscr{L} \otimes \bar{P})((I_N \otimes \bar{A} + \mu\mathscr{L} \otimes H)\delta + (M \otimes \mu H)z_e) \\
\leq &- \varepsilon\alpha\lambda_2\delta^T(I_N \otimes \bar{P})\delta - \sum_{i=1}^{N} \sigma_i \left\| S_{\delta_i}\delta + S_{e_i}z_e \right\|^2 + (1-\varepsilon)z_e^T(\Gamma_i^2 \otimes I_{2n})z_e.
\end{aligned}
\tag{5.31}
$$

According to (5.28) and (5.31), we have

$$
\begin{aligned}
&\langle \nabla V(\xi), F(\xi) \rangle \\
\leq &- \varepsilon\alpha\lambda_2\delta^T(I_N \otimes \bar{P})\delta - \varepsilon\sum_{i=1}^{N}\gamma_i^2 z_{ei}^T z_{ei} - \sum_{i=1}^{N}(\sqrt{\sigma_i}\left\|S_{\delta_i}\delta + S_{e_i}z_e\right\| - \frac{1}{\sqrt{\sigma_i}}\gamma_i\phi_i(\tau_i)\left\|z_{ei}\right\|)^2 \\
\leq &- \varepsilon\alpha\lambda_2\delta^T(I_N \otimes \bar{P})\delta - \varepsilon\sum_{i=1}^{N}\gamma_i^2 z_{ei}^T z_{ei}.
\end{aligned}
\tag{5.32}
$$

From (5.32), it is clear that the Lyapunov function decreases along the hybrid system's flow dynamics. Then, consider the change of $V(\xi)$ when ξ jumps. From the form of $V(\xi)$ in (5.27) and the form of $\mathscr{J}(\xi)$ in (5.16), it is known that δ has no jump dynamics, i.e., $\delta^+ = \delta$. τ_i and z_{ei} have jump dynamics at sampling instants of agent i, that indicates $\phi_i(\tau_i^+) = \phi_i(0)$ and $z_{ei}^+ = 0$, i.e., no matter what is the value of $\phi_i(0)$, the term $\phi_i(\tau_i)z_{ei}^T z_{ei}$ does not increase at any agents' sampling instants. Combined with the condition (5.32), we can conclude that the Lyapunov function $V(\xi)$ decreases along the hybrid system's dynamics, and δ and z_{ei} converge to the origin eventually. In other words, it implies that all agents' states reach the same value and the objective of consensus is achieved. □

Remark 5.1 According to Theorem 5.1, the proposed sampled-data based DOF controller (5.2) is feasible to control the consensus of MAS as long as the matrix condition (5.26) is satisfied and the sampling period h_i satisfies $h_i \in (0, \frac{\sqrt{\sigma_i}}{\gamma_i} \arctan \frac{\phi_i(0)}{\sqrt{\sigma_i}})$. For the matrix condition (5.26), one can always find a pair of satisfactory solution by carefully adjusting γ_i and σ_i. Observe that $-(1-\varepsilon)\alpha\lambda_2(I_N \otimes \bar{P})$ and $-(1-\varepsilon)\Gamma_i^2 \otimes I_{2n}$ are negative and γ_i and σ_i can be arbitrarily selected, therefore, we can guarantee the diagonal sub-matrices to be negative definite, namely, there always exist appropriate γ_i and σ_i such that the diagonal sub-matrices are negative. Furthermore, it can be seen that the off-diagonal matrices are independent of γ_i, this means that we can select sufficiently large γ_i such that the matrix inequality condition (5.26) is satisfied. As a result, the derived conditions in Theorem 5.1 can always be guaranteed. Besides, the implementation of the DOF controller does not depend on continuous communication, instead it only requires the sampled relative output information between agents. The sampling period is asynchronous, thus it is superior to the synchronous samplings.

Remark 5.2 The designed sampled-data based DOF consensus control has some distinctive features. First, the practical implementation is simpler in the sense that only output information rather than the full state information is needed to be transmitted through network, so that the volume of information transmissions is potentially reduced, and the proposed DOF control does not need observer and directly use the relative output information to design control law, which, compared with [7, 8, 10], where state observers were first designed to estimate the system states and then the estimated states still needed to be transmitted, is more easy to be implemented. Besides, compared with the existing DOF control methods in [12–14], the main advantage of the proposed method is that sampled rather than continuous relative output information is used, which implies that the proposed method is more suitable to be implemented in the context of network.

5.4 Extension to Leader-Following MAS

It is well known that MAS can be categorized into leader-following MAS and leaderless MAS according to if there exists leader or not. In previous section, the sampled-data DOF control has been applied to leaderless MAS. In this section, we will show that the proposed approach can also be applied to leader-following MAS.

Consider the leader-following MAS with N followers and one leader, the followers' dynamics is described by (5.1), and the leader's dynamics is given by

$$\begin{aligned} \dot{x}_0(t) &= Ax_0(t), \\ y_0(t) &= Cx_0(t). \end{aligned} \tag{5.33}$$

In leader-following case, $\bar{\mathcal{G}}$, defined in Chap. 1, is used to denote the communication graph. And it is assumed that the subgraph associated with the followers is connected and undirected, and the graph $\bar{\mathscr{G}}$ contains a directed spanning tree with the leader as the root.

The aim is to achieve consensus tracking, that is, all follower agents' states can track the leader's state asymptotically. To achieve this objective, the sampled-data DOF controller is designed as the following form. For $t \in (t_k^i, t_{k+1}^i]$,

$$
\begin{aligned}
u_i(t) &= Kv_i(t),\\
\dot{v}_i(t) &= (A+BK)v_i(t) + \mu G\left(\sum_{j\in\mathcal{N}_i}(y_{vi}(t_k^i) - y_{vj}(t_k^i))\right) + b_i y_{vi}(t_k^i))\\
&\quad - \mu G\left(\sum_{j\in\mathcal{N}_i}(y_i(t_k^i) - y_j(t_k^i))\right) + b_i(y_i(t_k^i) - y_0(t_k^i))),\\
y_{vi}(t) &= Cv_i(t),
\end{aligned}
\tag{5.34}
$$

where b_i denotes the connection relationship between leader and agent i. The other parameters are the same as in the leaderless case.

Denote the consensus error $\delta_i(t) = x_i(t) - x_0(t)$, and define

$$
\begin{aligned}
q_{xi}^l(t) &= \sum_{j\in\mathcal{N}_i}(x_i(t) - x_j(t)) + b_i(x_i(t) - x_0(t)),\\
q_{vi}^l(t) &= \sum_{j\in\mathcal{N}_i}(v_i(t) - v_j(t)) + b_i v_i(t),\\
\bar{q}_{xi}^l(t) &= q_{xi}^l(t_k^i),\ e_{xi}^l(t) = \bar{q}_{xi}^l(t) - q_{xi}^l(t),\ t \in (t_k^i, t_{k+1}^i],\\
\bar{q}_{vi}^l(t) &= q_{vi}^l(t_k^i),\ e_{vi}^l(t) = \bar{q}_{vi}^l(t) - q_{vi}^l(t),\ t \in (t_k^i, t_{k+1}^i].
\end{aligned}
\tag{5.35}
$$

By recalling that $\mathscr{H} = \mathscr{L} + \mathscr{B}$ with $\mathscr{B} = \text{diag}\{b_1, b_2, \ldots, b_N\}$, then, we have

$$
\begin{aligned}
q_{xi}^l(t) &= \sum_{j\in\mathcal{N}_i}(x_i(t) - x_j(t)) + b_i(x_i(t) - x_0(t)) = (\mathscr{H}_{row}^i \otimes I_n)\delta(t),\\
q_{vi}^l(t) &= \sum_{j\in\mathcal{N}_i}(v_i(t) - v_j(t)) + b_i v_i(t) = (\mathscr{H}_{row}^i \otimes I_n)v(t),
\end{aligned}
\tag{5.36}
$$

where $\delta(t) = [\delta_1^T(t)\ \delta_2^T(t)\ \cdots\ \delta_N^T(t)]^T$, and $v(t) = [v_1^T(t)\ v_2^T(t)\ \cdots\ v_N^T(t)]^T$.

Then, similar to the leaderless case, define $z_i = [\delta_i^T\ v_i^T]^T$, $z_{ei} = [e_{xi}^l\ e_{vi}^l]^T$, based on (5.1), and (5.33)–(5.36), it is derived that

$$
\begin{aligned}
\dot{z}_i &= \bar{A}z_i + \mu G(\mathscr{H}_{row}^i \otimes I_{2n})z + \mu G z_{ei},\\
\dot{z} &= (I_N \otimes \bar{A} + \mu\mathscr{H} \otimes G)z + (I_N \otimes \mu G)z_e,\\
\dot{z}_e &= -(\mathscr{H} \otimes \bar{A} + \mu\bar{\mathscr{L}}^2 \otimes G)z - (\bar{\mathscr{L}} \otimes \mu G)z_e.
\end{aligned}
\tag{5.37}
$$

Compared with (5.12) and (5.13) (leaderless MAS), the dynamics in (5.37) (leader-following MAS) has similar form, though the system matrices are different. Take z and z_e as closed-loop states, following the steps in Sect. 5.3.1, the same form of hybrid system model can be constructed. Furthermore, the main results obtained in Sect. 5.3.2 also hold for leader-following MAS with slightly change of system matrices. Thus, the proposed sampled-data based DOF control approach can be applied to both leaderless and leader-following MAS.

5.5 Simulation Study

Example 5.1 (*Leaderless Consensus*) Consider MAS with 6 agents, the communication topology is given in Fig. 5.1. This example is borrowed from [8], where observer based control was considered. The Laplacian matrix $\mathcal{L}$ can be obtained from the communication graph and the minimum positive eigenvalue of $\mathcal{L}$ is $\lambda_2 = 1$. The system matrices are $A = [0\ 1\ 0; -1\ 0\ 1; 0\ 0\ -1]$, $B = [0; 0; 1]$, $C = [1\ 0\ 0; 0\ 1\ 0]$. Select gain matrix $K = [0.5208\ \ -1.6152\ \ -1.3993]$ such that $A + BK$ is Hurwitz. Let the positive definite matrices $P = [1\ 0.2651\ 0.029; 0.2651\ 1.1192\ 0.3189; 0.029\ \ 0.3189\ \ 0.2371]$, $Q = [5.4012\ \ -0.3252\ \ -2.4108; -0.3252\ \ 3.5952\ \ -0.2316; -2.4108\ \ -0.2316\ \ 6.6096]$. Let $\kappa = 5$, then, $G = -\kappa Q^{-1} C^T = [-1.1173\ \ -0.1276; -0.1276\ \ -1.4085; -0.412\ \ -0.0959]$. Select $\mu = 0.2$, $\alpha = 0.15$, $\varepsilon = 0.001$, $\sigma_{1,2,3} = 0.3$, $\sigma_{4,5,6} = 0.2$, $\gamma_i = 15$, $\phi_i(0) = 20$. Based on these parameters, it can be verified that the stability conditions in Theorem 5.1 are satisfied. We can calculate that the MASP of agents 1–3 and agents 4–6 are 0.0704 and 0.0577, respectively. In this example, let the sampling periods of the agents are $h_1 = 0.07$, $h_2 = 0.06$, $h_3 = 0.05$, $h_4 = 0.04$, $h_5 = 0.03$, $h_6 = 0.02$. The initial conditions $v_i(0)$ are set to be $v_i(0) = [0; 0; 0]$ and the initial states $x_i(0)$ are randomly generated, e.g., $x_i(0) = 10\text{randn}(3, 1)$, where 'randn' is the random function in MATLAB.

The simulation results are presented in Figs. 5.2, 5.3. Figure 5.2 shows the state and input trajectories, all components of $x_i(t)$ converge to the same value, that is, the objective of consensus is achieved. As the agents' states tend to the same value, the internal controller state $v_i(t)$ converges to the origin, such that the control input $u_i(t) = K v_i(t)$ also converges to the origin. To show that the sampled relative information is used in the controller, Fig. 5.3 presents the sampled relative information of the agents. From the local enlarged figure, it can be seen that the sampling periods of the agents are different, i.e., asynchronous sampling.

Fig. 5.1 Communication topology without leader

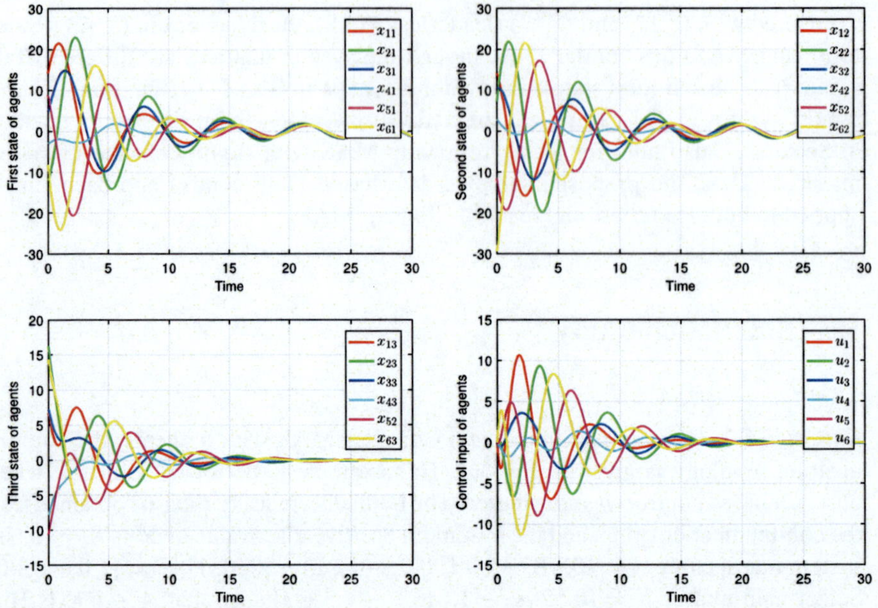

Fig. 5.2 State trajectories and control inputs of leaderless MAS (with the proposed method)

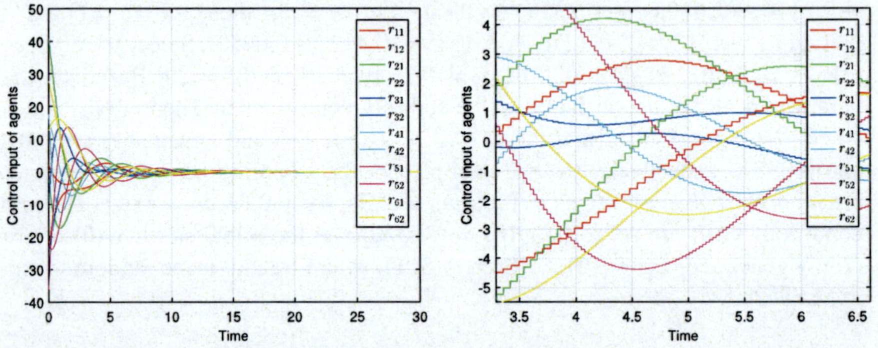

Fig. 5.3 Sampled relative information $r_i(t) = \bar{q}_{y_v i}(t) - \bar{q}_{yi}(t)$, $i \in [1, 6]$, with $r_i(t) = [r_{i1}(t); r_{i2}(t)]$, $\bar{q}_{y_v i}(t) = q_{y_v i}(t_k^i)$, $\bar{q}_{yi}(t) = q_{yi}(t_k^i)$, $t \in [t_k^i, t_{k+1}^i)$. The right figure is local enlargement of left figure

Moreover, in order to further illustrate the advantage of the proposed method, some comparison results and discussions are given in the following. Note that in recent work [8], observer-based output feedback consensus control of MAS was studied. Using the method and given parameters in [8], the simulation results are shown in Fig. 5.4. From Figs. 5.2 and 5.4, we can see that both the methods can achieve consensus with output-based control, but the proposed method is more superior in physical implementation. First, the control implementation is more simpler,

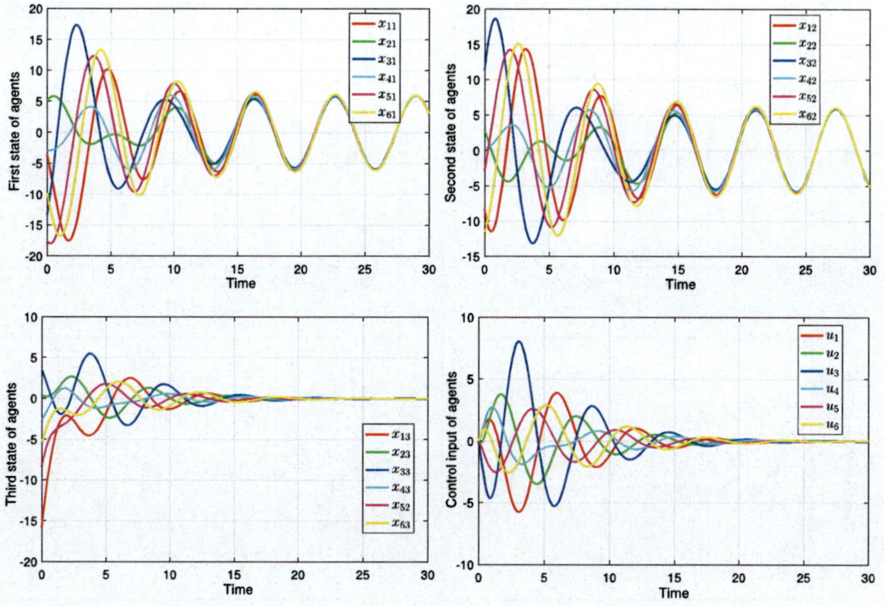

Fig. 5.4 State trajectories and control inputs of leaderless MAS (with the method in [8])

Fig. 5.5 Communication topology with leader

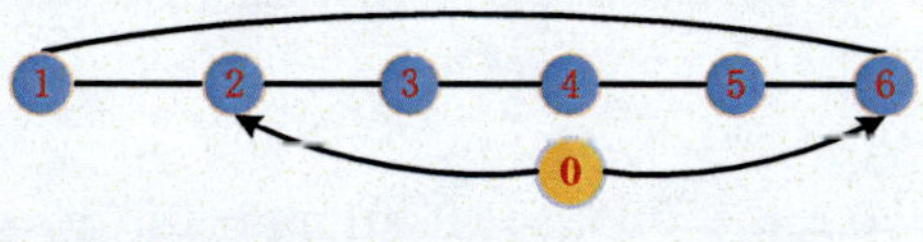

since observer design (as used in [8]) is not required, and sampled information rather than internal dynamic states (e.g., $\hat{x}_i$, $\hat{v}_i$) is used. Sampled information can be kept by using zero-order holder, but internal dynamic states need to be measured continuously. Second, with the proposed DOF control, output information rather than state information (e.g., $\hat{v}_i(t_k^i)$, $\tilde{x}_i(t_k^i)$, see [8]) is transmitted over network, it implies that the amount of information in each data transmission is usually less with the DOF control method, because the dimension of output is less than dimension of state (in this example, the output is 2-dimensional but the state is 3-dimensional). Therefore, it is illustrated that in the sense of similar consensus control performance, the proposed method can be easier implemented.

Example 5.2 (*Leader-Following Consensus*) Consider leader-following MAS, the communication topology is shown in Fig. 5.5, it is assumed that agent 2 and agent 6 can communicate with leader. The minimum positive eigenvalue of $\mathscr{H}$ is $\lambda_1 = 0.2434$. The system matrices and gain matrix K are the same as in Example 5.1. Let the positive definite matrices $P = [1.8036\ \ 0.4782\ \ 0.0524;\ 0.4782\ 2.0186 0.5752;\ 0.0524\ 0.5752\ \ 0.4277]$, $Q = [2.7006\ -0.1626\ -1.2054;\ -0.1626 1.7976\ -0.1158;\ -1.2054\ -0.1158\ 3.3048]$. Choose $\kappa = 5, G = -\kappa Q^{-1} C^T = [-2.2346\ \ -0.2552;\ -0.2552\ \ -2.8169;\ -0.8240\ \ -0.1918]$. Select $\mu =$

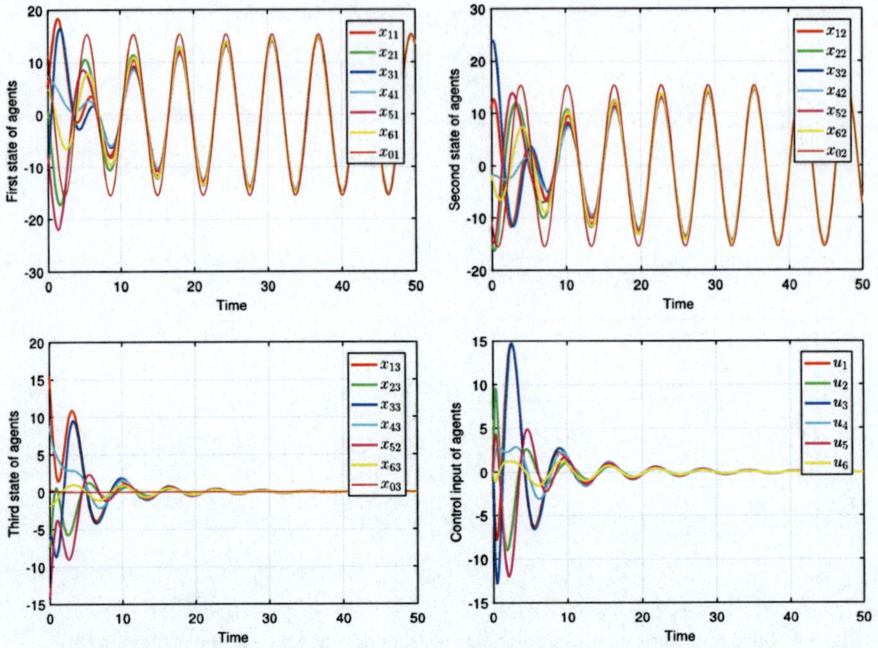

Fig. 5.6 State trajectories and control inputs of leader-following MAS (with the proposed method)

0.2, $\alpha = 0.15$, $\varepsilon = 0.001$, $\sigma_i = 0.1$, $\gamma_i = 15$, $\phi_i(0) = 20$. Based on these parameters, we can calculate that the MASP of agents are 0.0328. In this example, let the sampling periods of the agents be $h_i = 0.02$. The initial conditions $v_i(0)$ are set to be $v_i(0) = [0; 0; 0]$ and the initial states $x_i(0)$ are randomly generated. The simulation results are shown in Fig. 5.6, where it is clearly shown that the all the follower agents' states can track the leader's states asymptotically, i.e., leader-following consensus is achieved. For comparison, the simulation results with the method in [8] are shown in Fig. 5.7, it can be seen that both the proposed DOF control method and the observer-based control method in [8] can achieve leader-following consensus, but, as discussed in Example 5.1, the implementation of the proposed DOF control method is much simpler.

5.6 Conclusion

In this chapter, we study the consensus control problem of linear MAS with sampled-data DOF control. The DOF control law is designed based on sampled relative output information between neighboring agents, while continuous information exchange is avoided. For consensus analysis and determining the MASP, a hybrid model is con-

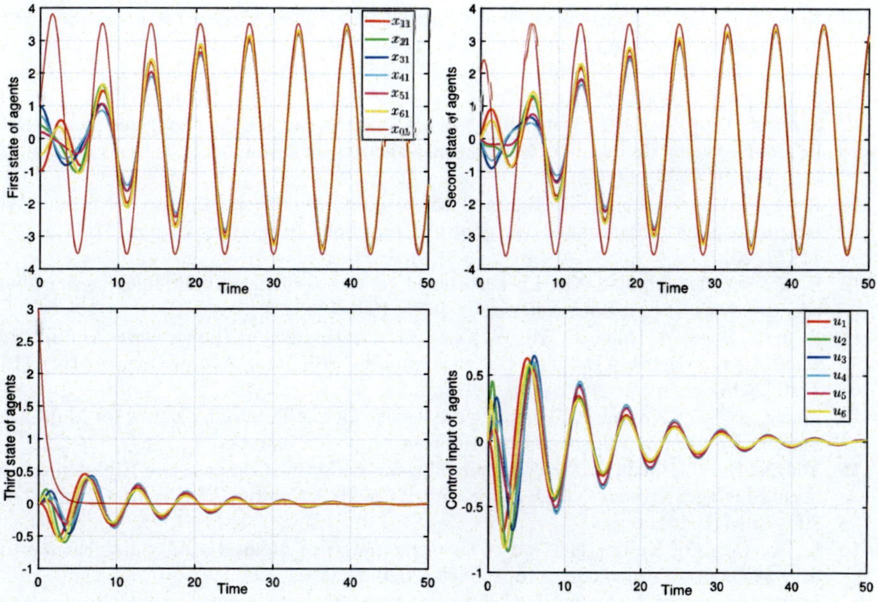

Fig. 5.7 State trajectories and control inputs of leader-following MAS (with the method in [8])

structed to completely describe the MAS dynamics and incorporate the sampling period. Based on this model, rigorous Lyapunov analysis is provided to prove stability, and the MASP can be explicitly calculated.

References

1. Xiao, F., Wang L.: Asynchronous consensus in continuous-time multi-agent systems with switching topology and time-varying delays. IEEE Trans. Autom. Control **53**(8), 1804-1816 (2008)
2. Zhang, W., Tang, Y., Han, Q., Liu, Y.: Sampled-Data Consensus of Linear Time-Varying Multiagent Networks With Time-Varying Topologies. IEEE Transactions on Cybernetics **52**(1), 128-137 (2022)
3. Xiao, F., Shi, Y., Ren, W.: Robustness Analysis of Asynchronous Sampled-Data Multiagent Networks With Time-Varying Delays. IEEE Trans. Autom. Control **63**(7), 2145-2152 (2018)
4. Ding, L., Han, Q., Guo, G.: Network-based leader-following consensus for distributed multi-agent systems. Automatica **49**(7), 2281-2286 (2013)
5. He, W., Zhang, B., Han, Q.: Leader-following consensus of nonlinear multiagent systems with stochastic sampling. IEEE Trans. Cybern. **47**(2), 327-338 (2017)
6. Liu, W., Huang, J.: An updated version of "Leader-following consensus for linear multi-agent systems via asynchronous sampled-data control. IEEE Trans. Autom. Control **65**(7), 3215-3222 (2020)
7. Zhang, J., Zhang, H., Zhang, K., Cai, Y.: Observer-Based Output Feedback Event-Triggered Adaptive Control for Linear Multiagent Systems Under Switching Topologies. IEEE Trans. Neural Netw. Learn. Syst. **33**(12), 7161-7171 (2022)

8. Koike, R., Endo, T., Matsuno, F.: Output-based dynamic event-triggered consensus control for linear multiagent systems, Automatica **133**, 109863 (2021)
9. Ma, Y., Che, W., Deng, C., Wu, Z.: Observer-Based Event-Triggered Containment Control for MASs Under DoS Attacks. IEEE Trans. Cybern. **52**(12), 13156-13167 (2022)
10. Ruan, X., Feng, J., Xu, C., Wang, J.: Observer-Based Dynamic Event-Triggered Strategies for Leader-Following Consensus of Multi-agent Systems with Disturbances. IEEE Trans. Net. Sci. Eng. **7**(4), 3148-3158 (2020)
11. Rong, L., Liu, X., Jiang, G., Xu,S.,: Observer-Based Multiagent Bipartite Consensus With Deterministic Disturbances and Antagonistic Interactions. IEEE Trans. Cybern. **52**(11), 11772-11779 (2022)
12. Li, Z., Ren, W., Liu, X., Xie, L.: Distributed consensus of linear multi-agent systems with adaptive dynamic protocols. Automatica **49**(7), 1986-1995 (2013)
13. Wen, G., Zhao, Y., Duan, Z., Yu, W., Chen, G.: Containment of Higher-Order Multi-Leader Multi-Agent Systems: A Dynamic Output Approach. IEEE Trans. Autom. Control **61**(4), 1135-1140 (2016)
14. Zhang, Y., Su, Y., Wang, X.: Distributed Adaptive Output Feedback Control for Multi-agent Systems with Unknown Dynamics. IEEE Trans. Autom. Control **66**(3), 1367-1374 (2020)
15. You, X., Hua, C., Guan, X.: Self-Triggered Leader-Following Consensus for High-Order Nonlinear Multiagent Systems via Dynamic Output Feedback Control. IEEE Trans. Cybern. **49**(6), 2002-2010 (2019)
16. Li, X., Tang, Y., Karimi, H.: Consensus of multi-agent systems via fully distributed event-triggered control. Automatica **116**, 108898 (2020)
17. Li, X., Soh, Y., Xie, L.: Output-feedback protocols without controller interaction for consensus of homogeneous multi-agent systems: A unified robust control view. Automatica **81**, 37-45 (2017)
18. Hu, W., Liu, L., Feng, G.: Consensus of linear multi-agent systems by distributed event-triggered strategy. IEEE Trans. Cybern. **46**(1), 148-157 (2016)
19. Smith, R., Hadaegh, F.: Control of deep-space formation-flying spacecraft; relative sensing and switched information. J. Guid. Control Dynam. **28**(1), 106-114 (2005)
20. Clarke, F.: Optimization and nonsmooth analysis, SIAM, Philadelphia, PA (1990)
21. Liu, J., Zhang, J., Zhang, H., He, M.: Stability and stabilization analysis for time-delay systems: a unified viewpoint of Finsler's lemma. IET Contr. Theory Appl. **28**(11), 1577-1582 (2011)

Part II
Event-Triggered Cooperative Control

Chapter 6
Hybrid Event-Triggered Bipartite Consensus of MAS

6.1 Introduction

As mentioned in Chap. 1, sampled-data control and event-triggered control are both suitable for the communication over network. Thus, by means of the proposed hybrid system approach, we have investigated the sampled-data based consensus control for MASs in Chaps. 2–5, and it is formally proved that sampled-data control is easy to implement in practical scenarios with asynchronous sampling periods, and has nice robustness to various network issues. Nevertheless, it should also be noticed that sampled-data control is a purely time-triggered method. It would always be triggered as long as a fixed time unit elapsed, as a result, even if there is no disturbance acting on the system and the system is operating smoothly [1], the periodic redundant communication still continues, making amount of communications resources be wasted. Besides, a clear periodicity in time on one hand makes the sampled-data control be predictable and easy to implement in the context of network, but on the other hand gives rise to network attacks in the sense that a more refined attack rule can be made by adversaries to destroy the target time points accurately. Alternatively, in this chapter, we would like to consider the event-triggered control for the networked MASs. With the event-triggered control protocol, the communication among agents occurs only when the state difference between real-time state and the sampled state exceeds a threshold. Thus, it has great potential to reduce the usage of communication resources (e.g., network bandwidth and node energy), and has received widespread research interests [2–5]. However, notice that although these results have shown that Zeno behavior (i.e., the phenomenon with infinite triggers in finite time period) is excluded, there still exist some limitations in the sense that the minimum inter-event time (MIET) is unknown, the triggering conditions depend on neighbors' information, and the exclusion of Zeno behavior depends on exponential decay term. In practice, it is desired that not only Zeno behavior is ruled out but also a strictly positive known MIET is guaranteed, which is more practical because hardware limitation always enforces a positive minimum time constraint between two triggering instants. Therefore, in this chapter, the hybrid system approach will

G. Zhao et al., *Cooperative Control of Multi-agent Systems*,
https://doi.org/10.1007/978-981-97-0968-7_6

be introduced to design the hybrid event-triggered control protocol. Benefited from the design, the timer variable used in former chapters (to generate sampling periods) can still be used to constitute the event-triggering mechanism, while the role of it is changed to record a minimum inter-event time, with which, the mentioned hardware limitations as well as the Zeno behavior can be addressed simultaneously.

In the context of consensus control of MAS, cooperative relationship is the most fundamental form in the sense that the interactions among agents are cooperative, and unsigned graph with nonnegative weights is used to describe the cooperative relationship. But, in real world applications, not only cooperation but also competition may exist in some scenarios, such as friend or foe relationships in social networks [6], bidirectional flying of unmanned air vehicles [4], two-party political systems [7], and ecological systems [5, 8]. In these scenarios, in order to appropriately describe both the cooperative and antagonistic interactions among agents, signed graph (instead of unsigned graph) with positive and negative edge weights is utilized, and bipartite consensus instead of consensus is considered, which means that all agents reach agreement with the same magnitude but with opposite signs. The concept of bipartite consensus was firstly proposed and studied in [9], where it was shown that bipartite consensus can be achieved if the signed graph is structurally balanced and connected. Then, around this research topic, considerable number of results have emerged in recent years. For example, [6] studied interval bipartite consensus of MAS with signed graph. The bipartite output tracking control and prescribed-time bipartite consensus control problems of MAS were investigated in [4, 10] respectively. Bipartite containment was achieved in [3] with adaptive observer, and output feedback based bipartite consensus problem with disturbances was considered in [11]. In [12], it was proved that bipartite consensus of MAS is equivalent to the stabilizability of each agent's state space model. In [13], extended state observer and distributed output feedback consensus control law were designed in order to achieve leader-following bipartite consensus. The observer-based bipartite formation of MAS was exploited in [14], and [15] studied finite-time and fixed-time bipartite tracking of MAS with antagonistic interactions. Note that, in these available results, only leaderless or leader-following bipartite consensus problem was considered, or they are studied separately, while few of them modeled the two cases into a unified framework. For the common control objective in the sense of driving some well defined error variables to the origin, it is suitable for representing the two cases in a unified form by means of the hybrid dynamical model, then the leaderless and leader-following bipartite consensus problems can be simultaneously transformed into the stability problem of a hybrid system.

In this chapter, we will study the event-triggered bipartite consensus problem of linear MAS, and aim to propose a hybrid ETM with determined MIET and a unified framework for leaderless and leader-following MAS. Firstly, a novel hybrid ETM is proposed to reduce usage of communication resources, the triggering instant is determined by joint action of an internal timer variable and an internal dynamic variable with hybrid dynamics. The minimum inter-event time of each agent is designable and can be pre-selected. Moreover, the designed triggering conditions only depend on a certain agent's own information, the neighbors' information is not required when implementing the ETM. Secondly, by exploiting features of leaderless and leader-following bipartite consensus of MAS, a unified hybrid model is constructed to describe closed-loop dynamics in these two cases, and a unified hybrid systems framework is proposed for stability analysis and ETM design.

6.2 Networked MAS and Hybrid Dynamic ETM

This section formulates the leaderless and leader-following bipartite consensus problems for the event-triggered MASs. The considered bipartite consensus is a more general case in the sense that both the cooperative and antagonistic interactions among agents are contained in a closed-loop MAS. The system dynamics of the considered MAS is depicted in the state space expression form

$$\dot{x}_i(t) = Ax_i(t) + Bu_i(t), \tag{6.1}$$

where $x_i(t) \in \mathbb{R}^n$ and $u_i(t) \in \mathbb{R}^m$ denote the state and control input of the follower agent i respectively. A, B are matrices with appropriate dimensions and it is assumed that (A, B) is stabilizable. If there is a leader in the MAS and its information can be received by at least one follower, then, the leader can be labeled as 0, and its dynamics is described by

$$\dot{x}_0(t) = Ax_0(t), \tag{6.2}$$

where $x_0(t) \in \mathbb{R}^n$ is the leader's state.

The preliminaries about the signed graph $\mathscr{G}$, which utilizes the positive and negative edge weights to describe the cooperative and antagonistic interaction relationship between agents, has been introduced in Chap. 1. Moreover, provided that the node set $\mathscr{V}$ of the signed graph $\mathscr{G}$ can be divided into two subsets $\mathscr{V}_1$ and $\mathscr{V}_2$ with $\mathscr{V}_1 \cup \mathscr{V}_2 = \mathscr{V}$ and $\mathscr{V}_1 \cap \mathscr{V}_2 = \emptyset$, if it is satisfied that for any $i, j \in \mathscr{V}_l (l \in \{1, 2\})$, $a_{ij} \geq 0$ and for $i \in \mathscr{V}_l, j \in \mathscr{V}_q, l \neq q (l, q \in \{1, 2\})$, $a_{ij} \leq 0$, then the signed graph $\mathscr{G}$ is said to be structurally balanced. Then, by means of the gauge transformation technique proposed in [9], a change of orthant order in $\mathbb{R}^n$ can be performed via transformation matrix $D = \text{diag}\{d_1, d_2, \ldots, d_N\}$ with $d_i \in \{\pm 1\}$, such that the modulus of the entries of matrix $\mathscr{L}$ and $\mathscr{H}$ keep unchanged, but the sign pattern is modified. For proceeding this chapter, we make the following Assumptions 6.1–6.2 to limit the communication connection among agents.

Assumption 6.1 The signed graph $\mathscr{G}$ is undirected, connected and structurally balanced.

Assumption 6.2 In leader-following case, at least one follower can receive information from the leader, i.e., there exists a spanning tree in the graph $\mathscr{G}$ with the leader agent 0 as the root.

Provided that Assumptions 6.1–6.2 are satisfied, then with respect to the Laplacian matrix $\mathscr{L}$, and the graph related matrices $\mathscr{A}$ and $\mathscr{H}$, the following lemma holds.

Lemma 6.1 ([9]) *Consider a structurally balanced signed graph $\mathscr{G}$, the following results hold:*

(1) There exists a diagonal matrix D such that the entries of the matrix $\mathscr{A}_D = D\mathscr{A}D$ are all nonnegative, the Laplacian matrix $\mathscr{L}_D = D\mathscr{L}D$ is nonnegative definite, and the matrix $\mathscr{H}_D = D\mathscr{H}D$ is positive definite;

(2) If a vector x satisfies $1^T x = 0$, we have $\min_{x\neq 0} \frac{x^T \hat{\mathscr{L}} x}{x^T x} = a(\hat{\mathscr{L}})$, where $\hat{\mathscr{L}} = \mathscr{L}_D$ or $\hat{\mathscr{L}} = \mathscr{H}_D$, $a(\hat{\mathscr{L}})$ denotes the minimum positive eigenvalue of $\hat{\mathscr{L}}$.

In network environment, a certain agent i's neighbors' state $x_j(t)$, $j \in \mathscr{N}_i$ cannot be directly used in controller design, since $x_j(t)$ is directly used in agent i indicates that continuous communication between agents is required, which is usually unpractical in applications. Thus, event-triggered control is used to avoid continuous communication, that is, the sampled information $x_i(t^i_{k_i})$ at triggering instant $t^i_{k_i}$ rather than $x_i(t)$ is used to design consensus controller. Specifically, for the leaderless MAS, the event-triggered bipartite consensus control protocol is designed as

$$u_i(t) = K \sum_{j\in\mathscr{N}_i} a_{ij}(\hat{x}_j(t) - \text{sgn}(a_{ij})\hat{x}_i(t)), \tag{6.3}$$

and, for the leader-following MAS,

$$u_i(t) = K\left(\sum_{j\in\mathscr{N}_i} a_{ij}(\hat{x}_j(t) - \text{sgn}(a_{ij})\hat{x}_i(t)) - b_i(x_i(t) - d_i x_0(t))\right), \tag{6.4}$$

where $\hat{x}_i(t) = e^{A(t-t^i_{k_i})}x_i(t^i_{k_i})$ denotes the open-loop estimate of state $x_i(t)$. From the definition of $\hat{x}_i(t)$, it can be seen that $\hat{x}_i(t)$ evolves according to $\dot{\hat{x}}_i(t) = A\hat{x}_i(t)$ when triggering event does not occur, and $\hat{x}_i(t)$ jumps to $x_i(t^i_{k_i})$ at triggering instant $t^i_{k_i}$. The gain matrix K is to be determined, and $\text{sgn}(\cdot)$ denotes the sign function. Note that continuous state $x_i(t)$ is used in the control protocol (6.4), since $x_i(t)$ is agent i's own state, it is always available for constructing $u_i(t)$.

We propose the following hybrid ETM to determine when a certain agent transmits information to its neighbors,

$$t^i_{k_i+1} := \inf\{t \geq t^i_{k_i} + \tau^i_{MIET} | \eta_i(t) \leq 0\}, \tag{6.5}$$

where $\tau^i_{MIET} > 0$ is a constant, which denotes the minimum inter-event time, and $\eta_i(t)$ is an internal dynamic variable in agent i, which has the following dynamics

$$\begin{cases} \dot{\eta}_i(t) = \Psi^i_F(o_i(t)), t \in (t^i_{k_i}, t^i_{k_i+1}] \\ \eta_i(t^+) = \Psi^i_J(o_i(t)), t = t^i_{k_i} \end{cases} \tag{6.6}$$

where $\Psi_F^i(o_i(t))$ and $\Psi_J^i(o_i(t))$ are flow and jump map functions that depend on local variable $o_i(t)$. How to design $\Psi_F^i(o_i(t))$ and $\Psi_J^i(o_i(t))$ will be given in Sect. 6.4.

Given the consensus control protocols (6.3), (6.4) and the hybrid ETM (6.5), if Assumptions 6.1–6.2 can be satisfied such that Lemma 6.1 holds, then the bipartite consensus is able to be achieved for the leaderless MASs in the sense that $\lim_{t\to\infty} \|x_i(t) - d_i d_j x_j(t)\| = 0$, namely, if both $i, j \in \mathscr{V}_1$ (or $i, j \in \mathscr{V}_2$), then, $d_i d_j = 1$ and $x_i(t) = x_j(t)$ as $t \to \infty$, and if $i \in \mathscr{V}_1$ and $j \in \mathscr{V}_2$, or vice versa, then, $d_i d_j = -1$ and $x_i(t) = -x_j(t)$ as $t \to \infty$, and for the leader-following MASs in the sense that $\lim_{t\to\infty} \|x_i(t) - d_i x_0(t)\| = 0$, namely, $x_i(t) = x_0(t)$ (if $d_i = 1$) as $t \to \infty$, or $x_i(t) = -x_0(t)$ (if $d_i = -1$) as $t \to \infty$. In what follows, the main concern of us will be focused on constituting a unified framework for leaderless and leader-following bipartite consensus problem by means of the hybrid dynamical model and then to determine the gain matrix K of control protocols (6.3), (6.4), and specify the flow and jump map functions $\Psi_F^i(o_i(t))$ and $\Psi_J^i(o_i(t))$ of the hybrid ETM (6.5).

6.3 Closed-Loop System Model

As discussed in the Introduction, the leaderless and leader-following bipartite consensus problems are suitable to be represented in a unified form by means of the hybrid dynamical model, because they have the common control objective, i.e., to drive some well defined error variables to the origin. Therefore, for the leaderless and leader-following MASs, a unified hybrid model is constructed in this section with using the flow dynamics and jump dynamics to completely describe the closed-loop system behaviors under event-triggered communication. For the sake of brevity, the arguments of state variables and functions will be omitted unless it is necessary.

6.3.1 Dynamics of Leaderless MAS

Recall the definition of $\hat{x}_i$, the measurement error between the real-time state and estimated state is defined as follows

$$e_i = \hat{x}_i - x_i. \tag{6.7}$$

Then, it follows from the dynamics of $\hat{x}_i$ that $\dot{e}_i = \dot{\hat{x}}_i - \dot{x}_i$ during the triggering intervals, and $e_i^+ = 0$ when a triggering event happens. By stacking the states in order, the system state vectors are given by $x = [x_1^T \; x_2^T \; \cdots \; x_N^T]^T$, $e = [e_1^T \; e_2^T \; \cdots \; e_N^T]^T$.

Note that $u_i = K \sum_{j\in\mathscr{N}_i} a_{ij}(\hat{x}_j - \text{sgn}(a_{ij})\hat{x}_i) = -K(\mathscr{L}_r^i \otimes I_n)(x + e)$, where $\mathscr{L}_r^i$ denotes ith row of Laplacian matrix $\mathscr{L}$. Then, it follows from (6.1) and (6.3) that the dynamics of x can be represented as a compact form

$$\dot{x} = (I_N \otimes A)x - (\mathscr{L} \otimes BK)x - (\mathscr{L} \otimes BK)e. \tag{6.8}$$

To achieve bipartite consensus, we introduce intermediate variable $z_i = d_i x_i$, and define $z = [z_1^T \ z_2^T \ \cdots z_N^T]^T$. Based on $D = \mathrm{diag}\{d_1, d_2, \ldots, d_N\}$, while noting that $D\mathscr{L}D = \mathscr{L}_D$, we have $z = (D \otimes I_n)x$ and

$$\dot{z} = (I_N \otimes A)z - (\mathscr{L}_D \otimes BK)z - (D\mathscr{L} \otimes BK)e. \tag{6.9}$$

For the purpose of consensus analysis, define a combinational measurement variable $\delta_i = \sum_{j\in\mathscr{N}_i} \bar{a}_{ij}(z_j - z_i)$ with $\bar{a}_{ij} = d_i a_{ij} d_j$. Let $\delta = [\delta_1^T \ \delta_2^T \ \cdots \delta_N^T]^T$, it can be derived that $\delta_i = \sum_{j\in\mathscr{N}_i} d_i a_{ij} d_j (z_j - z_i) = \sum_{j\in\mathscr{N}_i} d_i a_{ij}(d_j z_j - d_j z_i) = \sum_{j\in\mathscr{N}_i} d_i a_{ij}(d_j z_j - \mathrm{sgn}(a_{ij}) d_i z_i) = -d_i(\mathscr{L}_r^i \otimes I_n)(D \otimes I_n)z$, and it follows that $\delta = -(\mathscr{L}_D \otimes I_n)z$. Define $\bar{e}_i = d_i e_i$, it follows that $\bar{e} = (D \otimes I_n)e$. Then, according to (6.9) and the definition of e, the flow dynamics of δ and $\bar{e} = [\bar{e}_1^T \ \bar{e}_2^T \ \cdots \bar{e}_N^T]^T$ can be represented as

$$\begin{aligned} \dot{\delta} &= (I_N \otimes A - \mathscr{L}_D \otimes BK)\delta + (\mathscr{L}_D^2 \otimes BK)\bar{e}, \\ \dot{\bar{e}} &= (I_N \otimes A + \mathscr{L}_D \otimes BK)\bar{e} - (I_N \otimes BK)\delta. \end{aligned} \tag{6.10}$$

Since the error variable e_i has jump dynamics at triggering instants, the closed-loop system states also have jump dynamics that can be represented as

$$\delta(t^+) = \delta(t), \ \ \bar{e}(t^+) \in \cup_{i\in[1,N]} T_i(\bar{e}(t)) \tag{6.11}$$

where $T_i(\bar{e}(t)) = (I_{nN} - \Gamma_i \otimes I_n)\bar{e}(t)$ when $t = t_{k_i}^i$, Γ_i is a diagonal matrix with the iith entry being 1 and all the other elements are zeros.

6.3.2 Dynamics of Leader-Following MAS

In this case, the error e_i is defined as the same form as (6.7). Let $\delta_{di} = x_i - d_i x_0$, it indicates that $\hat{x}_i = e_i + \delta_{di} + d_i x_0$. The control law u_i in (6.4) can be represented as

$$u_i = -K(\mathscr{L}_r^i \otimes I_n)\delta_d - K(\mathscr{L}_r^i \otimes I_n)e - Kb_i\delta_{di}, \tag{6.12}$$

where $\delta_d = [\delta_1^T \ \delta_2^T \ \cdots \ \delta_N^T]^T$. According to (6.1), (6.2), and (6.4), by some calculations, we obtain

$$\begin{aligned} \dot{\delta}_d &= (I_N \otimes A - \mathscr{H} \otimes BK)\delta_d - (\mathscr{L} \otimes BK)e, \\ \dot{e} &= (I_N \otimes A + \mathscr{L} \otimes BK)e + (\mathscr{H} \otimes BK)\delta_d. \end{aligned} \tag{6.13}$$

Note that $\mathscr{L}$ is the Laplacian matrix of signed graph $\mathscr{G}$, $\mathscr{L}$ is generally not negative semi-definite. Thus, for the purpose of consensus analysis, we further define

$\delta_i = d_i\delta_{di}$ and $\bar{e}_i = d_i e_i$, that have the same modulus with δ_{di} and e_i, but may have different signs depending on the value of d_i. Based on $\delta = (D \otimes I_n)\delta_d$ and $\bar{e} = (D \otimes I_n)e$, the flow dynamics of δ and $\bar{e}$ can be represented as

$$\begin{aligned}\dot{\delta} &= (I_N \otimes A - \mathscr{H}_D \otimes BK)\delta - (\mathscr{L}_D \otimes BK)\bar{e},\\ \dot{\bar{e}} &= (I_N \otimes A + \mathscr{L}_D \otimes BK)\bar{e} + (\mathscr{H}_D \otimes BK)\delta.\end{aligned} \tag{6.14}$$

According to the definition of δ_i and $\bar{e}_i$, the jump dynamics of δ and $\bar{e}$ have the same form as (6.11). Thus, from (6.10), (6.11), and (6.14), we can see that the closed-loop dynamics of the MAS in both leaderless and leader-following cases have the same form with different system matrices. However, the aforementioned closed-loop form is not appropriate for consensus analysis and ETM design, we next construct a hybrid system model for the closed-loop MAS.

6.3.3 A Unified Hybrid Model

First, we introduce an internal timer variable $\tau_i \in \mathbb{R}$ with dynamics $\dot{\tau}_i = 1$ when the triggering event does not occur, and $\tau_i^+ = 0$ at triggering instants. The role of τ_i is to record the time elapsed since the last triggering instant, and it is useful to describe the minimum inter-event time condition (i.e., $t \geq t^i_{k_i} + \tau^i_{MIET}$ in the ETM (6.5), which is equivalent to $\tau_i \geq \tau^i_{MIET}$) in the closed-loop model. A positive integer κ_i is introduced to record the number of triggering events. Define

$$\begin{aligned}\tau &= [\tau_1\ \tau_2\ \cdots\ \tau_N]^T, \kappa = [\kappa_1\ \kappa_2\ \cdots\ \kappa_N]^T,\\ \eta &= [\eta_1\ \eta_2\ \cdots\ \eta_N]^T, \xi = [\delta^T\ \bar{e}^T\ \tau^T\ \kappa^T\ \eta^T]^T,\end{aligned} \tag{6.15}$$

where ξ denotes the closed-loop system state. Then, according to (6.5), (6.6), (6.10), (6.11), and (6.14), the flow dynamics of ξ can be represented as

$$\dot{\xi} = \mathscr{F}(\xi), \text{ If } \xi \in \mathbb{F}, \tag{6.16}$$

where

$$\mathscr{F}(\xi) = \begin{bmatrix} A^{11}_{cl}\delta + A^{12}_{cl}\bar{e} \\ A^{21}_{cl}\delta + A^{22}_{cl}\bar{e} \\ 1_N \\ 0_N \\ \Psi_F(o) \end{bmatrix}, \tag{6.17}$$

where $\Psi_F(o) = [\Psi_F^1(o_1)\ \Psi_F^2(o_2)\ \cdots\ \Psi_F^N(o_N)]^T$, and the system matrices have the following form

$$\begin{aligned} A_{cl}^{11} &= I_N \otimes A - \mathscr{L}_D \otimes BK, A_{cl}^{12} = \mathscr{L}_D^2 \otimes BK, \\ A_{cl}^{21} &= -(I_N \otimes BK), A_{cl}^{22} = I_N \otimes A + \mathscr{L}_D \otimes BK, \end{aligned} \tag{6.18}$$

for leaderless MAS, and have the following form

$$\begin{aligned} A_{cl}^{11} &= I_N \otimes A - \mathscr{H}_D \otimes BK, A_{cl}^{12} = -\mathscr{L}_D \otimes BK, \\ A_{cl}^{21} &= \mathscr{H}_D \otimes BK, A_{cl}^{22} = I_N \otimes A + \mathscr{L}_D \otimes BK, \end{aligned} \tag{6.19}$$

for leader-following MAS. The flow set $\mathbb{F}$ is defined as $\mathbb{F} := \bigcap_{i \in [1,N]} \mathbb{F}_i$, where

$$\mathbb{F}_i := \{\xi \in \mathbb{X} | \tau_i < \tau_{MIET}^i \text{ or } \eta_i > 0\},$$

where $\mathbb{X} := \{(\delta, \bar{e}, \tau, \kappa, \eta) \in \mathbb{R}^{nN} \times \mathbb{R}^{nN} \times \mathbb{R}_{\geq 0}^N \times \mathbb{N}_{>0}^N \times \mathbb{R}_{\geq 0}^N\}$. Besides, the jump dynamics of ξ can be represented as

$$\xi^+ \in \bigcup_{i \in [1,N]} \mathscr{J}_i(\xi), \text{ If } \xi \in \mathbb{J}, \tag{6.20}$$

where

$$\mathscr{J}_i(\xi) = \begin{bmatrix} \delta \\ (I_{nN} - \Gamma_i \otimes I_n)\bar{e} \\ (I_N - \Gamma_i)\tau \\ (I_N + \Gamma_i)\kappa \\ \Psi_J(o) \end{bmatrix},$$

where $\Psi_J(o) = [\Psi_J^1(o_1)\ \Psi_J^2(o_2)\ \cdots\ \Psi_J^N(o_N)]^T$. The jump set $\mathbb{J}$ is defined as $\mathbb{J} := \bigcup_{i \in [1,N]} \mathbb{J}_i$, where

$$\mathbb{J}_i := \{\xi \in \mathbb{X} | \tau_i \geq \tau_{MIET}^i \text{ and } \eta_i \leq 0\}.$$

By introducing internal timer variable τ_i and integer variable κ_i, the leaderless MAS and leader-following MAS with signed graph are represented in a unified hybrid systems form (6.16) and (6.20), and the event-triggering conditions are also integrated into the hybrid model. Based on the constructed hybrid model, the leaderless and leader-following bipartite consensus of MAS is equivalent to stability of the hybrid system (6.16) and (6.20), which is stated as follows.

Definition 6.1 [16] Consider the MAS represented by (6.16) and (6.20), the set $\{\xi \in \mathbb{X} | \delta = 0_{nN}, \bar{e} = 0_{nN}\}$ is said to be uniformly globally asymptotically stable (UGAS) if there exists a function $\beta \in \mathscr{KL}$, such that for any initial condition $\xi(0) \in \mathbb{X}$, the corresponding solution ξ satisfies

$$\|(\delta(t), \bar{e}(t))\| \leq \beta(\|(\delta(0), \bar{e}(0))\|, t). \tag{6.21}$$

6.4 Stability Analysis and ETM Design

In this section, we develop conditions such that the dynamics $(\delta, \bar{e})$ in hybrid system (6.16) and (6.20) is UGAS, which indicates that leaderless and leader-following bipartite consensus is achieved. In addition, how to design functions $\Psi_F^i(o_i)$ and $\Psi_J^i(o_i)$ are also presented. The stability analysis is based on an appropriately constructed Lyapunov function $U(\xi) : \mathbb{X} \to \mathbb{R}_{\geq 0}$, which is locally Lipschitz in its first two arguments, and there exist positive constants $\underline{\alpha}_U, \bar{\alpha}_U, \varepsilon_U$ such that

$$\begin{aligned} &\underline{\alpha}_U \left\|\xi_1\right\|^2 \leq U(\xi) \leq \bar{\alpha}_U \left\|\xi_1\right\|^2, \\ &\langle \nabla U(\xi), \mathscr{F}(\xi) \rangle \leq -\varepsilon_U \left\|\xi_1\right\|^2, \tau_i \in [0, t^i_{k_i+1} - t^i_{k_i}) \\ &U(\mathscr{J}_i(\xi)) - U(\xi) \leq 0, \tau_i = t^i_{k_i+1} - t^i_{k_i} \end{aligned} \tag{6.22}$$

where $\xi_1 = [\delta^T \ \bar{e}^T]^T$. The above conditions imply that the Lyapunov function $U(\xi)$ is radially unbounded, it is decreasing when there is no triggering event, and it is non-increasing at triggering instants. In fact, these conditions can be satisfied by selecting Lyapunov functions for δ and $\bar{e}$ obeying the following assumption.

Assumption 6.3 There exist function $W_i(\kappa_i, \bar{e}_i)$, $i \in [1, N]$, that is locally Lipschitz in their second arguments, function $V(\delta) : \mathbb{R}^{nN} \to \mathbb{R}_{\geq 0}$, continuous functions $H_i(\delta, \bar{e})$, constants $0 \leq \lambda_i < 1$, $L_i \geq 0$, $\varepsilon_q > 0$, $\upsilon_i > 0$, $0 < \varepsilon_{\upsilon i} \leq 1$, $0 < \alpha_0 < 1$, $\epsilon > 0$ and positive constants $\underline{\alpha}_{Wi}, \bar{\alpha}_{Wi}, \underline{\alpha}_V, \bar{\alpha}_V$, such that

(1) For $\kappa_i \in \mathbb{N}$ and $\bar{e}_i \in \mathbb{R}^n$,

$$\begin{aligned} \underline{\alpha}_{Wi} \left\|\bar{e}_i\right\| \leq W_i(\kappa_i, \bar{e}_i) \leq \bar{\alpha}_{Wi} \left\|\bar{e}_i\right\|, \\ W_i(\kappa_i + 1, \bar{e}_i^+) \leq \lambda_i W_i(\kappa_i, \bar{e}_i). \end{aligned} \tag{6.23}$$

(2) For all $\kappa_i \in \mathbb{N}$, $\delta \in \mathbb{R}^{nN}$ and $\bar{e} \in \mathbb{R}^{nN}$,

$$\left\langle \frac{\partial W_i(\kappa_i, \bar{e}_i)}{\partial \bar{e}_i}, \dot{\bar{e}}_i \right\rangle \leq L_i W_i(\kappa_i, \bar{e}_i) + H_i(\delta, \bar{e}). \tag{6.24}$$

(3) For all $\delta \in \mathbb{R}^{nN}$,

$$\underline{\alpha}_V \left\|\delta\right\|^2 \leq V(\delta) \leq \bar{\alpha}_V \left\|\delta\right\|^2. \tag{6.25}$$

(4) For all $\kappa_i \in \mathbb{N}$, $\bar{e} \in \mathbb{R}^{nN}$, and almost all $\delta \in \mathbb{R}^{nN}$,

$$\langle \nabla V(\delta), A_{cl}^{11}\delta + A_{cl}^{12}\bar{e} \rangle \leq -\varepsilon_q \left\|\delta\right\|^2 - \sum_{i=1}^{N} \upsilon_i H_i^2(\delta, \bar{e}) + \sum_{i=1}^{N} (\alpha_0 \gamma_i^2 \upsilon_i \varepsilon_{\upsilon i} - \epsilon) W_i^2(\kappa_i, \bar{e}_i). \tag{6.26}$$

The conditions in Assumption 6.3 are similar to the conditions used in [17, 18], where networked control systems rather than consensus control of MAS were considered. In fact, the conditions in Assumption 6.3 can recover the conditions used in [17] for $N = 1, \upsilon_i = 1, \varepsilon_{\upsilon i} = 1, \alpha_0 = 1$, and can recover the conditions used in [18] for $\upsilon_i = 1, \varepsilon_{\upsilon i} = 1, \alpha_0 = 1$, so, these conditions form a generalization for MAS. Although these conditions seem to be difficult to obtain, this is not the case, in order to not interrupt the presentation of main results, how to construct explicit form of functions $W_i(\kappa_i, \bar{e}_i)$ and $V(\delta)$ to satisfy these conditions will be given in Sect. 6.5. Next, for stability analysis and ETM design, we introduce internal functions $\phi_i(\tau_i) : \mathbb{R}_{\geq 0} \to \mathbb{R}_{\geq 0}$ with the following piecewise dynamics

$$\frac{d\phi_i(\tau_i)}{d\tau_i} = \begin{cases} -2L_i\phi_i(\tau_i) - \gamma_i(\upsilon_i\varepsilon_{\upsilon i} + \upsilon_i^{-1}\phi_i^2(\tau_i)), & 0 \leq \tau_i \leq \tau_{MIET}^i \\ 0, & \tau_i > \tau_{MIET}^i \end{cases} \tag{6.27}$$

where L_i, γ_i, $\varepsilon_{\upsilon i}$, and υ_i are constant parameters in Assumption 6.3, the initial conditions $\phi_i(0)$ need to be pre-given.

According to the form of dynamics in (6.27), it can be seen that the derivative of $\phi_i(\tau_i)$ with respect to τ_i is strictly negative for $0 \leq \tau_i \leq \tau_{MIET}^i$, which means that $\phi_i(\tau_i)$ is strictly decreasing for $0 \leq \tau_i \leq \tau_{MIET}^i$. So, if τ_{MIET}^i is sufficiently large, $\phi_i(\tau_i)$ will become negative eventually. The function $\phi_i(\tau_i)$ is useful for stability analysis and will be used to construct Lyapunov function, so, it is usually required that $\phi_i(\tau_i)$ is guaranteed to be positive, i.e., the value of τ_{MIET}^i should be upper bounded. To ensure $\phi_i(\tau_i)$ to be positive, τ_{MIET}^i can be selected according to the following lemma.

Lemma 6.2 *Consider the dynamics of $\phi_i(\tau_i)$ in (6.27), with initial conditions $\phi_i(0) > 1$ for $i \in [1, N]$, the parameters $\upsilon_i > 0$, $0 < \varepsilon_{\upsilon i} \leq 1$, $L_i > 0$, $\gamma_i > 0$. Select the minimum event-triggering interval τ_{MIET}^i to satisfy*

$$\tau_{MIET}^i \leq \tau_{MAX}^i, \tag{6.28}$$

where τ_{MAX}^i is such that $\phi_i(\tau_{MAX}^i) = \phi_i^{-1}(0)$, and τ_{MAX}^i can be explicitly calculated by

$$\tau_{MAX}^i = \begin{cases} \dfrac{1}{L_i p_i} \arctan \dfrac{p_i p_{1i} p_{2i}}{p_{3i} + p_{2i}^2}, & \gamma_i^2 \varepsilon_{\upsilon i}^2 > L_i^2, \\ L_i^{-1} \upsilon_i^{-0.5} p_{1i} p_{2i}^{-1}, & \gamma_i^2 \varepsilon_{\upsilon i}^2 = L_i^2, \\ \dfrac{1}{L_i p_i} \text{arctanh} \dfrac{p_i p_{1i} p_{2i}}{p_{3i} + p_{2i}^2}, & \gamma_i^2 \varepsilon_{\upsilon i}^2 < L_i^2, \end{cases} \tag{6.29}$$

where $p_i = \left\| (\gamma_i/L_i)^2 \varepsilon_{\upsilon i} - 1 \right\|^{0.5}$, $p_{1i} = 1 - \phi_i^{-1}(0)$, $p_{2i} = 1 + \phi_i^{-1}(0)$, $p_{3i} = 2\phi_i^{-1}(0)(\gamma_i \upsilon_i^{-0.5} L_i^{-1} - 1)$. Then, $\phi_i(\tau_i)$ is ensured to be positive.

Proof Based on $\phi_i(\tau_i)$'s dynamics with $\phi_i(0) > 1$ and $\phi_i(\tau_{MAX}^i) = \phi_i^{-1}(0)$, we have

$$\tau_{MAX}^i = -\int_{\phi_i(0)}^{\phi_i^{-1}(0)} \frac{d\phi_i}{2L_i\phi_i + \gamma_i\varepsilon_{\upsilon i}\upsilon_i + \upsilon_i^{-1}\gamma_i\phi_i^2}$$

Define $p_i = \left\|(\gamma_i/L_i)^2\varepsilon_{vi} - 1\right\|^{0.5}$ and $s_i = \phi_i v_i^{-0.5} + v_i^{0.5}L_i\gamma_i^{-1}$, the aforementioned integral can be rewritten as

$$\tau^i_{MAX} = -\int_{s_{1i}}^{s_{2i}} \frac{ds_i}{s_i^2 - \mathrm{sgn}(L_i^2 - \gamma_i^2\varepsilon_{vi})(L_i p_i v_i^{-0.5}\gamma_i^{-1})^2}$$

where $s_{1i} = \phi_i(0)v_i^{-0.5} + v_i^{0.5}L_i\gamma_i^{-1}, s_{2i} = \phi_i^{-1}(0)v_i^{-0.5} + v_i^{0.5}L_i\gamma_i^{-1}$. First, consider the case $\gamma_i^2\varepsilon_{vi}^2 > L_i^2$, utilizing differentiation formulas of arctangent function and by some routine calculations, we obtain

$$\tau^i_{MAX} = \frac{1}{L_i p_i}\arctan\left(\frac{p_i p_{1i} p_{2i}}{\phi_i^{-1}(0)(\frac{\gamma_i}{L_i}(v_i\varepsilon_{vi} + \frac{1}{v_i}) - 2) + p_{2i}^2}\right)$$

where the value $v_i\varepsilon_{vi} + \frac{1}{v_i}$ can be minimized by setting $v_i = \varepsilon_{vi}^{-0.5}$, which leads to $\tau^i_{MAX} = \frac{1}{L_i p_i}\arctan\frac{p_i p_{1i} p_{2i}}{p_{3i} + p_{2i}^2}$. The rest two cases with $\gamma_i^2\varepsilon_{vi}^2 = L_i^2$ and $\gamma_i^2\varepsilon_{vi}^2 < L_i^2$ can be calculated by similar procedures. Thus, with the obtained τ^i_{MAX} in (6.29), it is ensured that $\phi_i(\tau_i) \in [\phi_i^{-1}(0), \phi_i(0)]$. □

Theorem 6.1 *Consider the hybrid systems composed of* (6.16) *and* (6.20). *The conditions in Assumption* 6.3 *hold, the minimum event-triggering interval* τ^i_{MIET} *is selected according to Lemma* 6.2, *and the following conditions are satisfied for* $i \in [1, N]$.

(1).*The flow map function* $\Psi^i_F(o_i)$, *with local variables* $o_i = (\bar{e}_i, \tau_i, \kappa_i, \eta_i)$, *is such that for all* $\xi \in \mathbb{X}$

$$\Psi^i_F(o_i) = T_i(\xi) - \pi_i\eta_i, \tag{6.30}$$

where $\pi_i > 0$ *is a constant, and*

$$T_i(\xi) = \begin{cases} T_{1i}(\xi), & 0 \le \tau_i \le \tau^i_{MIET}, \\ T_{2i}(\xi), & \tau_i > \tau^i_{MIET}, \end{cases} \tag{6.31}$$

with

$$\begin{aligned} T_{1i}(\xi) &= (1-\alpha_0)\gamma_i^2\varepsilon_{vi}v_i W_i^2(\kappa_i, \bar{e}_i), \\ T_{2i}(\xi) &= -\gamma_i\phi_i(\tau_i)(2L_i + v_i^{-1}\gamma_i\phi_i(\tau_i))W_i^2(\kappa_i, \bar{e}_i) - \alpha_0\gamma_i^2 v_i\varepsilon_{vi}W_i^2(\kappa_i, \bar{e}_i). \end{aligned} \tag{6.32}$$

(2).*The jump map function* $\Psi^i_J(o_i)$ *satisfies*

$$\Psi^i_J(o_i) = \gamma_i(\phi_i(\tau^i_{MIET}) - \phi_i(0)\lambda_i^2)W_i^2(\kappa_i, \bar{e}_i). \tag{6.33}$$

Then, with the hybrid ETM designed in (6.5), *the set as defined in Definition* 6.1 *is UGAS. In other words,* δ *converges to the origin means that the leaderless and* *leader-following bipartite consensus is achieved, meanwhile, the triggering intervals $t^i_{k_i+1} - t^i_{k_i}$ is lower bounded by $\tau^i_{MIET} > 0$, i.e., Zeno-freeness is guaranteed.*

Proof Choose the candidate Lyapunov function

$$U(\xi) = V(\delta) + \sum_{i=1}^{N} \left\{ \gamma_i \phi_i(\tau_i) W_i^2(\kappa_i, \bar{e}_i) + \eta_i \right\}, \tag{6.34}$$

where $V(\delta)$ satisfies conditions (6.25) and (6.26).

First, consider the evolution of $U(\xi)$ along ξ's flow dynamics, represented as $\langle \nabla U(\xi), \mathscr{F}(\xi) \rangle$ with $\mathscr{F}(\xi)$ given by (6.17). Based on conditions in Assumption 6.3, (6.17), and (6.27), for all $\tau \in \mathbb{R}^N$ and almost all $(\delta, \bar{e}, \eta) \in \mathbb{R}^{nN} \times \mathbb{R}^{nN} \times \mathbb{R}^N$

$$\begin{aligned}
&\langle \nabla U(\xi), \mathscr{F}(\xi) \rangle \\
\leq & - \sum_{i=1}^{N} v_i H_i^2(\delta, \bar{e}) + \sum_{i=1}^{N} (\alpha_0 \gamma_i^2 v_i \varepsilon_{vi} - \epsilon) W_i^2(\kappa_i, \bar{e}_i) - \varepsilon_q \|\delta\|^2 + \sum_{i=1}^{N} \Psi_F^i(o_i) \\
& + \sum_{i=1}^{N} b(\tau_i)(-2L_i \gamma_i \phi_i(\tau_i) - \gamma_i^2 (\varepsilon_{vi} v_i + v_i^{-1} \phi_i^2(\tau_i))) W_i^2(\kappa_i, \bar{e}_i) \\
& + \sum_{i=1}^{N} 2\gamma_i \phi_i(\tau_i) W_i(\kappa_i, \bar{e}_i)(L_i W_i(\kappa_i, \bar{e}_i) + H_i(\delta, \bar{e})),
\end{aligned} \tag{6.35}$$

where $b(\tau_i) = 1$ for $0 \leq \tau_i \leq \tau^i_{MIET}$, and $b(\tau_i) = 0$ for $\tau_i > \tau^i_{MIET}$. Notice that $2\gamma_i \phi_i(\tau_i) W_i(\kappa_i, \bar{e}_i) H_i(\delta, \bar{e}) \leq v_i H_i^2(\delta, \bar{e}) + v_i^{-1} \gamma_i^2 \phi_i^2(\tau_i) W_i^2(\kappa_i, \bar{e}_i)$, and according to (6.30), (6.35) can be rewritten as

$$\begin{aligned}
&\langle \nabla U(\xi), \mathscr{F}(\xi) \rangle \\
\leq & - \varepsilon_q \|\delta\|^2 + \sum_{i=1}^{N} (\alpha_0 \gamma_i^2 v_i \varepsilon_{vi} - \epsilon) W_i^2(\kappa_i, \bar{e}_i) + \sum_{i=1}^{N} (T_i(\xi) - \pi_i \eta_i) \\
& + \sum_{i=1}^{N} -b(\tau_i) \gamma_i^2 \varepsilon_{vi} v_i W_i^2(\kappa_i, \bar{e}_i) + \sum_{i=1}^{N} (1 - b(\tau_i)) \gamma_i \phi_i(\tau_i)(2L_i + v_i^{-1} \gamma_i \phi_i(\tau_i)) W_i^2(\kappa_i, \bar{e}_i).
\end{aligned} \tag{6.36}$$

Based on (6.31) and (6.32), $T_i(\xi)$ can be expressed as

$$\begin{aligned}
T_i(\xi) = & - \alpha_0 \gamma_i^2 \varepsilon_{vi} v_i W_i^2(\kappa_i, \bar{e}_i) + b(\tau_i) \gamma_i^2 \varepsilon_{vi} v_i W_i^2(\kappa_i, \bar{e}_i) \\
& - (1 - b(\tau_i)) \gamma_i \phi_i(\tau_i)(2L_i + v_i^{-1} \gamma_i \phi_i(\tau_i)) W_i^2(\kappa_i, \bar{e}_i),
\end{aligned} \tag{6.37}$$

which combined with (6.36), yields

$$
\begin{aligned}
&\langle \nabla U(\xi), \mathscr{F}(\xi) \rangle \\
&\leq -\varepsilon_q \|\delta\|^2 - \sum_{i=1}^{N} \epsilon W_i^2(\kappa_i, \bar{e}_i) - \sum_{i=1}^{N} \pi_i \eta_i \\
&\leq -\min\left\{ \varepsilon_q \bar{\alpha}_V^{-1}, \min_{i\in[1,N]} \left\{ \frac{\epsilon}{\gamma_i \phi_i(\tau_i)} \right\}, \min_{i\in[1,N]} \pi_i \right\} U(\xi).
\end{aligned} \tag{6.38}
$$

Then, consider the case $\xi \in \mathbb{J}$, i.e., triggering event occurs and ξ has jump dynamics. For a certain agent i, when $\xi \in \mathbb{J}_i$, according to (6.20), we obtain

$$
\begin{aligned}
&U(\xi^+) - U(\xi) \\
&= \sum_{i=1}^{N} (\gamma_i \phi_i(0) \lambda_i^2 W_i^2(\kappa_i, \bar{e}_i) + \psi_J^i(o_i)) - \sum_{i=1}^{N} (\gamma_i \phi_i(\tau_{MIET}^i) W_i^2(\kappa_i, \bar{e}_i) + 0) \leq 0,
\end{aligned} \tag{6.39}
$$

the last inequality comes from (6.33). The obtained inequality conditions in (6.38) and (6.39) show that the required stability conditions in (6.22) are satisfied, thus, we can conclude that the closed-loop MAS is UGAS, which indicates that the leaderless and leader-following bipartite consensus is achieved. □

6.5 Construction of Functions $V(\delta)$ and $W_i(\kappa_i, \bar{e}_i)$

In this section, the explicit form of functions $V(\delta)$ and $W_i(\kappa_i, \bar{e}_i)$ is presented, then it is verifiable that all the conditions in Assumption 6.3 are satisfied, and the stability results obtained in Theorem 6.1 is reliable.

First, from the definition of $\bar{e}_i = d_i e_i$, it is known that $\bar{e}_i$ jumps to zero at agent i's triggering instants $t_{k_i}^i$. Hence, to satisfy the conditions in (6.23), we can select $W_i(\kappa_i, \bar{e}_i) = \|\bar{e}_i\|$, in this case, we have $\underline{\alpha}_{Wi} \|\bar{e}_i\| \leq W_i(\kappa_i, \bar{e}_i) \leq \bar{\alpha}_{Wi} \|\bar{e}_i\|$ with $\underline{\alpha}_{Wi} \leq 1$ and $\bar{\alpha}_{Wi} \geq 1$, and $\bar{e}_i^+ = h_i(\kappa_i, \bar{e}_i) = 0$ such that $W_i(\kappa_i + 1, h(\kappa_i, \bar{e}_i)) \leq \lambda_i W_i(\kappa_i, \bar{e}_i)$ holds for $0 \leq \lambda_i < 1$. With the selected $W_i(\kappa_i, \bar{e}_i)$, we next need to verify if the condition (6.24) holds and give explicit form of L_i and $H_i(\delta, \bar{e})$. From the condition (6.24), we have

$$
\left\langle \frac{\partial W_i(\kappa_i, \bar{e}_i)}{\partial \bar{e}_i}, \dot{\bar{e}}_i \right\rangle \leq \left\| \frac{\partial (\bar{e}_i^T \bar{e}_i)^{\frac{1}{2}}}{\partial \bar{e}_i} \right\| \|\dot{\bar{e}}_i\| = \|\dot{\bar{e}}_i\|. \tag{6.40}
$$

Define $E_i = [0 \ \cdots \ I_n \ \cdots \ 0] \in \mathbb{R}^{n \times nN}$, which is a matrix composed of n blocks, with the ith block being I_n and all the rest blocks being zero matrices. Based on the flow dynamics of $\bar{e}$ (see (6.16)), $\|\dot{\bar{e}}_i\|$ satisfies

$$\begin{aligned} \|\dot{\bar{e}}_i\| &= \|E_i A_{cl}^{21}\delta + E_i A_{cl}^{22}\bar{e}\| \\ &= \|E_i A_{cl}^{21}\delta + E_i A_{cl}^{22}(I_{nN} - \Gamma_i \otimes I_n)\bar{e} + E_i A_{cl}^{22}\bar{e}_i\| \\ &\leq L_i W_i(\kappa_i, \bar{e}_i) + H_i(\delta, \bar{e}), \end{aligned} \tag{6.41}$$

where $L_i = \|E_i A_{cl}^{22}\|$, and $H_i(\delta, \bar{e}) = \|A_{\delta i}\delta + A_{\bar{e}i}\bar{e}\|$ with $A_{\delta i} = E_i A_{cl}^{21}$ and $A_{\bar{e}i} = E_i A_{cl}^{22}(I_{nN} - \Gamma_i \otimes I_n)$. Thus, conditions (6.23) and (6.24) in Assumption 6.3 can be satisfied by selecting $W_i(\kappa_i, \bar{e}_i) = \|\bar{e}_i\|$.

Then, we select $V(\delta) = \delta^T(I_N \otimes P)\delta$ with a positive definite matrix P that satisfies the following Riccati equation

$$PA + A^T P - 2\mu a(\hat{\mathscr{L}})PBB^T P + Q = 0, \tag{6.42}$$

where $\mu > 0$, matrix Q is positive definite, $\hat{\mathscr{L}} = \mathscr{L}_D$ in the leaderless case and $\hat{\mathscr{L}} = \mathscr{H}_D$ in the leader-following case, $a(\hat{\mathscr{L}})$ denotes the minimum positive eigenvalue of matrix $\hat{\mathscr{L}}$. It is easy to verify that the condition (6.25) holds with $\underline{\alpha}_V = \lambda_{\min}(I_N \otimes P)$ and $\bar{\alpha}_V = \lambda_{\max}(I_N \otimes P)$.

The next result presents LMI conditions to guarantee satisfaction of the condition (6.26).

Proposition 6.1 *Consider the hybrid systems (6.16) and (6.20). Let $V(\delta) = \delta^T(I_N \otimes P)\delta$ and $K = \mu B^T P$, where the constant $\mu > 0$ and matrix $P > 0$ are obtained from (6.42). If there exist constants $0 < \sigma < 1$, $\upsilon_i > 0$, $0 < \varepsilon_{\upsilon i} \leq 1$, $0 < \alpha_0 < 1$, $\gamma_i > 0$, and $\epsilon > 0$, such that the following linear matrix inequality holds*

$$\begin{bmatrix} \Omega_{11} & \Omega_{12} \\ * & \Omega_{22} \end{bmatrix} \leq 0, \tag{6.43}$$

where

$$\Omega_{11} = -(1-\sigma)(I_N \otimes Q) + \sum_{i=1}^{N} \upsilon_i A_{\delta i}^T A_{\delta i},$$

$$\Omega_{12} = (I_N \otimes P)A_{cl}^{12} + \sum_{i=1}^{N} \upsilon_i A_{\delta i}^T A_{\bar{e}i},$$

$$\Omega_{22} = -\alpha_0 \mathrm{diag}\{\gamma_1^2 \upsilon_1 \varepsilon_{\upsilon 1}, \ldots, \gamma_N^2 \upsilon_N \varepsilon_{\upsilon N}\} + \sum_{i=1}^{N} \upsilon_i A_{\bar{e}i}^T A_{\bar{e}i} + \epsilon I_{nN}.$$

Then, the condition (6.26) in Assumption 6.3 is satisfied by defining $\varepsilon_q = \sigma\lambda_{\min}(I_N \otimes Q)$ and $H_i(\delta, \bar{e}) = \|A_{\delta i}\delta + A_{\bar{e}i}\bar{e}\|$.

Proof Due to δ is continuous and evolves according to the flow dynamics (6.17), thus, based on (6.17), $V(\delta) = \delta^T(I_N \otimes P)\delta$, $K = \mu B^T P$ and the definition of $\hat{\mathscr{L}}$ in Lemma 6.1, we have

$$\begin{aligned}
&\langle \nabla V(\delta), A_{cl}^{11}\delta + A_{cl}^{12}\bar{e}\rangle \\
=&\delta^T((I_N \otimes P)(I_N \otimes A - \hat{\mathscr{L}} \otimes BK) + (I_N \otimes A - \hat{\mathscr{L}} \otimes BK)^T(I_N \otimes P))\delta + 2\delta^T(I_N \otimes P)A_{cl}^{12}\bar{e} \\
=&\delta^T(I_N \otimes (PA + A^T P) - 2\mu\hat{\mathscr{L}} \otimes PBB^T P)\delta + 2\delta^T(I_N \otimes P)A_{cl}^{12}\bar{e} \\
\le& - \delta^T(I_N \otimes Q)\delta + 2\delta^T(I_N \otimes P)A_{cl}^{12}\bar{e},
\end{aligned} \tag{6.44}$$

where the last inequality is obtained based on Lemma 6.1 and Riccati equation (6.42). Specifically, from item (2) of Lemma 6.1 and the fact $(\mathbf{1}^T \otimes I_n)\delta = -(\mathbf{1}^T \otimes I_n)(\mathscr{L}_D \otimes I_n)z = -(\mathbf{1}^T\mathscr{L}_D \otimes I_n)z = 0$, we have $-2\mu\delta^T(\hat{\mathscr{L}} \otimes PBB^T P)\delta \le -2\mu a(\hat{\mathscr{L}})\delta^T(I_N \otimes PBB^T P)\delta$, then, from (6.42), the last inequality is obtained. Next, let the matrix in (6.43) be left-multiplied by $[\delta^T \ \bar{e}^T]$ and right-multiplied by $[\delta^T \ \bar{e}^T]^T$, it follows that

$$\begin{aligned}
&- \delta^T(1-\sigma)(I_N \otimes Q)\delta + \sum_{i=1}^{N} \upsilon_i\delta^T A_{\delta i}^T A_{\delta i}\delta + 2\delta^T(I_N \otimes P)A_{cl}^{12}\bar{e} + 2\sum_{i=1}^{N} \upsilon_i\delta^T A_{\delta i}^T A_{\bar{e}i}\bar{e} \\
&- \alpha_0\bar{e}^T(\mathrm{diag}\{\gamma_1^2\upsilon_1\varepsilon_{\upsilon 1}, \ldots, \gamma_N^2\upsilon_N\varepsilon_{\upsilon N}\} - \epsilon I_{nN})\bar{e} + \sum_{i=1}^{N} \upsilon_i\bar{e}^T A_{\bar{e}i}^T A_{\bar{e}i}\bar{e} \le 0.
\end{aligned} \tag{6.45}$$

Based on the above inequality, it can be derived that

$$\begin{aligned}
&- \delta^T(I_N \otimes Q)\delta + 2\delta^T(I_N \otimes P)A_{cl}^{12}\bar{e} \\
\le& - \sigma\delta^T(I_N \otimes Q)\delta - \sum_{i=1}^{N} \upsilon_i(A_{\delta i}\delta + A_{\bar{e}i}\bar{e})^T(A_{\delta i}\delta + A_{\bar{e}i}\bar{e}) + \sum_{i=1}^{N} \alpha_0\gamma_i^2\upsilon_i\varepsilon_{\upsilon i}\bar{e}_i^T\bar{e}_i - \sum_{i=1}^{N} \epsilon\bar{e}_i^T\bar{e}_i.
\end{aligned}$$

Then, note that $\delta^T(I_N \otimes Q)\delta \ge \lambda_{\min}(I_N \otimes Q)\,\|\delta\|^2$, $(A_{\delta i}\delta + A_{\bar{e}i}\bar{e})^T(A_{\delta i}\delta + A_{\bar{e}i}\bar{e}) = \|A_{\delta i}\delta + A_{\bar{e}i}\bar{e}\|^2$, $\bar{e}_i^T\bar{e}_i = \|\bar{e}_i\|^2$, and according to the definitions $\varepsilon_q = \sigma\lambda_{\min}(I_N \otimes Q)$, $H_i(\delta, \bar{e}) = \|A_{\delta i}\delta + A_{\bar{e}i}\bar{e}\|$ and $W_i(\kappa_i, \bar{e}_i) = \|\bar{e}_i\|$, we have

$$\begin{aligned}
&- \delta^T(I_N \otimes Q)\delta + 2\delta^T(I_N \otimes P)A_{cl}^{12}\bar{e} \\
\le& - \varepsilon\,\|\delta\|^2 - \sum_{i=1}^{N} \upsilon_i H_i^2(\delta, \bar{e}) + \sum_{i=1}^{N} \alpha_0\gamma_i^2\upsilon_i\varepsilon_{\upsilon i}W_i^2(\kappa_i, \bar{e}_i) - \sum_{i=1}^{N} \epsilon W_i^2(\kappa_i, \bar{e}_i),
\end{aligned}$$

which has the same form as (6.26), i.e., the condition (6.26) holds. This completes the proof. □

According to the aforementioned results, all the conditions in Assumption 6.3 hold, and the form of Lyapunov function (6.34) can be explicitly obtained based on the constructed $W_i(\kappa_i, \bar{e}_i) = \|\bar{e}_i\|$ and $V(\delta)$. Moreover, the flow map functions $\Psi_F^i(o_i)$ and jump map functions $\Psi_J^i(o_i)$ in the proposed hybrid ETM (6.5) can be represented by

$$\begin{aligned}\Psi_F^i(o_i) &= -\pi_i\eta_i + ((b(\tau_i) - \alpha_0)\gamma_i^2\varepsilon_{vi}v_i - (1 - b(\tau_i))\gamma_i\phi_i(\tau_i)(2L_i + v_i^{-1}\gamma_i\phi_i(\tau_i)))\,\|\bar{e}_i\|^2\,,\\ \Psi_J^i(o_i) &= \gamma_i(\phi_i(\tau_{MIET}^i) - \phi_i(0)\lambda_i^2)\,\|\bar{e}_i\|^2\,,\end{aligned} \tag{6.46}$$

where $b(\tau_i) = 1$ ($b(\tau_i) = 0$) for $0 \le \tau_i \le \tau_{MIET}^i$ ($\tau_i > \tau_{MIET}^i$). □

Remark 6.1 Observe the form of ETM (6.5), we know it is the combination of periodic event detection and continuous event detection that constitutes the final event detection mechanism (6.5), for this reason, it is called hybrid ETM. Benefited from the design, immediate event detection is no more needed after a new triggering event, instead there always exists a period of sleeping time $(t_{k_i}^i, t_{k_i}^i + \tau_{MIET}^i)$ provided by the periodic event detection mechanism, during which any triggered event is not allowed while the closed-loop system performance is guaranteed. Then, from the time instant $t_{k_i}^i + \tau_{MIET}^i$, agent i begins to detect the condition $\eta_i(t) \le 0$ continuously. The proposed hybrid ETM is able to achieve exclusion of Zeno behavior by the aid of periodic event detection, and achieve effectively using system information by the aid of continuous event detection.

Remark 6.2 According to the hybrid ETM (6.5), it is known that the next triggered event happens only after a time period τ_{MIET}^i has elapsed, therefore, a minimum inter-event time (MIET) is always guaranteed in between successive events. It implies that Zeno-triggering behavior can be absolutely excluded, thus, for the hybrid ETM (6.5) Zeno-freeness analysis is not needed. Note that, the hybrid ETM (6.5) has two main advantages. To be specific, (1) the MIET τ_{MIET}^i is a pre-selected and adjustable positive constant, and the upper bound of τ_{MIET}^i can be explicitly calculated; (2) the proposed ETM is in decentralized form, that is, the triggering condition only depends on agent i's own information. Thus, its physical implementation is much simple.

Remark 6.3 According to whether an internal variable is used to memory past system's information, there are two types of ETM: static ETM and dynamic ETM. The dynamic ETM was proposed in [19] and it has been shown that dynamic ETM has potential to further reduce usage of communication resources compared with static ETM, see [20–22] as examples, just to name a few. Since an internal dynamic variable $\eta_i(t)$ is used in the proposed hybrid ETM, it also belongs to dynamic ETM, but, the hybrid ETM (6.5) has some differences with the dynamic ETMs in the existing results. First, the form of dynamics of $\eta_i(t)$ is different, and $\eta_i(t)$ is allowed to reach zero at triggering instants, this provides potential to achieve larger inter-event times, since it generally takes longer time for $\eta_i(t)$ to reach zero than reach a positive value. Second, the condition $t \ge t_{k_i}^i + \tau_{MIET}^i$ in (6.5) enforces a strictly positive MIET τ_{MIET}^i to triggering intervals, and τ_{MIET}^i can be pre-specified. By contrast, the MIET in the existing dynamic ETMs is usually unknown. Third, from the ETM design results in Theorem 6.1, the proposed hybrid ETM is in decentralized form, i.e., only agent i's own information is needed in implementation, but the dynamic ETM in the aforementioned results are distributed, which indicates that the implementation of the proposed hybrid ETM is more convenient. Fourth, the internal variable $\eta_i(t)$ is allowed to have jump dynamics, which provides extra freedom to get larger inter-event times.

6.6 Application to Satellite Formation

Consider a satellite formation problem. Each satellite flying in the low Earth orbit can be viewed as an agent, suppose that there are 12 agents and one leader, the communication topology is shown in Fig. 6.1, for simplicity, the weight values of edges denoted by black lines all equal one. The linearized equations of the relative dynamics of the ith satellite with respect to the virtual satellite are described as follows

$$\begin{aligned} \ddot{\bar{x}}_i - 2\omega_0\dot{\bar{y}}_i &= u_{xi} \\ \ddot{\bar{y}}_i + 2\omega_0\dot{\bar{x}}_i - 3\omega_0^2\bar{y}_i &= u_{yi} \\ \ddot{\bar{z}}_i + \omega_0^2\bar{z}_i &= u_{zi} \end{aligned} \tag{6.47}$$

where $(\bar{x}_i, \bar{y}_i, \bar{z}_i)$ denote the position of the ith satellite in the rotating coordinate, (u_{xi}, u_{yi}, u_{zi}) are the control inputs, and ω_0 is the virtual satellite's angular rate. The physical units of the variables $\bar{x}_i/\bar{y}_i/\bar{z}_i$, $u_{xi}/u_{yi}/u_{zi}$, and ω_0 are m (meter), Nm (Newton metre), and rad/s (radian/second) respectively. Define the position vector $r_i = \begin{bmatrix}\bar{x}_i & \bar{y}_i & \bar{z}_i\end{bmatrix}^T$, velocity vector $\dot{r}_i = \begin{bmatrix}v_{xi} & v_{yi} & v_{zi}\end{bmatrix}^T$, control vector $u_i = \begin{bmatrix}u_{xi} & u_{yi} & u_{zi}\end{bmatrix}^T$. The satellite formation can be achieved if their velocity vectors converge to the same value and their positions keep a prescribed distance from each other, i.e., $r_i - h_i \to r_j - h_j, \dot{r}_i - \dot{r}_j$ for $i, j \in [1, N]$, where $h_i - h_j \in \mathbb{R}^3$ denotes the prescribed distance between satellite i and j. Let $\bar{r}_i = r_i - h_i$ and define $x_i = \begin{bmatrix}\bar{r}_i^T & \dot{r}_i^T\end{bmatrix}^T$, it can be derived from (6.47) that $\dot{x}_i = Ax_i + B_1u_i$ with $A = [0\ I_3;\ A_1\ A_2]$, $B_1 = [0;\ I_3]$, and $A_1 = [0\,0\,0;\ 0\,3\omega_0^2\,0;\ 0\,0\ -\omega_0^2]$, $A_2 = [0\ 2\omega_0^2\ 0;\ -2\omega_0^2\ 0\ 0;\ 0\,0\,0]$ where $\omega_0 = 0.001$ is the angular rate of the satellite.

First, consider the leaderless case. Select $Q = 50I$, $\mu = 0.1$, solving Riccati equation (6.42) leads to $K = [2.6665\ \ -0.002\ 0\ 3.8333\ 0\ 0;\ 0.002\ 2.6665\ 0\ 0\ 3.8333\ 0;\ 0\ 0\ 2.6665\ 0\ 0\ 3.8333]$. The rest parameters are selected as $\phi_i(0) = 5$, $\alpha_0 = 0.2$, $\gamma_i = 10$. $v_i = 1$, $\varepsilon_{vi} = 1$. It can be calculated that $\tau_{MAX}^{i\in[1,12]} = 0.098, 0.063, 0.079, 0.089, 0.082, 0.077, 0.052, 0.142, 0.082, 0.073, 0.072, 0.095$. For simplicity, select $\tau_{MIET}^{i} = 0.05$ for $i \in [1, 12]$. With these parameters and let the initial state $x_i(0) = i \times [1; 2; 3; 4; 5; 6]$ for $i \in [1, 6]$ and $x_i(0) = -i \times [1; 2; 3; 4; 5; 6]$ for $i \in [7, 12]$, the simulation results are shown in Figs. 6.2 and 6.3. Figure 6.2 shows that the bipartite consensus is achieved. Figure 6.3 presents the triggering instants, the triggering

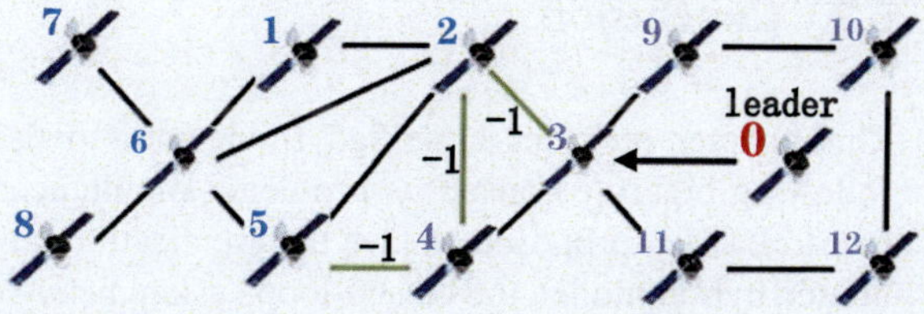

Fig. 6.1 Communication topology of multiple satellite systems

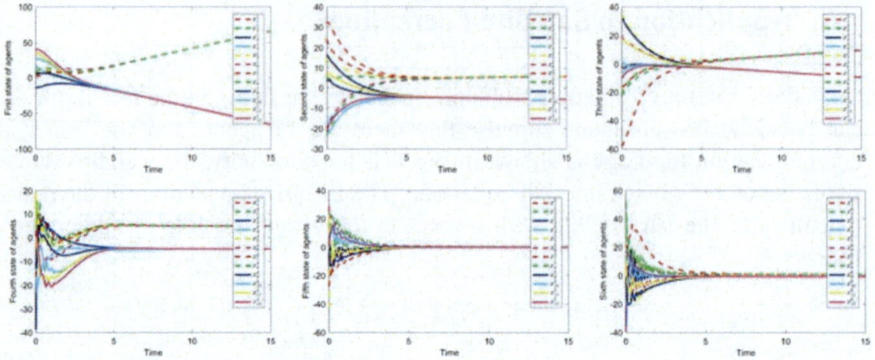

Fig. 6.2 Agents' states in leaderless case

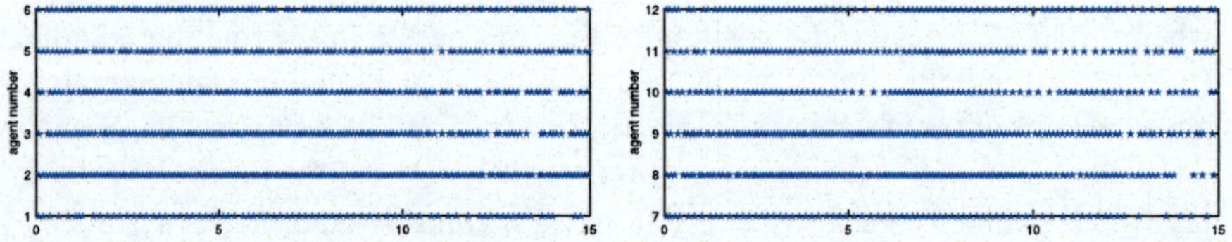

Fig. 6.3 Triggering instants in leaderless case

number of agents are 60, 151, 124, 97, 97, 99, 52, 104, 112, 83, 71, 75. The lower bound of triggering intervals is always greater than or equal to 0.05.

Then, consider the leader-following case. In this case, suppose that agents 3 can receive information from leader, as shown in Fig. 6.1. For simplicity, the parameters are selected to be the same as in leaderless case. Let $x_0(0) = [6; -8; 10; -6; 8; -10]$, the simulation results are shown in Figs. 6.4 and 6.5. Figure 6.4 shows that the leader-following bipartite consensus is achieved. Figure 6.5 presents the triggering instants, the simulation time is 0–25 s and the triggering number of the agents are 149, 262, 260, 177, 179, 175, 166, 283, 286, 192, 189, 188. The simulation results show that the proposed framework and hybrid ETM can be applied to both leaderless and leader-following consensus of MAS, and the lower bound of triggering intervals can be pre-selected such that Zeno-freeness can be guaranteed.

6.7 Conclusion

This chapter provides a unified framework to deal with leaderless and leader-following bipartite consensus problems simultaneously, and a novel decentralized hybrid ETM is proposed to save communication resources. Through constructing a unified hybrid model, the closed-loop system behaviors of the leaderless and leader-following MASs are completely described with the flow dynamics and jump dynam-

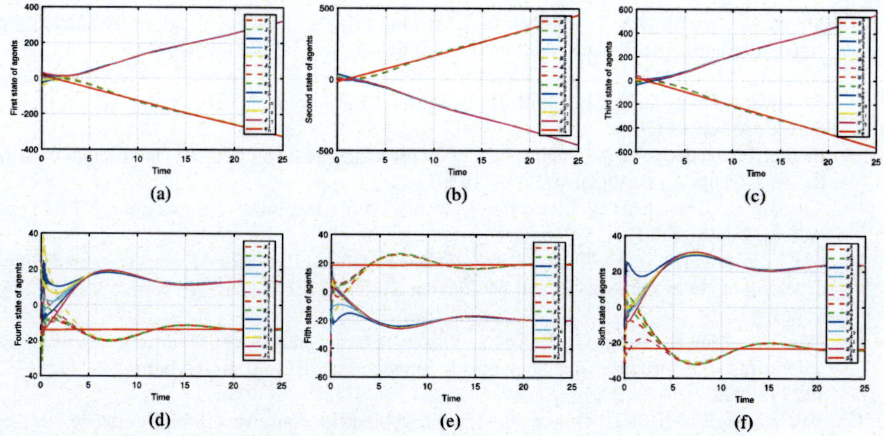

Fig. 6.4 Agents' states in leader-following case

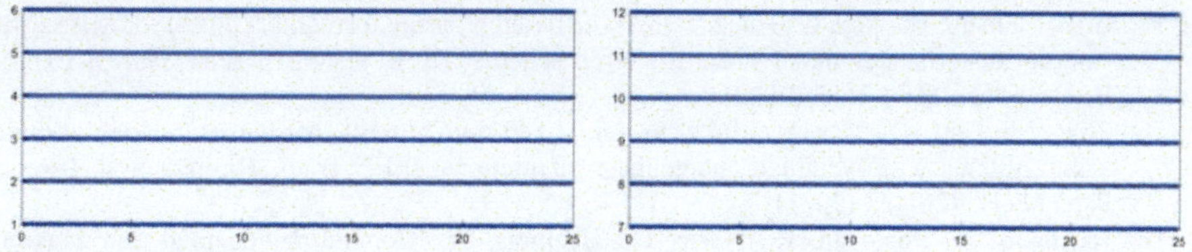

Fig. 6.5 Triggering instants in leader-following case

ics. The designed triggering condition does not depend on any neighbors' information, including the information at their triggering instants, and the minimum inter-event time can always be pre-selected to be strictly positive to exclude Zeno behavior. In addition, a satellite formation example is provided to testify the effectiveness of the obtained results.

References

1. Heemels, W., Donkers, M., Teel, A.: Periodic Event-Triggered Control for Linear Systems. IEEE Trans. Autom. Control **58**(4), 847-861 (2013)
2. Cai, Y., Zhang, H. Duan, J., Zhang, J.: Distributed Bipartite Consensus of Linear Multiagent Systems Based on Event-Triggered Output Feedback Control Scheme. IEEE Trans. Syst., Man, Cybern: Syst. **51**(11), 6743-6756 (2021)
3. Zhang, H., Zhou, Y., Liu, Y., Sun, J.: Cooperative Bipartite Containment Control for Multiagent Systems Based on Adaptive Distributed Observer. IEEE Trans. Cybern. **52**(6), 5432-5440 (2022)
4. Chen, X., Yu, H., Hao, F.: Prescribed-Time Event-Triggered Bipartite Consensus of Multiagent Systems. IEEE Trans. Cybern. **52**(4), 2589-2598 (2022)
5. Yu, H., Chen, X., Chen, T., Hao, F.: Event-triggered bipartite consensus for multi-agent systems: A Zeno-free analysis. IEEE Trans. Autom. Control **65**(11), 4866-4873 (2020)

6. Meng, D., Du, M., Jia, Y.: Interval bipartite consensus of networked agents associated with signed digraphs. IEEE Trans. Autom. Control **61**(12), 3755-3770 (2016)
7. Ning, B., Han, Q.-L., Zuo, Z.: Bipartite Consensus Tracking for Second-Order Multiagent Systems: A Time-Varying Function-Based Preset-Time Approach. IEEE Trans. Autom. Control **66**(6), 2739-2745 (2021)
8. Fiedler, B., Gedeon, T.: A Lyapunov function for tridiagonal competitive-cooperative systems. SIAM J. Math. Anal. **30**(3), 469-478 (1999)
9. Altafini, C.: Consensus Problems on Networks With Antagonistic Interactions. IEEE Trans. Autom. Control **58**(4), 935-946 (2013)
10. Cai, Y., Zhang, H., Su, H., Zhang, J., He, Q.: The Bipartite Edge-Based Event-Triggered Output Tracking of Heterogeneous Linear Multiagent Systems. IEEE Trans. Cybern. **53**(2), 967-978 (2023)
11. Rong, L., Liu, X., Jiang, G.-P., Xu, S.: Observer-based multiagent bipartite consensus with deterministic disturbances and antagonistic interactions. IEEE Trans. Cybern., **52**(11), 11772-11779 (2022)
12. Valcher, M. E., Misra, P.: On the consensus and bipartite consensus in high-order multi-agent dynamical systems with antagonistic interactions. Syst. Control Lett. **66**(4), 94-103 (2014)
13. Bhowmick, S., Panja, S.: Leader-follower bipartite consensus of linear multiagent systems over a signed directed graph. IEEE Trans. Circuits Syst. II-Express Briefs, **58**(4), 935-946 (2013)
14. Li, M., Zhang, W., Yan, C., Hu, Z.: Observer-based bipartite formation control for MASs with external disturbances under event-triggered scheme. IEEE Trans. Circuits Syst. II-Express Briefs. **69**(3), 1178-1182 (2022)
15. Zhao, M., Peng, C., Tian, E.: Finite-time and fixed-time bipartite consensus tracking of multi-agent systems with weighted antagonistic interactions. IEEE Trans. Circuits Syst. I-Regul. Pap., **68**(1), 426-433 (2021)
16. Heijmans, S., Postoyan, R., Nesic, D., Heemels, W.: Computing Minimal and Maximal Allowable Transmission Intervals for Networked Control Systems Using the Hybrid Systems Approach. IEEE Control Systems Letters **1**(1), 56-61 (2017)
17. Heemels, W., Teel, A., van de Wouw, D.: Networked control systems with communication constraints: Tradeoffs between transmission intervals, delays, and performance. IEEE Trans. Autom. Control **55**(8), 1781-1796 (2010)
18. Dolk, V., Borgers, D., Heemels, W.: Output-Based and Decentralized Dynamic Event-Triggered Control With Guaranteed $\mathscr{L}_p$-Gain Performance and Zeno-Freeness. IEEE Trans. Autom. Control **62**(1), 34-49 (2017)
19. Girard, A.: Dynamic triggering mechanisms for event-triggered control. IEEE Trans. Autom. Control **60**(7), 1992-1997 (2015)
20. Hu, S., Chen, X., Qiu, J., Zhao, F., Jiang, X., Du, Y.: Dynamic Event-triggered Bipartite Consensus of Multi-agent Systems with Estimator and Cooperative-Competitive Interactions. IEEE Trans. Circuits Syst. II, Express Briefs **69**(7), 3309-3313 (2022)
21. Hu, W., Yang, C., Huang, T., Gui, W.: A Distributed Dynamic Event-Triggered Control Approach to Consensus of Linear Multiagent Systems With Directed Networks. IEEE Trans. Cybern. **50**(2), 869-874(2020)
22. Du, S., Liu, T., Ho, D.: Dynamic event-triggered control for leader-following consensus of multiagent systems. IEEE Trans. Syst. Man, Cybern.: Syst. **23**(6), 602-619 (2018)

Chapter 7
Hybrid Event-Triggered Consensus of MAS with Time Delay

7.1 Introduction

In previous Chap. 6, it is clarified that the event-triggered control scheme has great potential to reduce unnecessary utilization of network resources and maintain desired performance, then, by means of the proposed hybrid system approach, a novel hybrid event-triggered control (ETC) scheme is designed for the bipartite consensus control of MASs, which possesses the following superiorities:

(i) The minimum inter-event time (MIET) is strictly positive;
(ii) Asymptotic consensus is achieved;
(iii) Continuous communication is not required for both control protocols and event-triggering conditions.

Clearly, all of these properties are of high relevance to realize real-life implementations. But in the literature, few results have achieved them at the same time, despite some of them are able to guarantee property (ii) or property (iii), or both of them. Nevertheless, it should also be noted that being able to guarantee property (i) is very important from both theoretical and practical viewpoints. Because, for the design of event-triggering mechanisms (ETM), there always exists a nontrivial problem, i.e., Zeno behavior, which implies that there exists an accumulation time $T > 0$ by which an infinite number of events will be triggered. In recent results, to rule out Zeno behavior, inter-event times were set to be larger than zero, but this may be not always sufficient for the implementation of ETMs, since the arbitrarily small inter-event time may suggest the 'worst' case $\lim_{k\to\infty} t_{k+1} - t_k = 0$. Several examples satisfying the 'worst' case presented in [1] show that by the arbitrarily small inter-event time, Zeno behavior may be not truly precluded or although Zeno behavior can be precluded, as time goes by, the triggering interval even tends to zero. Besides, the 'worst' case also implies a high frequency of information transmission, which would require the hardware to be responded infinitely fast. But it is problematic for physical implementation on real-time system due to hardware constraints. Therefore, the

G. Zhao et al., *Cooperative Control of Multi-agent Systems*,
https://doi.org/10.1007/978-981-97-0968-7_7

proposed hybrid ETC scheme does provide a feasible solution to the MAS consensus problem.

However, to only consider the ideal network is not adequate in general as it is well known that even very small communication delays may destroy or degrade the stability and performance of the closed-loop systems. Therefore, in the presence of communication delays, how to achieve asymptotic event-triggered consensus of MAS without requiring continuous communication among agents while guaranteeing Zeno-freeness is not a trivial problem. To the best of the authors' knowledge, few results on MAS with general linear dynamics and communication delays have been reported that can achieve the above mentioned properties simultaneously. For example, distributed periodic event-triggered control (ETC) was proposed in [2] for MAS with communication delays and event-detecting delays, and in [3] for MAS with multiple time-varying delays, but the results are restricted to MAS with single integrator dynamics. In [4], the event-based leader-following consensus was investigated for MAS with constant input time-delay, but the neighbors' information at triggering instants is required, which causes excessive communication. To address this issue, in [5], a periodic ETC scheme was proposed for MAS with time-varying delays, the agent broadcasts information to its neighbors without requiring its neighbors's information at triggering instants, however, only bounded consensus can be achieved.

In order to achieve the properties (i)–(iii) under communication delays, this chapter presents a model-based hybrid ETC scheme to force the MIET to be strictly positive such that Zeno behavior is excluded, and develops the consensus analysis and ETC scheme design under hybrid systems framework [6, 7]. In the constructed closed-loop system state, error dynamics are contained for a full consideration, since the errors have jump dynamics at triggering instants and at the instants when the delayed-states are received by the agent's neighbors, the closed-loop system model is described by a hybrid system. Then, Lyapunov-based consensus analysis and event-triggering condition design results are presented.

7.2 Networked MAS and Problem Formulation

This section formulates the leader-following consensus problem for the event-triggered MASs with communication delays. In the presence of communication delays, how to achieve asymptotic event-triggered consensus of MAS without requiring continuous communication among agents while guaranteeing Zeno-freeness performance poses the main technical challenges. Consider the MAS with one leader and N followers. The leader's dynamics is represented as

$$\dot{x}_0(t) = Ax_0(t) \tag{7.1}$$

and the dynamics of each follower i is represented as

$$\dot{x}_i(t) = Ax_i(t) + Bu_i(t) \tag{7.2}$$

where $x_0(t) \in \mathbb{R}^n$ and $x_i(t) \in \mathbb{R}^n$, $i \in [1, N]$ are the states of the leader and followers respectively, $u_i(t) \in \mathbb{R}^m$ denotes the control input of follower i. The matrices A, B are system matrices and assume that (A, B) is stabilizable.

The preliminaries about communication graph have been introduced in Chap. 1. For proceeding this chapter, we make the following Assumption on the communication graph.

Assumption 7.1 The graph $\bar{\mathscr{G}}$ is undirected and contains a spanning tree with the root node being the leader node 0.

The communication between agent i and its neighbors is performed via network in event-triggered fashion. The agent i's state x_i is sampled and transmitted to its neighbors at t_k^i, $k \in \mathbb{N}$, with $0 \leq t_0^i < t_1^i < \cdots$, however, due to communication delays, the measurement $x_i(t_k^i)$ arrives at agent j $(j \in \mathscr{N}_i)$ at $t_k^i + \tau_k^i$, where τ_k^i denotes the time delays from agent i to its out-neighbor. Notice that it is assumed the time delay τ_k^i is the same for all the out-neighbors of agent i, and satisfies $0 \leq \tau_k^i \leq \tau_{MAD} \leq \tau_{MIET}$, for $k \in \mathbb{N}$, where τ_{MAD} denotes the maximum allowable delay (MAD) of the network, and τ_{MIET} denotes the lower bound of the minimum inter-event time (MIET) with the ETC scheme, i.e., $\tau_{MIET} = \inf_{k \in \mathbb{N}, i \in [1,N]}(t_{k+1}^i - t_k^i)$. The assumption implies that the transmitted information from agent i always arrives at its neighbors before the next transmission begins, thus it belongs to the so-called short-delay case. t_k^i denotes the transmission instant, and when the next transmission instant comes is judged according to the following designed event-triggering mechanism (ETM)

$$t_{k+1}^i := \inf\{t \geq t_k^i + \tau_{MIET} | O_i(o_i) \geq 0\}, \tag{7.3}$$

where $O_i(\cdot)$ denotes a function depending on local variable o_i, and the explicit form of $O_i(o_i)$ will be given later.

Remark 7.1 The local variable o_i contains the state of agent i and the model states of its neighbors, it may also contain internal auxiliary variables, such that the event-triggering condition (7.3) only depends on local information. Moreover, observe that by selecting $\tau_{MIET} > 0$, it is imposed that the next triggering event can only occur after at least τ_{MIET} time has elapsed from t_k, therefore, it means that Zeno behavior is naturally excluded.

In view of the nature of ETC, agent i cannot continuously get access to its neighbors' states and only receives information at instants $t_k^j + \tau_k^j$, $j \in \mathscr{N}_i$. Thus, to construct control protocol for agent i, the estimation of agent j's $(j \in \mathscr{N}_i)$ states is necessary. Each agent needs to implement $d_i + 1$ models of itself and its neighbors, these models are used to estimate its own and neighbors' states based on the state information at latest triggering instants. Due to communication delays, agent i's models as seen by agent i and as seen by agent j, $j \in \mathcal{N}_i$ are different. In order to distinguish the models as seen by agent i and its neighbors, let $\hat{x}_{ii}(t)$ denote the model state of agent i as seen by agent i, and its dynamics is given by

$$\begin{cases} \dot{\hat{x}}_{ii}(t) = A\hat{x}_{ii}(t), & t \in (t_k^i, t_{k+1}^i] \\ \hat{x}_{ii}^+(t) = x_i(t), & t = t_k^i \end{cases} \tag{7.4}$$

where t_k^i ($k \in \mathbb{N}$) denotes the triggering instants of agent i. Here, with slightly abuse of notation, we use $\hat{x}_{ii}^+(t)$ to denote the state of $\hat{x}_{ii}(t)$ right after jump, i.e., $\hat{x}_{ii}^+(t) = \hat{x}_{ii}(t^+) = \lim_{\epsilon \to 0^+} \hat{x}_{ii}(t + \epsilon)$.

As mentioned before, due to communication delays, the state measurement $x(t_k^i)$ transmitted by agent i at triggering instant t_k^i is received by agent j ($j \in \mathcal{N}_i$) at instant $t_k^i + \tau_k^i$, then, $x(t_k^i)$ is used to update the model state of agent i as seen by agent j, and denote this model state as $\hat{x}_{ij}(t)$, its dynamics is given by

$$\begin{cases} \dot{\hat{x}}_{ij}(t) = A\hat{x}_{ij}(t), & t \in (t_k^i + \tau_k^i, t_{k+1}^i + \tau_{k+1}^i] \\ \hat{x}_{ij}^+(t) = f_d(t_k^i, \tau_k^i), & t = t_k^i + \tau_k^i \end{cases} \tag{7.5}$$

where $f_d(t_k^i, \tau_k^i) = G(\tau_k^i)x_i(t_k^i)$ and $G(\tau_k^i) = \exp(A\tau_k^i)$. It is implied by (7.5) that τ_k^i is necessarily known to the receiving agents, this can be realized by using timestamp when transmitting data packets via network.

With the defined model states of the follower agents, the control protocol $u_i(t)$ can be designed as follows

$$u_i(t) = K\left[\sum_{j \in \mathcal{N}_i} \left(\hat{x}_{ji}(t) - \hat{x}_{ii}(t)\right) - b_i(x_i(t) - x_0(t))\right]. \tag{7.6}$$

Since the follower agents cannot always get access to their neighbors' states, the model state $\hat{x}_{ji}(t)$ instead of $x_j(t)$ is used in the control protocol $u_i(t)$. Moreover, note that the leader's information in $u_i(t)$ is $x_0(t)$ other than $\hat{x}_0(t)$, this is because the leader has no control input, when the leader transmits the current state to its neighbors at $t_0 = 0$, and updates the model state as seen by its neighbors at $t_0 + \tau_0^0$, we can obtain that $x_0(t) = \hat{x}_0(t)$ ($t \geq t_0 + \tau_0^0$) holds all the time.

Define the state errors

$$\begin{aligned} e_{ii}(t) &= \hat{x}_{ii}(t) - x_i(t), \\ e_{ij}(t) &= \hat{x}_{ij}(t) - x_i(t). \end{aligned} \tag{7.7}$$

Based on the dynamics of $\hat{x}_{ii}(t)$ and $x_i(t)$, $e_{ii}(t)$ has jump dynamics at time t_k^i in the form of $e_{ii}^+(t_k^i) = 0$. Similarly, according to (7.5), we can obtain the jump dynamics of $e_{ij}(t)$ at instant $t_k^i + \tau_k^i$ as

$$
\begin{aligned}
e_{ij}^{+}(t_k^i+\tau_k^i) &= f_d(t_k^i,\tau_k^i) - x_i(t_k^i+\tau_k^i) \\
&= f_d(t_k^i,\tau_k^i) - \hat{x}_{ij}(t_k^i+\tau_k^i) + \hat{x}_{ij}(t_k^i+\tau_k^i) - x_i(t_k^i+\tau_k^i) \\
&= -G(\tau_k^i)e_{ij}(t_k^i) + e_{ij}(t_k^i+\tau_k^i).
\end{aligned} \tag{7.8}
$$

Define the consensus error between the follower agent i and the leader as $\delta_i(t) = x_i(t) - x_0(t)$. Then, from (7.1), (7.2), (7.6) and (7.7), the dynamics of $\delta_i(t)$ can be described by

$$
\dot{\delta}_i(t) = A\delta_i(t) + BK\left[\sum_{j\in\mathcal{N}_i}(e_{ji}(t)+\delta_j(t)-e_{ii}(t)-\delta_i(t)) - b_i\delta_i(t)\right]. \tag{7.9}
$$

To derive the dynamics of the overall MAS, note that at triggering instant t_k^i, agent i transmits the current state to its neighbors such that all agents $j \in \mathcal{N}_i$ update their model states $\hat{x}_{ij}(t)$ at the same time $t_k^i + \tau_k^i$. In this manner, we have $\hat{x}_{ij_1}(t) = \hat{x}_{ij_2}(t)$ for any $j_1, j_2 \in \mathcal{N}_i$, then, for simplicity, we can define a unified error $e_{iJ} = e_{ij}$ for $j \in \mathcal{N}_i$.

Define

$$
\begin{aligned}
\delta(t) &= \left[\delta_1^T(t)\ \delta_2^T(t)\ \cdots\ \delta_N^T(t)\right]^T, \\
e_I(t) &= \left[e_{11}^T(t)\ e_{22}^T(t)\ \cdots\ e_{NN}^T(t)\right]^T, \\
e_J(t) &= \left[e_{1J}^T(t)\ e_{2J}^T(t)\ \cdots\ e_{NJ}^T(t)\right]^T.
\end{aligned} \tag{7.10}
$$

Based on (7.7), (7.9) and (7.10), the flow dynamics of the MAS can be described by

$$
\begin{bmatrix}\dot{\delta}(t)\\ \dot{e}(t)\end{bmatrix} = \begin{bmatrix}\bar{A}_{11} & \bar{A}_{12}\\ \bar{A}_{21} & \bar{A}_{22}\end{bmatrix}\begin{bmatrix}\delta(t)\\ e(t)\end{bmatrix}, \tag{7.11}
$$

where

$$
\begin{aligned}
&e(t) = \left[e_I^T(t)\ e_J^T(t)\right]^T,\ \bar{A}_{11} = I_N \otimes A - \mathscr{H} \otimes BK,\ \bar{A}_{12} = \left[-\mathscr{D} \otimes BK\ \ \mathscr{A} \otimes BK\right], \\
&\bar{A}_{21} = -\begin{bmatrix}1\\1\end{bmatrix} \otimes (\mathscr{H} \otimes BK),\ \bar{A}_{22} = \begin{bmatrix} I_N \otimes A - \mathscr{D} \otimes BK & \mathscr{A} \otimes BK \\ -\mathscr{D} \otimes BK & I_N \otimes A + \mathscr{A} \otimes BK\end{bmatrix}.
\end{aligned} \tag{7.12}
$$

In the following, *transmission event* means that at triggering instant t_k^i, agent i transmits information to its neighbors, *update event* means that at instant $t_k^i + \tau_k^i$, agent j receives information from agent i and updates the model state $\hat{x}_{ij}(t)$. The error dynamics $e_I(t)$, $e_J(t)$ experience jumps when transmission event occurs or update event occurs, and the jump dynamics of the MAS can be described by

$$\delta^+(t) = \delta(t),\ e_I^+(t) \in \bigcup_{i\in[1,N]} (I - \Gamma_i \otimes I_n)e_I(t),\ t \in \bigcup_{i\in[i,N]} t = t_k^i,$$
$$e_J^+(t) \in \bigcup_{i\in[1,N]} (\Gamma_i \otimes G_i(\tau_k^i))e_J(t - \tau_k^i) + e_J(t),\ t \in \bigcup_{i\in[i,N]} t = t_k^i + \tau_k^i, \tag{7.13}$$

where $\Gamma_i \in \mathbb{R}^{N\times N}$ denotes a matrix with the iith diagonal entry is equal to one and all the other entries are equal to zeros.

According to (7.13), it can be seen that the transmission event and update event always make the error dynamics experience jumps, as a result, the state errors cannot be explicitly handled as closed-loop system states in most of the existing literature. Thus, a hybrid system consisting of both flow and jump dynamics is needed to fully describe the system dynamics of the MAS. For the purpose, in the following, first we will construct a suitable closed-loop state by introducing some necessary internal variables, and describe it as a hybrid system. Then, stability analysis with respect to the constructed hybrid system will be proceeded, and it is proved that the stability of the hybrid system is equivalent to the consensus of MAS (7.1) and (7.2), that is, by means of the proposed hybrid ETC protocol (7.6) and (7.3), the leader-following consensus of MAS (7.1) and (7.2) can be achieved in the sense that $\lim_{t\to\infty} |x_i(t) - x_0(t)| = 0,\ i \in [1, N]$, despite there exist communication delays.

7.3 Reformulation of Closed-Loop MAS

The form of the constructed MAS model (7.11) and (7.13) is not suitable for stability analysis, since there are two classes of error jump dynamics that caused by transmission event and update event respectively. In this section, we transform the constructed MAS model into a hybrid system. In view of the general hybrid model given in Chap. 1, in this chapter, it is represented as

$$\mathbb{H}: \begin{cases} \dot{\xi} = \mathscr{F}(\xi), & \text{If } \xi \in \mathbb{F}, \\ \xi^+ \in \bigcup\limits_{i\in[1,N]} \mathscr{J}_i(\xi), & \text{If } \xi \in \mathbb{J}, \end{cases} \tag{7.14}$$

where $\mathscr{F}(\xi)$ describes the flow dynamics, $\mathscr{J}(\xi)$ describes the jump dynamics, $\mathbb{F}$ and $\mathbb{J}$ denote the flow set and jump set respectively. In the following, we transform the MAS model into the form of a hybrid system.

In order to transform the MAS model into hybrid systems form, we introduce auxiliary variables $r_i \in \{0, 1\}, s_i \in \mathbb{R}^n, \kappa_i \in \mathbb{N}, \tau_i \in \mathbb{R}_{\geq 0}$, for $i \in [1, N]$. The variable r_i is a two-value number, which indicates whether the next event is a transmission event ($r_i = 0$) or an update event ($r_i = 1$). The variable s_i is used to memory the value $e_{iJ}(t_k^i)$ at transmission instant t_k^i and is reset to zero at update instant $t_k^i + \tau_k^i$. The integer variable κ_i keeps track of the number of triggering events. The timer variable τ_i can capture the time elapsed since the last triggering instant and also be used to guarantee Zeno-freeness.

Consider the combined state variable $\xi := (\delta, e, \tau, \kappa, s, r) \in \mathbb{X}$ with the set $\mathbb{X} := \{(\delta, e, \tau, \kappa, s, r) \in \mathbb{R}^{nN} \times \mathbb{R}^{2nN} \times \mathbb{R}^{N}_{\geq 0} \times \mathbb{N}^{N} \times \mathbb{R}^{nN} \times \{0, 1\}^{N}\}$, where $\tau = \left[\tau_1\ \tau_2 \cdots \tau_N\right]^T$, $\kappa = \left[\kappa_1\ \kappa_2 \cdots \kappa_N\right]^T$, $s = \left[s_1^T\ s_2^T \cdots s_N^T\right]^T$, $r = \left[r_1\ r_2 \cdots r_N\right]^T$. Based on the above defined state variables, the flow dynamics of ξ (including the flow dynamics (7.11) of the MAS) can be described as

$$\mathscr{F}(\xi) := \left[\bar{A}_{11}\delta + \bar{A}_{12}e\ \bar{A}_{21}\delta + \bar{A}_{22}e\ \mathbf{1}_N\ \mathbf{0}_N\ \mathbf{0}_{nN}\ \mathbf{0}_N\right]. \tag{7.15}$$

The flow set $\mathbb{F}$ is given by

$$\mathbb{F} := \bigcap_{i \in [1,N]} \mathbb{F}_i,$$

with

$$\mathbb{F}_i := \left\{\xi \in \mathbb{X} | [(\tau_i \leq \tau_{MIET} \text{ or } O_i \leq 0) \text{ and } r_i = 0] \text{ or } [\text{event } \mathscr{E}^i_{pkt} \text{ does not occur and } r_i = 1]\right\},$$

where the event $\mathscr{E}^i_{pkt}$ denotes that the data transmitted by agent i arrives at its neighbors, thus, event $\mathscr{E}^i_{pkt}$ does not occur means that the data has not been received by its neighbors yet.

Recall the definition of $\Gamma_i \in \mathbb{R}^{N \times N}$ and expression (7.14), then, the jump dynamics of $\mathscr{J}_i(\xi)$ (including the jump dynamics (7.13) of the MAS) can be represented as

$$\mathscr{J}_i(\xi) := \begin{cases} \mathscr{J}_{0i}(\xi), & \text{If } \xi \in \mathbb{J}_i \text{ and } r_i = 0, \\ \mathscr{J}_{1i}(\xi), & \text{If } \xi \in \mathbb{J}_i \text{ and } r_i = 1, \end{cases} \tag{7.16}$$

$$\mathscr{J}_{0i}(\xi) = \begin{bmatrix} \delta \\ \left(I_{2nN} - \begin{bmatrix} \Gamma_i \otimes I_n & 0 \\ 0 & 0_{nN \times nN} \end{bmatrix}\right) e \\ (I_N - \Gamma_i)\tau \\ \kappa + \Gamma_i \mathbf{1}_N \\ -(\Gamma_i \otimes I_n) e_J \\ r + \Gamma_i \mathbf{1}_N \end{bmatrix}, \quad \mathscr{J}_{1i}(\xi) = \begin{bmatrix} \delta \\ \begin{bmatrix} 0_{nN \times nN} \\ \Gamma_i \otimes G_i \end{bmatrix} s + e \\ \tau \\ \kappa \\ 0 \\ r - \Gamma_i \mathbf{1}_N \end{bmatrix}.$$

The jump set $\mathbb{J}$ is given by

$$\mathbb{J} := \bigcup_{i \in [1,N]} \mathbb{J}_i,$$

with

$$\mathbb{J}_i := \{\xi \in \mathbb{X} | [(\tau_i > \tau_{MIET} \text{ and } O_i > 0) \text{ and } r_i = 0] \text{ or } [\text{event } \mathscr{E}^i_{pkt} \text{ occur and } r_i = 1] \Big\}.$$

The function $\mathscr{J}_{0i}(\xi)$ describes how the state ξ jumps when transmission event in agent i occurs, i.e., the event-triggering condition is satisfied and agent i transmits data to its neighbors. The error e_{ii} is reset to zero and e_{iJ} remains unchanged. The variable τ_i is reset to zero and the variable κ_i is updated to be $\kappa_i = \kappa_i + 1$. The memory state s_i is set to be $-e_{iJ}$, which will be used to update the error dynamics when update event occurs. Moreover, r_i jumps to 1 indicates that the next event for agent i as well as its neighbors is an update event (induced by short-delay assumption). The function $\mathscr{J}_{1i}(\xi)$ describes how ξ jumps when an update event occurs, namely after agent i transmits data, the delayed data is received by agent i's neighbors and used to update model state of agent i as seen by its neighbors. The jump dynamics of e corresponds to e_{ii} remains unchanged and e_{iJ} jumps according to (7.8). r_i toggles back to 0 means that the next event in agent i is a transmission event. There is no information needs to be stored in s_i, so s_i is reset to zero and this is also necessary as will be seen in Appendix when constructing Lyapunov function.

Under hybrid systems framework, the MAS model (7.11), (7.13) has been transformed into hybrid systems form (7.14) with introducing some auxiliary variables. It implies that the leader-following consensus can be achieved if the obtained hybrid system is stable. The stability property of the hybrid system is defined as follows.

Definition 7.1 For the constructed hybrid system model as described by (7.14). The set given by $\{\xi \in \mathbb{X} | \delta = \mathbf{0}_{nN}, e = \mathbf{0}_{2nN}, s = \mathbf{0}_{nN}\}$ is said to be uniformly globally asymptotically stable (UGAS) if there exists a function $\beta \in \mathscr{KL}$ such that, for any initial condition $\xi(0) \in \mathbb{X}_0$ with $\mathbb{X}_0 = \{\xi \in \mathbb{X} | r = \mathbf{0}_N\}$, the following condition holds

$$|(\delta(t), e(t), s(t))| \leq \beta(|(x(0), e(0), s(0))|, t). \tag{7.17}$$

7.4 Consensus Analysis and ETM Design Under Time Delay

In this section, we present conditions such that the (δ, e, s) dynamics of the hybrid system (7.14) is UGAS (i.e., the leader-following consensus is achieved), meanwhile, the explicit form of the event-triggering condition (7.3) (i.e., the function O_i which depends on local variables) is developed. First, we suppose that the following Condition I is satisfied for the error dynamics, similar conditions were employed in [8–10] for delay-free networked control systems. Then, based on Condition I, we construct a function relating to the errors that will decrease even if the errors have jumps, which is useful for constructing Lyapunov function. Furthermore, the main results including consensus analysis and event-triggering condition design are presented.

For the sake of brevity, the arguments of state variables and functions will be omitted according to the context.

Condition I: For $i \in [1, N]$, there exits a function $W_i : \mathbb{N} \times \mathbb{R}^{2n} \to \mathbb{R}_{\geq 0}$ that is locally Lipschitz in its second argument such that for all $\kappa_i \in \mathbb{N}$ and all $e_i \in \mathbb{R}^{2n}$, the following conditions

$$\underline{\alpha}_{Wi}\,|e_i| \leq W_i(\kappa_i, e_i) \leq \bar{\alpha}_{Wi}\,|e_i|, \tag{7.18a}$$

$$W_i(\kappa_i + 1, 0) \leq \lambda_i W_i(\kappa_i, e_i), \tag{7.18b}$$

$$W_i(\kappa_i + 1, e_i) \leq \lambda_{Wi} W_i(\kappa_i, e_i) \tag{7.18c}$$

hold for constants $0 < \underline{\alpha}_{Wi} \leq \bar{\alpha}_{Wi}, 0 < \lambda_i < 1$ and $\lambda_{Wi} \geq 1$. Additionally, for almost all $e_i \in \mathbb{R}^{2n}$ and all $\kappa_i \in \mathbb{N}$,

$$\left|\frac{\partial W_i}{\partial e_i}(\kappa_i, e_i)\right| \leq c_i \tag{7.19}$$

holds for some constant $c_i > 0$.

In Condition I, the state variable $e_i = \left[e_{ii}^T \; e_{iJ}^T\right]^T$, based on the definition of e_I, e_J, e in (7.10) and (7.12), the relation between e_i and e can be represented as

$$e_i = \begin{bmatrix} R_i & 0 \\ 0 & R_i \end{bmatrix} e, \; e = \sum_{i=1}^{N} \begin{bmatrix} R_i^T \begin{bmatrix} I & 0 \end{bmatrix} \\ R_i^T \begin{bmatrix} 0 & I \end{bmatrix} \end{bmatrix} e_i, \tag{7.20}$$

where $R_i \in \mathbb{R}^{n \times nN}$ is composed of N sub-matrices, each with $n \times n$ dimension, and the ith sub-matrix is an identity matrix I_n and all the other sub-matrices are zero matrices.

Based on Condition I, the next result presents how to construct functions for variable communication delays case.

Proposition 7.1 *Consider the MAS with hybrid system model (7.14). If Condition I holds, define the function* $\bar{W}_i(\kappa_i, r_i, e_i, s_i) : \mathbb{N} \times \{0, 1\} \times \mathbb{R}^{2n} \times \mathbb{R}^n \to \mathbb{R}_{\geq 0}$ *as follows*

$$\begin{aligned} &\bar{W}_i(\kappa_i, 0, e_i, s_i) = W_i(\kappa_i, e_i), \\ &\bar{W}_i(\kappa_i, 1, e_i, s_i) = \max\left\{\frac{\lambda_i}{\lambda_{Wi}} W_i(\kappa_i, e_i), W_i\left(\kappa_i, e_i + \begin{bmatrix} 0_{n\times n} \\ G_i s_i \end{bmatrix}\right)\right\}. \end{aligned} \tag{7.21}$$

where $G_i = G(\tau_k^i)$, $\bar{W}_i$ *is locally Lipschitz for all fixed* $\kappa_i \in \mathbb{N}$ *and* $r_i \in \{0, 1\}$, *and there exist* $\mathcal{K}_\infty$ *functions* $\underline{\alpha}_{\bar{W}i}$ *and* $\bar{\alpha}_{\bar{W}i}$, *and* $0 < \lambda_i < 1$ *such that for all* $\kappa_i \in \mathbb{N}$, $r_i \in \{0, 1\}$, $e_i \in \mathbb{R}^{2n}$ *and* $s_i \in \mathbb{R}^n$, $i \in [1, N]$, *if the maximal allowable delay* τ_{MAD} *satisfies*

$$\tau_{MAD} \leq \min\left\{\frac{\ln(1+\lambda_i)}{|A|}, -\frac{\ln(1-\lambda_i)}{|A|}\right\}, \tag{7.22}$$

then, the following conditions hold

$$\bar{W}_i\left(\kappa_i + 1, 1, \begin{bmatrix} 0_n \\ e_{iJ} \end{bmatrix}, -e_{iJ}\right) \leq \lambda_i \bar{W}_i(\kappa_i, 0, e_i, s_i), \tag{7.23a}$$

$$\bar{W}_i\left(\kappa_i, 0, e_i + \begin{bmatrix} 0_n \\ G_i s_i \end{bmatrix}, 0\right) \leq \bar{W}_i(\kappa_i, 1, e_i, s_i), \tag{7.23b}$$

$$\underline{\alpha}_{\bar{W}i}(|(e_i, s_i)|) \leq \bar{W}_i(\kappa_i, r_i, e_i, s_i) \leq \bar{\alpha}_{\bar{W}i}(|(e_i, s_i)|). \tag{7.23c}$$

Proposition 7.2 *Consider the flow dynamics of the error e in (7.15) and (7.16). If the inequalities (7.18a) and (7.19) in Condition I hold for $i \in [1, N]$, then there always exist $\bar{R}_i \in \mathbb{R}^{2nN}$, $\bar{A}_{Ri} \in \mathbb{R}^{2nN}$, constants $M_{ei\geq 0}$ and $L_{r_i i} \geq 0$ with $r_i \in \{0, 1\}$, such that for all $\kappa_i \in \mathbb{N}$, $r_i \in \{0, 1\}$, $s_i \in \mathbb{R}^{nN}$, $\delta \in \mathbb{R}^{nN}$, and almost all $e_i \in \mathbb{R}^{2n}$, the following condition*

$$\left\langle \frac{\partial \bar{W}_i}{\partial e_i}, \dot{e}_i \right\rangle \leq c_i \left|\bar{R}_i \bar{A}_{21}\delta + \bar{A}_{Ri} e\right| + L_{r_i i} \bar{W}_i \tag{7.24}$$

holds with

$$L_{r_i i} = \begin{cases} \dfrac{c_i M_{ei}}{\underline{\alpha}_{Wi}}, & \text{If } r_i = 0 \\[2ex] \dfrac{c_i M_{ei} \lambda_{Wi}}{\underline{\alpha}_{Wi} \lambda_i}, & \text{If } r_i = 1 \end{cases}$$

where

$$M_{ei} = \left|\bar{R}_i \bar{A}_{22}\right|, \ \bar{A}_{Ri} = \bar{R}_i \bar{A}_{22}\left(I_{2nN} - \begin{bmatrix} \Gamma_i \otimes I_n & 0 \\ 0 & \Gamma_i \otimes I_n \end{bmatrix}\right), \ \bar{R}_i = \begin{bmatrix} R_i & 0 \\ 0 & R_i \end{bmatrix},$$

in which R_i is defined in (7.20).

The proofs for Propositions 7.1–7.2 are given in the Appendix. Notice that the conditions in (7.23) describe how the value of function $\bar{W}_i$ changes when the error e_i has jumps. Proposition 7.2 describes how the function $\bar{W}_i$ evolves when the error e_i flows, and the upper bound on $\left\langle \frac{\partial \bar{W}_i}{\partial e_i}, \dot{e}_i \right\rangle$ is presented. These properties are useful for consensus analysis.

Remark 7.2 In order to guarantee that the conditions in Condition I hold, we can choose $W_i(\kappa_i, e_i) = |e_i|$ with parameters $\underline{\alpha}_{Wi} = \bar{\alpha}_{Wi} = 1$, $0 < \lambda_i < 1$, $\lambda_{Wi} = 1$ and $c_i = 1$ for $i \in [1, N]$, in this case, the bound condition (7.24) can be reduced to

$$\left\langle \frac{\partial \bar{W}_i}{\partial e_i}, \dot{e}_i \right\rangle \leq \left|\bar{R}_i \bar{A}_{21}\delta + \bar{A}_{Ri} e\right| + L_{r_i i} \bar{W}_i$$

with

$$L_{r_i i} = \begin{cases} M_{ei}, & \text{If } r_i = 0, \\ M_{ei}/\lambda_i, & \text{If } r_i = 1. \end{cases}$$

Before presenting the main results, first, consider the function $\phi_{r_i i} : \mathbb{R}_{\geq 0} \to \mathbb{R}_{\geq 0}$ with $i \in [1, N]$, $r_i \in \{0, 1\}$, which has the following form

$$\dot{\phi}_{r_i i}(t) = b(t)(-2L_{r_i i}\phi_{r_i i}(t) - \gamma_{r_i i}(\phi^2_{r_i i}(t) + 1)), \tag{7.25}$$

where $L_{r_i i}$ comes from condition (7.24) in Proposition 7.2, $\gamma_{r_i i}$ is constant that will be given later and the initial condition $\phi_{r_i i}(0)$ needs to be given, and the function $\theta(t) : \mathbb{R}_{\geq 0} \to \{0, 1\}$ is defined as

$$\theta(t) = \begin{cases} 1, & \text{if } 0 \leq t \leq \tau_{MIET}, \\ 0, & \text{if } t > \tau_{MIET}. \end{cases} \tag{7.26}$$

Theorem 7.1 *Consider the MAS with hybrid system model (7.14). Suppose that the conditions (7.23a) and (7.23b) in Proposition 7.1 and the condition (7.24) in Proposition 7.2 hold, let* $\gamma_{0i} = \gamma_i$, $\gamma_{1i} = \gamma_i \lambda_{Wi}/\lambda_i$ *for the given constant* $\gamma_i > 0$, *and the following conditions hold for each* $i \in [1, N]$*:*

(1) There exist constants τ_{MIET} *and* τ_{MAD} *with* $\tau_{MIET} \geq \tau_{MAD}$ *such that*

$$\begin{aligned} &\gamma_{0i}\phi_{0i}(\tau_i) \geq \lambda_i^2 \gamma_{1i}\phi_{1i}(0), \ \textit{for } \tau_i \geq \tau_{MIET} \\ &\gamma_{1i}\phi_{1i}(\tau_i) \geq \gamma_{0i}\phi_{0i}(\tau_i), \ \textit{for } 0 \leq \tau_i \leq \tau_{MAD} \end{aligned} \tag{7.27}$$

hold for any solution $\phi_{r_i i}$ *of (7.25) with the given initial conditions* $\phi_{r_i i}(0) > 0$ *satisfying* $\gamma_{1i}\phi_{1i}(0) \geq \gamma_{0i}\phi_{0i}(0) \geq \lambda_i^2 \gamma_{1i}\phi_{1i}(0)$.

(2) There exist positive definite matrices $P \in \mathbb{R}^{nN \times nN}$, $Q \in \mathbb{R}^{nN \times nN}$, *constants* $0 < \rho_0 < \rho < 1$, $0 < \varepsilon < \gamma_i^2$, $a > 0$, $\rho_1 > 0$, $\rho_2 > 0$ *such that the following matrix inequalities hold*

$$P\bar{A}_{11} + \bar{A}_{11}^T P \leq -Q \tag{7.28}$$

$$\begin{bmatrix} -(1-\rho)Q + \tilde{A}_{21}^T \tilde{A}_{21} & P\bar{A}_{12} + \tilde{A}_{21}^T \bar{A}_R \\ * & -\sum_{i \in [1,N]}(\gamma_i^2 - \varepsilon)\bar{R}_i^T \bar{R}_i + \bar{A}_R^T \bar{A}_R \end{bmatrix} \leq 0 \tag{7.29}$$

$$\rho_0 Q > \frac{\max_{i \in [1,N]}\{\alpha_i\}((1+a)H_1^T H_1 + \rho_2/\rho_1)}{\min_{i \in [1,N]}\{\alpha_i\}/\rho_1 - (1/a + 1)\left|H_2^T H_2\right|}, \tag{7.30}$$

where

$$H_1 = -\mathscr{L} \otimes I_n, \ H_2 = -(\mathscr{D} \otimes I_n)R_{eI} + (\mathscr{A} \otimes I_n)R_{eJ}, \ \alpha_i = \gamma_{0i}(2L_{0i}\phi_{0i} + \gamma_{0i}(\phi_{0i}^2 + 1)),$$

$$\tilde{A}_{21} = \begin{bmatrix} \bar{R}_1\bar{A}_{21} \\ \bar{R}_2\bar{A}_{21} \\ \vdots \\ \bar{R}_N\bar{A}_{21} \end{bmatrix}, \ \bar{A}_R = \begin{bmatrix} \bar{A}_{R1} \\ \bar{A}_{R2} \\ \vdots \\ \bar{A}_{RN} \end{bmatrix}, \ R_{eI} = \begin{bmatrix} R_1 & 0 \\ \vdots & \vdots \\ R_N & 0 \end{bmatrix}, \ R_{eJ} = \begin{bmatrix} 0 & R_1 \\ \vdots & \vdots \\ 0 & R_N \end{bmatrix}.$$

(3) The hybrid event-triggering condition is designed as the following form

$$t_{k+1}^i = \inf\{t > t_k^i | \tau_i > \tau_{MIET} \text{ and } O_i \geq 0\} \tag{7.31}$$

with

$$O_i = -\rho_1 \left| \sum_{j \in \mathcal{N}_i} (\hat{x}_{ji} - \hat{x}_{ii}) \right|^2 - \rho_2 b_i |\delta_i|^2 + (1 - \theta(\tau_i))\gamma_{0i}(2L_{0i}\phi_{0i} + \gamma_{0i}(\phi_{0i}^2 + 1))\bar{W}_i^2. \tag{7.32}$$

Then, the (δ, e, s) dynamics of the hybrid system (7.14) is UGAS with the event-triggering condition in (7.31). In other words, δ converges to zero indicates that the leader-following consensus is achieved with Zeno-freeness. Moreover, the lower bound of MIET can be selected to be any positive constant that satisfies

$$\tau_{MIET} \leq \min_{i \in [1,N]} \tau_M^i \tag{7.33}$$

with

$$\tau_M^i = \begin{cases} -\dfrac{1}{L_{0i}p_i} \arctan \dfrac{\bar{\phi}_{0i}}{(L_{0i}p_i)^2 + \bar{\phi}_{1i}\bar{\phi}_{2i}}, & \gamma_{0i} > L_{0i}, \\ \dfrac{1}{\bar{\phi}_{1i}} - \dfrac{1}{\bar{\phi}_{2i}}, & \gamma_{0i} = L_{0i}, \\ \dfrac{1}{L_{0i}p_i} \operatorname{arctanh} \dfrac{\bar{\phi}_{0i}}{(L_{0i}p_i)^2 - \bar{\phi}_{1i}\bar{\phi}_{2i}}, & \gamma_{0i} < L_{0i}, \end{cases} \tag{7.34}$$

where

$$p_i = \left|(\gamma_{0i}/L_{0i})^2 - 1\right|^{0.5}, \bar{\phi}_{0i} = L_{0i}p_i\gamma_{0i}(\lambda_i\phi_{1i}(0) - \phi_{0i}(0)),$$
$$\bar{\phi}_{1i} = \lambda_i\gamma_{0i}\phi_{1i}(0) + L_{0i}, \bar{\phi}_{2i} = \gamma_{0i}\phi_{0i}(0) + L_{0i}.$$

Proof Consider the candidate Lyapunov function

$$U(\xi) = \delta^T P \delta + \sum_{i=1}^{N} \left[\gamma_{r_i i}\phi_{r_i i}(\tau_i)\bar{W}_i^2(\kappa_i, r_i, e_i, s_i)\right]. \tag{7.35}$$

Due to $\gamma_{r_i i} > 0$ and $\phi_{r_i i} \geq 0$ for $\tau_i \geq 0$, from the radial unboundedness of functions $\delta^T P\delta$ and $\bar{W}_i$, we can see that $U(\xi)$ is also radially unbounded such that it is an appropriate candidate Lyapunov function.

According to hybrid systems theory [7], for the MAS with hybrid model (7.14), if we can show that the following inequalities are satisfied

$$\begin{aligned} &\langle \nabla U(\xi), \mathscr{F}(\xi)\rangle \leq -\sigma_\delta(|\delta|) - \sum_{i=1}^{N} \sigma_W(\bar{W}_i), \text{ for } \xi \in \mathbb{F} \\ &U(\xi^+) - U(\xi) \leq 0, \text{ for } \xi \in \mathbb{J} \end{aligned} \tag{7.36}$$

for some positive definite functions σ_δ, σ_W, then, the (δ, e, s) dynamics in ξ converge to zeros asymptotically, namely the leader-following consensus is achieved as $\delta_i = x_i - x_0 \to 0$.

First, with slightly abuse of notation, consider the quantity $\langle \nabla U(\xi), \mathscr{F}(\xi)\rangle$ with $\mathscr{F}(\xi)$ in (7.15), even though $\bar{W}_i$ is not differentiable with respect to κ_i and r_i, however, due to $\dot{\kappa}_i = \dot{r}_i = 0$ in the flow set, this does not lead to any problem. Hence, for all (τ, κ, r) and almost all (δ, e, s), we have

$$\begin{aligned} &\langle \nabla U(\xi), \mathscr{F}(\xi)\rangle \\ =&\delta^T(P\bar{A}_{11} + \bar{A}_{11}^T P)\delta + 2\delta^T P\bar{A}_{12}e + \sum_{i=1}^{N}\Big[-\theta(\tau_i)\gamma_{r_i i}(2L_{r_i i}\phi_{r_i i} + \gamma_{r_i i}(\phi_{r_i i}^2 + 1))\bar{W}_i^2 \\ &+ 2\gamma_{r_i i}\phi_{r_i i}(\tau_i)\bar{W}_i\Big\langle \frac{\partial \bar{W}_i}{\partial e_i}, \bar{R}_i\bar{A}_{21}\delta + \bar{R}_i\bar{A}_{22}e\Big\rangle\Big]. \end{aligned} \tag{7.37}$$

Let the matrix at the left side of inequality (7.29) left-multiplied by $\begin{bmatrix}\delta^T & e^T\end{bmatrix}$ and right-multiplied by $\begin{bmatrix}\delta^T & e^T\end{bmatrix}^T$, we have

$$\begin{aligned} &-\delta^T Q\delta + 2\delta^T P\bar{A}_{12}e \\ \leq& -\rho\delta^T Q\delta - \sum_{i=1}^{N}\Big[\delta^T\bar{R}_i\bar{A}_{21}^T\bar{A}_{21}\bar{R}_i\delta + e^T\bar{A}_{Ri}^T\bar{A}_{Ri}e + 2\delta^T\bar{R}_i^T\bar{A}_{21}^T\bar{A}_{Ri}e\Big] + \sum_{i=1}^{N}(\gamma_i^2 - \varepsilon)e_i^T e_i \\ =& -\rho\delta^T Q\delta - \sum_{i=1}^{N}\Big[|\bar{R}_i\bar{A}_{21}\delta + \bar{A}_{Ri}e|^2 - (\gamma_i^2 - \varepsilon)e_i^T e_i\Big] \end{aligned} \tag{7.38}$$

For the sake of brevity, suppose that $c_i = 1, i \in [1, N]$ as discussed in Remark 7.2 (if $c_i > 1$, the matrix inequality (7.29) can be slightly modified without changing the stability analysis framework), then, combined with (7.28) and based on the bound condition in Proposition 7.2, we have

$$\begin{aligned} &\langle \nabla U(\xi), \mathscr{F}(\xi)\rangle \\ \leq& -\rho\delta^T Q\delta - \sum_{i=1}^{N}\Big[|\bar{R}_i\bar{A}_{21}\delta + \bar{A}_{Ri}e|^2 - (\gamma_i^2 - \varepsilon)e_i^T e_i\Big] \\ &+ \sum_{i=1}^{N}\big[-\theta(\tau_i)\gamma_{r_i i}(2L_{r_i i}\phi_{r_i i} + \gamma_{r_i i}(\phi_{r_i i}^2 + 1))\bar{W}_i^2 + 2\gamma_{r_i i}\phi_{r_i i}(\tau_i)\bar{W}_i\left(|\bar{R}_i\bar{A}_{21}\delta + \bar{A}_{Ri}e| + L_{r_i i}\bar{W}_i\right)\big]. \end{aligned} \tag{7.39}$$

Note that $\theta(\tau_i)$ is a piecewise constant function, in the case of $0 \leq \tau_i \leq \tau_{MIET}$, based on (7.21) and Remark 7.2, the following inequalities

$$\begin{aligned}
&\langle \nabla U(\xi), \mathscr{F}(\xi) \rangle \\
\leq & -\rho\delta^T Q\delta - \sum_{i=1}^{N} \Big[|\bar{R}_i \bar{A}_{21}\delta + \bar{A}_{Ri} e|^2 - (\gamma_i^2 - \varepsilon) e_i^T e_i - 2L_{r_i i}\gamma_{r_i i}\phi_{r_i i}\bar{W}_i^2 - \gamma_{r_i i}^2\phi_{r_i i}^2 \bar{W}_i^2 - \gamma_{r_i i}^2 \bar{W}_i^2 \\
& + 2\gamma_{r_i i}\phi_{r_i i} |\bar{R}_i \bar{A}_{21}\delta + \bar{A}_{Ri} e| \bar{W}_i + 2\gamma_{r_i i}\phi_{r_i i} L_{r_i i}\bar{W}_i^2 \Big] \\
\leq & -\rho\delta^T Q\delta - \sum_{i=1}^{N} (|\bar{R}_i \bar{A}_{21}\delta + \bar{A}_{Ri} e| - \gamma_{r_i i}\phi_{r_i i}\bar{W}_i)^2 + (\gamma_i^2 - \varepsilon) e_i^T e_i - \gamma_{r_i i}^2 \bar{W}_i^2 \\
\leq & -\rho\delta^T Q\delta - \sum_{i=1}^{N} \varepsilon \bar{W}_i^2 \leq -\sigma_\delta(|\delta|) - \sum_{i=1}^{N} \sigma_W(\bar{W}_i)
\end{aligned} \tag{7.40}$$

hold with positive definite functions $\sigma_\delta(x) = \rho\lambda_{\min}(Q)|x|^2$ and $\sigma_W(x) = \varepsilon x^2$. Subsequently, from $\gamma_{r_i i} \geq \gamma_i \lambda_{Wi}/\lambda_i$ and consider the case of $\tau_i > \tau_{MIET}$, it follows that $r_i = 0$ and

$$\begin{aligned}
&\langle \nabla U(\xi), \mathscr{F}(\xi) \rangle \\
\leq & -\rho\delta^T Q\delta - \sum_{i=1}^{N} [-\gamma_{0i}^2\phi_{0i}^2(\tau_i)\bar{W}_i^2 - \gamma_{0i}^2 \bar{W}_i^2 + \varepsilon \bar{W}_i^2 - 2\gamma_{0i}\phi_{0i}(\tau_i) L_{r_i i}\bar{W}_i^2] \\
= & -\rho\delta^T Q\delta + \sum_{i=1}^{N} \alpha_i \bar{W}_i^2 - \sum_{i=1}^{N} \varepsilon \bar{W}_i^2.
\end{aligned} \tag{7.41}$$

From (7.31), we know that $O_i \leq 0$ when the triggering event does not occur. Hence, summing both size of $O_i \leq 0$ with respect to i, and note from Proposition 7.1 and Remark 7.2 that $\bar{W}_i = |e_i|$, then, we have

$$\begin{aligned}
\min_{i\in[1,N]}\{\alpha_i\}|e_i|^2 &\leq \rho_2\delta^T(\mathscr{B}\otimes I_n)\delta + \rho_1|-(\mathscr{D}\otimes I_n)e_I + (\mathscr{A}\otimes I_n)e_J - (\mathscr{L}\otimes I_n)\delta|^2 \\
&\leq \rho_1|H_1\delta + H_2 e|^2 + \rho_2\delta^T\delta \\
&\leq \rho_1\left((1+a)\delta^T H_1^T H_1\delta + \left(\frac{1}{a}+1\right)e^T H_2^T H_2 e\right) + \rho_2\delta^T\delta,
\end{aligned}$$

where the Young's inequality $2x^T y \leq a x^T x + (1/a) y^T y$ is used in the above last inequality. From the inequality, we can derive that

$$|e|^2 \leq \frac{\delta^T((1+a)H_1^T H_1 + \rho_2/\rho_1)\delta}{\min_{i\in[1,N]}\{\alpha_i\}/\rho_1 - (1/a+1)\left|H_2^T H_2\right|},$$

it follows from (7.30) that $\max_{i\in[1,N]}\{\alpha_i\}|e|^2 < \rho_0\delta^T Q\delta$. Then, we obtain from (7.41) that

$$
\begin{aligned}
&\langle \nabla U(\xi), \mathscr{F}(\xi)\rangle \\
&\leq -(\rho-\rho_0)\delta^T Q\delta - \rho_0\delta^T Q\delta + \max_{i\in[1,N]}\{\alpha_i\}\bar{W}_i^2 - \sum_{i=1}^{N}\varepsilon\bar{W}_i^2 \\
&\leq -(\rho-\rho_0)\delta^T Q\delta - \sum_{i=1}^{N}\varepsilon\bar{W}_i^2 \\
&\leq -\sigma_\delta(|\delta|) - \sum_{i=1}^{N}\sigma_W(\bar{W}_i)
\end{aligned}
$$

holds by selecting $\sigma_\delta(x) = (\rho-\rho_0)\lambda_{\min}(Q)\,|x|^2$ and $\sigma_W(x) = \varepsilon x^2$. Thus, it is proved that $\langle \nabla U(\xi), \mathscr{F}(\xi)\rangle \leq -\sigma_\delta(|\delta|) - \sum_{i=1}^{N}\sigma_W(\bar{W}_i)$, for $\xi \in \mathbb{F}$ is satisfied as in (7.36).

Next, consider the jump dynamics when $\xi \in \mathbb{J}$, there exist two cases.

Case 1: $\xi \in \mathbb{J}_i$ and $r_i = 0$.

In this case, $r_i = 0$ indicates that the next event for agent i is a transmission event, so $\tau_i \geq \tau_{MIET}$. For each $i \in [1, N]$, we have

$$
\begin{aligned}
&U(\xi^+) - U(\xi) \\
=&\delta^T P\delta + \sum_{j=1, j\neq i}^{N}(\gamma_{r_j j}\phi_{r_j j}(\tau_i)\bar{W}_j^2(\kappa_j, r_j, e_j, s_j)) + \gamma_{1i}\phi_{1i}(0)\bar{W}_i^2\left(\kappa_i+1, 1, \begin{bmatrix}0\\ e_{iJ}\end{bmatrix}, -e_{iJ}\right) \\
&-\delta^T P\delta + \sum_{j=1, j\neq i}^{N}(\gamma_{r_j j}\phi_{r_j j}(\tau_i)\bar{W}_j^2(\kappa_j, r_j, e_j, s_j)) - \gamma_{0i}\phi_{0i}(\tau_{MIET})\bar{W}_i^2(\kappa_i, 0, e_i, s_i) \\
\leq& \ 0,
\end{aligned}
\tag{7.42}
$$

where the last inequality is from condition (7.27) and condition (7.23a).

Case 2: $\xi \in \mathbb{J}_i$ and $r_i = 1$.

In this case, $r_i = 1$ indicates that the next event for agent i is an update event, so $0 \leq \tau_i \leq \tau_{MIET}$. For each $i \in [1, N]$, using the condition (7.27) and condition (7.23b) in Proposition 7.1, it follows that

$$
\begin{aligned}
&U(\xi^+) - U(\xi) \\
=&\gamma_{0i}\phi_{0i}(\tau_i)\bar{W}_i^2\left(\kappa_i, 0, e_i + \begin{bmatrix}0\\ G_i s_i\end{bmatrix}, 0\right) - \gamma_{1i}\phi_{1i}(\tau_i)\bar{W}_i^2(\kappa_i, 1, e_i, s_i) \\
\leq& \ 0.
\end{aligned}
\tag{7.43}
$$

Both (7.42) and (7.43) ensure that the condition $U(\xi^+) - U(\xi) \leq 0$ holds for $\xi \in \mathbb{J}$. Combined with (7.40) and (7.41), it is known that the conditions (7.36) are guaranteed. Hence, we can conclude that the states (δ, e, s) converge to zeros asymptotically, which indicates that $\delta_i, i \in [1, N]$ converges to zeros and the leader-following consensus is achieved.

To prove the conditions (7.33) for the lower bound of the MIET, first, from (7.27) and $\gamma_{0i} = \gamma_i, \gamma_{1i} = \gamma_i \lambda_{Wi}/\lambda_i$, by selecting the initial conditions $\phi_{0i}(0)$ and $\phi_{1i}(0)$ such that $\phi_{1i}(0) > \lambda_i \phi_{0i}(0)$ is satisfied, the second inequality in (7.27) can be guaranteed for some $\tau_{MAD} \geq 0$. Then, to guarantee that the first inequality in (7.27) holds, which is equivalent to $\phi_{0i}(\tau_{MIET}) \geq \lambda_i \phi_{1i}(0)$ holds, we note from (7.25) and (7.26) that $\phi_{r_i i}(t)$ is monotone decreasing for $0 \leq t \leq \tau_{MIET}$ and remains unchanged for $t > \tau_{MIET}$, hence, if we can calculate a τ_M^i such that $\phi_{0i}(\tau_M^i) = \lambda_i \phi_{1i}(0)$ and then by selecting $\tau_{MIET} \leq \tau_M^i$, the first inequality in (7.27) can be satisfied. In the following, we briefly show how to calculate τ_M^i based on (7.25).

According to (7.25), we have $dt = d\phi_{0i}/(-2L_{r_i i}\phi_{r_i i}(t) - \gamma_{r_i i}(\phi_{r_i i}^2(t) + 1))$ for $0 \leq t \leq \tau_M^i$, thus, τ_M^i can be calculated by

$$\begin{aligned}\tau_M^i &= \int_{\phi_{0i}(0)}^{\lambda_i \phi_{1i}(0)} \frac{d\phi_{0i}}{(-2L_{r_i i}\phi_{r_i i}(t) - \gamma_{r_i i}(\phi_{r_i i}^2(t) + 1))}\\ &= -\frac{1}{\gamma_{0i}} \int_{\phi_{0i}(0)}^{\lambda_i \phi_{1i}(0)} \frac{d\phi_{0i}}{(\phi_{0i} + L_{0i}/\gamma_{0i})^2 - (L_{0i}/\gamma_{0i})^2 + 1}.\end{aligned}$$

Based on the definition $p_i = \left|(\gamma_{0i}/L_{0i})^2 - 1\right|^{0.5}$, when $\gamma_{0i} > L_{0i}$, we have

$$\tau_M^i = -\frac{1}{Lp_i} \arctan\left(\frac{\gamma_{0i}\phi_{0i} + L_{0i}}{L_{0i} p_i}\right)\Bigg|_{\phi_{0i}=\phi_{0i}(0)}^{\phi_{0i}=\lambda_i \phi_{1i}(0)}.$$

When $\gamma_{0i} = L_{0i}$, we obtain

$$\tau_M^i = \frac{1}{\gamma_{0i}} \frac{1}{\phi_{0i} + L_{0i}/\gamma_{0i}}\Bigg|_{\phi_{0i}=\phi_{0i}(0)}^{\phi_{0i}=\lambda_i \phi_{1i}(0)}.$$

When $\gamma_{0i} < L_{0i}$, we obtain

$$\tau_M^i = \frac{1}{Lp_i} \text{arctanh}\left(\frac{\gamma_{0i}\phi_{0i} + L_{0i}}{L_{0i} p_i}\right)\Bigg|_{\phi_{0i}=\phi_{0i}(0)}^{\phi_{0i}=\lambda_i \phi_{1i}(0)}.$$

Then, using inverse trigonometric function formulas $\arctan a - \arctan b = \arctan((a-b)/(1+ab))$ and $\text{arctanh}a - \text{arctanh}b = \arctan((a-b)/(1-ab))$, the conditions in (7.34) can be obtained, and by selecting $\tau_{MIET} \leq \tau_M^i$, for $i \in [1, N]$, the first inequality in (7.27) holds. □

Remark 7.3 Notice that the condition (7.28) in Theorem 7.1 implies that the system matrix $\bar{A}_{11}$ is required to be Hurwitz. From (7.12), $\bar{A}_{11}$ depends on the controller gain K, the following Lemma 7.1 shows how to construct K such that $\bar{A}_{11}$ is Hurwitz.

Lemma 7.1 *Suppose that the graph $\bar{\mathcal{G}}$ is undirected and connected. Denote $\alpha = Eig_{\min}(\mathcal{H})$ as the minimum eigenvalue of $\mathcal{H}$, since (A, B) is stabilizable, then, there exists a solution $P_1 > 0$ to the following Riccati inequality [11]*

$$P_1 A + A^T P_1 - 2\alpha P_1 B B^T P_1 + \alpha I_n < 0. \tag{7.44}$$

The controller gain K in (7.6) is constructed as $K = B^T P_1$. Then, the closed-loop system matrix $\bar{A}_{11}$ is Hurwitz.

Proof Since the graph $\bar{\mathscr{G}}$ is undirected and connected, it is known that the matrix $\mathscr{H}$ is positive definite [11, Lemma 4]. From $K = B^T P_1$ and (7.44), we have $P_1(A - \alpha BK) + (A - \alpha BK)^T P_1 < -\alpha I_n$, which implies that $A - \alpha BK$ is Hurwitz. Due to $\mathscr{H}$ is positive definite, there exists a nonsingular matrix T such that $\mathscr{H} = T^{-1}\Lambda_H T$, where Λ_H is a diagonal matrix with the diagonal elements being the eigenvalues of $\mathscr{H}$. Then, based on $\bar{A}_{11} = I_N \otimes A - \mathscr{H} \otimes BK$, we have $\bar{A}_{11} = (T^{-1} \otimes I_n)(I_N \otimes A - \Lambda_H \otimes BK)(T \otimes I_n)$. Assume that the eigenvalues of $\mathscr{H}$ are $\lambda_{Hi}, i \in [1, N]$, it follows that $\lambda_{Hi} \geq \alpha$. Then, from (7.44), we have $P_1 A + A^T P_1 - 2\lambda_{Hi} P_1 B B^T P_1 < -\alpha I_n$, which implies that $A - \lambda_{Hi} BK$ is Hurwitz. Therefore, $\bar{A}_{11}$ is also Hurwitz. □

Remark 7.4 From (7.27), it implies that the following inequalities should be satisfied

$$\underbrace{\phi_{1i}(\tau_i) \geq \lambda_i \phi_{0i}(\tau_i)}_{0 \leq \tau_i \leq \tau_{MAD}} \geq \lambda_i \phi_{0i}(\tau_{MIET}) \geq \lambda_i^2 \phi_{1i}(0).$$

The third inequality holds by setting $\tau_{MIET} \leq \tau_M^i$ for $i \in [1, N]$. The second inequality holds with $\tau_{MAD} \leq \tau_{MIET}$. Consider the first inequality, by selecting the initial conditions $\phi_{0i}(0)$ and $\phi_{1i}(0)$ such that $\phi_{1i}(0) > \lambda_i \phi_{0i}(0)$, it is possible that $\phi_{1i}(\tau_i) \geq \lambda_i \phi_{0i}(\tau_i)$ holds for any $0 \leq \tau_i \leq \tau_{MIET}$, in this case, the MAD is equal to the MIET. On the other hand, there may exist a $t_m < \tau_{MIET}$ such that $\phi_{1i}(\tau_i) < \lambda_i \phi_{0i}(\tau_i)$ when $\tau_i > t_m$, in this case, the MAD is equal to t_m. Therefore, there exists tradeoff between τ_{MIET} and τ_{MAD} for various parameters $L_{r_i i}, \lambda_i, \phi_{0i}(0)$ and $\phi_{1i}(0)$. For given parameters, the specific value of the maximum allowable MIET can be calculated based on (7.33), and then, the specific value of the MAD can also be calculated based on (7.27).

Remark 7.5 The condition (7.33) for the lower bound of the MIET is actually a little conservative, since a common τ_{MIET} is adopted for all agents, so, if the initial conditions and parameters of each agent are different, τ_{MIET} is required to be less than the minimum $\tau_M^i, i \in [1, N]$. However, this condition can be relaxed by selecting independent τ_{MIET}^i for each agent i instead of a common τ_{MIET}, and meanwhile, by defining the function $\theta_i(t) = 1$ for $0 \leq t \leq \tau_{MIET}^i$ and $\theta_i(t) = 0$ for $t > \tau_{MIET}^i$. In this way, each τ_{MIET}^i only depends on the local upper bound τ_M^i and the conservativeness is relaxed. Note that using independent lower bound of the MIET for each agent instead of a common one does not change the proof framework.

Remark 7.6 Let N_i and n_i denote the number of agent i's neighbors and triggering events respectively, and suppose that the communication is performed via broadcasting. If the control protocol is updated only at agent i's own triggering instant, but neighbors' information is needed at this instant, in this case, the number of communication is $N_i + 1$ because agent i needs to inform its neighbors to transmit their information, thus the total number of communication is $\sum_{i=1}^{N} n_i(N_i + 1)$ and the total number of controller update is $\sum_{i=1}^{N} n_i$ for the MAS. If the control protocol is also updated at neighbors' triggering instants, but the agent does not need its neighbors' information at triggering instants, in this case, the total number of communication is $\sum_{i=1}^{N} n_i$ and the total number of controller update is $\sum_{i=1}^{N} n_i(N_i + 1)$ for the MAS. In contrast with the latter control protocol, the former control protocol leads to less number of controller update but more number of communication, namely each control protocol has its own advantages and disadvantages. However, how to qualitatively analyze that which control protocol leads to less usage of resources is not a trivial problem and needs further investigation.

7.5 Simulation Study

Consider a spacecraft formation problem [12]. Each spacecraft can be viewed as an agent, suppose that there are one leader agent and five follower agents, and the communication topology is shown in Fig. 7.1. The linearized equations of the dynamics of the ith spacecraft are described as follows

$$\begin{aligned} \ddot{\bar{x}}_i - 2\omega_0 \dot{\bar{y}}_i &= u_{xi} \\ \ddot{\bar{y}}_i + 2\omega_0 \dot{\bar{x}}_i - 3\omega_0^2 y_i &= u_{yi} \\ \ddot{\bar{z}}_i + \omega_0^2 \bar{z}_i &= u_{zi} \end{aligned} \tag{7.45}$$

where $(\bar{x}_i, \bar{y}_i, \bar{z}_i)$ denote the position of the ith spacecraft in the rotating coordinate, (u_{xi}, u_{yi}, u_{zi}) are the control inputs, and ω_0 is the spacecraft's angular rate. Define the position vector $r_i = \left[\bar{x}_i\ \bar{y}_i\ \bar{z}_i\right]^T$, velocity vector $\dot{r}_i = \left[v_{xi}\ v_{yi}\ v_{zi}\right]^T$, control vector $u_i = \left[u_{xi}\ u_{yi}\ u_{zi}\right]^T$. The spacecraft formation can be achieved if their velocity vectors converge to the same value and their positions keep a prescribed distance from each other, i.e., $r_i - h_i \to r_j - h_j, \dot{r}_i - \dot{r}_j$ for $i, j \in [1, N]$, where $h_i - h_j \in \mathbb{R}^3$ denotes the prescribed distance between spacecraft i and j. Let $\bar{r}_i = r_i - h_i$ and define $x_i = \left[\bar{r}_i^T\ \dot{r}_i^T\right]^T$, the MAS dynamics can be described by

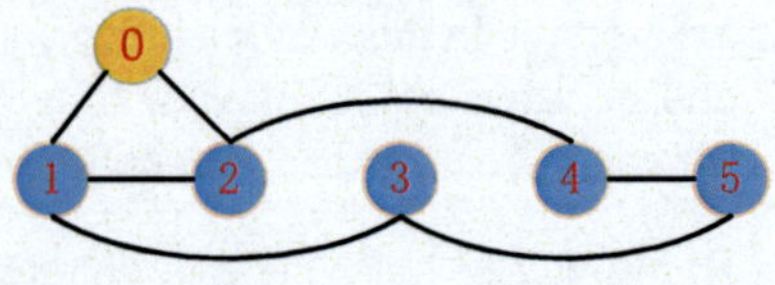

Fig. 7.1 Communication topology

$$\text{Leader: } \dot{x}_0 = Ax_0,$$
$$\text{Follower: } \dot{x}_i = Ax_i + Bu_i, \ i \in [1, 5]$$

with

$$A = \begin{bmatrix} 0 & 0 & 0 & 1 & 0 & 0 \\ 0 & 0 & 0 & 0 & 1 & 0 \\ 0 & 0 & 0 & 0 & 0 & 1 \\ 0 & 0 & 0 & 0 & 2\omega_0 & 0 \\ 0 & 3\omega_0^2 & 0 & -2\omega_0 & 0 & 0 \\ 0 & 0 & -\omega_0^2 & 0 & 0 & 0 \end{bmatrix}, B = \begin{bmatrix} 0 & 0 & 0 \\ 0 & 0 & 0 \\ 0 & 0 & 0 \\ 1 & 0 & 0 \\ 0 & 1 & 0 \\ 0 & 0 & 1 \end{bmatrix}$$

where $\omega_0 = 0.001$ is the angular rate of the spacecraft. Therefore, the spacecraft formation problem is converted to consensus problem of MASs with agent 0 as the leader and the other agents as followers. The objective is to achieve leader-following consensus such that $\delta_i = x_i - x_0 \rightarrow 0$.

Based on the communication topology, the augmented Laplacian matrix $\mathscr{H}$ can be easily obtained. The minimum eigenvalue of $\mathscr{H}$ is 0.2679, and the gain matrix K can be solved by the Riccati equation (7.44) as

$$K = \begin{bmatrix} 0.7071 & -0.0015 & 0 & 1.7717 & 0 & 0 \\ 0.0015 & 0.7071 & 0 & 0 & 1.7717 & 0 \\ 0 & 0 & 0.7071 & 0 & 0 & 1.7717 \end{bmatrix}$$

According to Theorem 7.1, we select parameters $L_{r_i i} = 0.2, \gamma_{r_i i} = 10, \phi_{0i} = 3, \phi_{1i} = 2, \lambda_i = 0.2$ for $i \in [1, N]$, then, the maximum allowable MIET can be calculated from (7.33) as MIET $= 0.086$s. To verify whether the conditions in (7.28)–(7.30) are satisfied, we select parameters $\rho = 0.1, \rho_1 = 0.3, \rho_2 = 0.1, \varepsilon = 10^{-3}, a = 1$, then, there exist matrices P, Q such that inequalities (7.28), (7.29) hold by using LMI ToolBox.

In the simulation, select the minimum inter-event time as $\tau_{MIET} = 0.04$s and maximum allowable delay as $\tau_{MAD} = 0.03$s, the communication delays are time-varying and generated randomly within the pre-specified bound. The initial conditions are $x_0(0) = \begin{bmatrix} 1 & 0 & -1 & 2 & 0 & -2 \end{bmatrix}^T$ and $x_i(0) = \mathrm{i} \times \begin{bmatrix} 1 & 2 & 3 & 4 & 5 & 6 \end{bmatrix}^T$ for $i \in [1, 5]$. The simulation results are shown in Fig. 7.2, where δ_{ij} denotes the j-th sub-state of $\delta_i(t) = x_i(t) - x_0(t)$. It is shown that the leader-following consensus is achieved with the proposed ETC design under hybrid systems framework. Actually, due to asymptotic consensus can be achieved with the proposed consensus protocol, in the simulation, we can suppose that the system achieve consensus if all the state trajectories of $\delta(t)$ converge to a neighborhood of the origin with an upper bound 10^{-3}, under this index, when the simulation is performed after 42 s, this index is satisfied and the simulation is stopped. Figure 7.3 shows the triggering instants and intervals.

Note from (7.31) that the parameters ρ_1, ρ_2 are relevant to the event-triggering condition, after the other parameters are determined, we can show the role of ρ_1, ρ_2 by selecting different values, especially ρ_1 is important because if some agent i

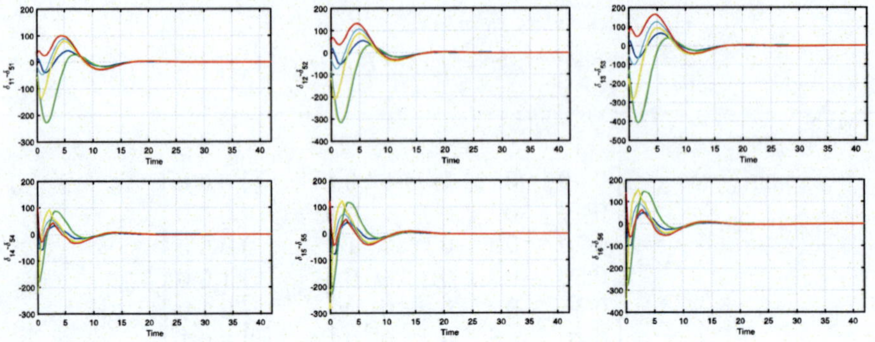

Fig. 7.2 The trajectories of state $\delta(t)$

Fig. 7.3 Triggering instants and intervals ($\rho_1 = 0.3$)

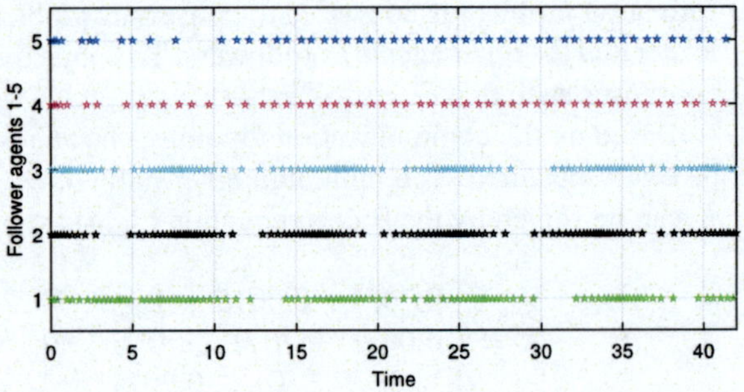

Table 7.1 The number of triggering events (NTE) and the average triggering interval (ATI) of the follower agents

	Agent1	Agent2	Agent3	Agent4	Agent5
NTE($\rho_1 = 0.1$)	156	168	172	141	143
ATI($\rho_1 = 0.1$)	0.2692	0.2500	0.2442	0.2979	0.2937
NTE($\rho_1 = 0.3$)	112	117	123	104	106
ATI($\rho_1 = 0.3$)	0.3750	0.3589	0.3415	0.4038	0.3962
NTE($\rho_1 = 0.5$)	86	91	98	83	78
ATI($\rho_1 = 0.5$)	0.4884	0.4615	0.4286	0.5060	0.5385

does not connect to the leader directly, $b_i = 0$ implies that ρ_2 has no effect on the triggering event. Therefore, we select $\rho_2 = 0.1$ and several different ρ_1 (i.e., $\rho_1 = 0.1$, $\rho_1 = 0.3$ and $\rho_1 = 0.5$) to investigate the role of ρ_1, the triggering instants and average triggering intervals are given in Table 7.1, it is shown that smaller ρ_1 leads to more number of triggering events and vice versa. However, ρ_1 is upper bounded from the condition (7.30) (i.e., the number of triggering events is lower bounded), where $\min_{i\in[1,N]}\{\alpha_i\}/\rho_1 - (1/a+1)\left|H_2^T H_2\right| > 0$ should be ensured, in this example, the upper bound of ρ_1 is 0.52.

7.6 Conclusion

In this chapter, the leader-following consensus problem is addressed for MAS with communication delays, and ETC scheme without requiring continuous communication is designed to reduce the usage of communication resources. By means of the hybrid system approach, the error states, which would experience jumps when transmission event occurs or update event occurs, have been taken into account as closed-loop system state to fully describe the closed-loop MAS. Then, under hybrid systems framework, Lyapunov based consensus analysis is proceeded in the sense that sufficient conditions for leader-following consensus are proposed, and the hybrid event-triggering condition is also developed.

Appendix

Proof of Proposition 7.1 From (7.21), the condition (7.23a) is equivalent to

$$\max\left\{\frac{\lambda_i}{\lambda_{Wi}} W_i\left(\kappa_i+1, \begin{bmatrix}0_n\\ e_{iJ}\end{bmatrix}\right) W_i\left(\kappa_i+1, \begin{bmatrix}0_n\\ e_{iJ}\end{bmatrix} - \begin{bmatrix}0_n\\ G_i e_{iJ}\end{bmatrix}\right)\right\} \leq \lambda_i W_i(\kappa_i, e_i). \quad (7.46)$$

As mentioned before, we can take $W_i(\kappa_i, e_i) = |e_i|$, and $0 < \lambda_i < 1, \lambda_{Wi} \geq 1$, it is obviously that

$$\frac{\lambda_i}{\lambda_{Wi}} W_i\left(\kappa_i+1, \begin{bmatrix}0_n\\ e_{iJ}\end{bmatrix}\right) \leq \lambda_i W_i(\kappa_i, e_i).$$

Note that $G_i = G(\tau_k^i) = \exp(A\tau_k^i)$, we have

$$\begin{aligned}\left\|\begin{bmatrix}0_n\\ e_{iJ}\end{bmatrix} - \begin{bmatrix}0_n\\ G_i e_{iJ}\end{bmatrix}\right\| &= \left\|\begin{bmatrix}0_n\\ (I_n - \exp(A\tau_k^i))e_{iJ}\end{bmatrix}\right\| \\ &\leq \left|I_n - \exp(A\tau_k^i)\right| |e_i| .\end{aligned}$$

To prove the condition $W_i\left(\kappa_i+1, \left[\begin{smallmatrix}0_n\\ e_{iJ}\end{smallmatrix}\right] - \left[\begin{smallmatrix}0_n\\ G_i e_{iJ}\end{smallmatrix}\right]\right) \leq \lambda_i W_i(\kappa_i, e_i)$ is equivalent to show that

$$\begin{aligned}&\left|I_n - \exp(A\tau_k^i)\right| \leq \lambda_i \\ \Leftrightarrow &-\lambda_i I_n \leq I_n - \exp(A\tau_k^i) \leq \lambda_i I_n \\ \Leftrightarrow &(1-\lambda_i)I_n \leq \exp(A\tau_k^i) \leq (1+\lambda_i)I_n .\end{aligned}$$

Due to $-|A|\, I_n \leq A \leq |A|\, I_n$, it follows that

$$\exp(A\tau_k^i) \le \exp(|A|\,\tau_{MAD} I_n)$$
$$\overset{(7.22)}{\le} \exp\left(|A|\,\frac{\ln(1+\lambda_i)}{|A|} I_n\right) \le (1+\lambda_i) I_n.$$

Similarly, it can be shown that $\exp(A\tau_k^i) \ge (1-\lambda_i)I_n$. Hence, the condition (7.23a) holds. Moreover, condition (7.23b) is equivalent to

$$W_i\left(\kappa_i, e_i + \begin{bmatrix} 0_n \\ G_i s_i \end{bmatrix}\right) \max\left\{\frac{\lambda_i}{\lambda_{Wi}} W_i(\kappa_i, e_i), W_i\left(\kappa_i, e_i + \begin{bmatrix} 0_n \\ G_i s_i \end{bmatrix}\right)\right\},$$

which holds trivially.

Proof of Proposition 7.2 Note from (7.20) that $e_i = \bar{R}_i e$, thus, we have

$$\begin{aligned} \dot{e}_i &= \bar{R}_i \bar{A}_{21}\delta + \bar{R}_i \bar{A}_{22} e \\ &= \bar{R}_i \bar{A}_{21}\delta + \bar{R}_i \bar{A}_{22}\left(I_{2nN} - \begin{bmatrix} \Lambda_i & 0 \\ 0 & \Lambda_i \end{bmatrix}\right) e + \bar{R}_i \bar{A}_{22} \begin{bmatrix} \Lambda_i & 0 \\ 0 & \Lambda_i \end{bmatrix} e \\ &\le \left|\bar{R}_i \bar{A}_{21}\delta + \bar{A}_{Ri} e\right| + M_{ei}\,|e_i|. \end{aligned} \tag{7.47}$$

Then, according to the form of $\bar{W}_i$, three cases are obtained.

Case 1: $r_i = 0$, $\bar{W}_i = W_i(\kappa_i, e_i)$.

Based on (7.18a) and (7.19), we have

$$\begin{aligned} \left\langle \frac{\partial \bar{W}_i}{\partial e_i}, \dot{e}_i \right\rangle &\le \left|\frac{\partial \bar{W}_i}{\partial e_i}\right| \dot{e} \\ &\le c_i \left|\bar{R}_i \bar{A}_{21}\delta + \bar{A}_{Ri} e\right| + \frac{c_i M_{ei}}{\underline{\alpha}_{Wi}} \bar{W}_i \\ &= c_i \left|\bar{R}_i \bar{A}_{21}\delta + \bar{A}_{Ri} e\right| + L_{0i} \bar{W}_i. \end{aligned}$$

Case 2: $r_i = 1$, $\frac{\lambda_i}{\lambda_{Wi}} W_i(\kappa_i, e_i) \ge W_i\left(\kappa_i, e_i + \begin{bmatrix} 0_{n\times n} \\ G_i s_i \end{bmatrix}\right)$.

Similarly as in Case 1, it follows that

$$\begin{aligned} \left\langle \frac{\partial \bar{W}_i}{\partial e_i}, \dot{e}_i \right\rangle &\le \left|\frac{\lambda_i}{\lambda_{Wi}} \frac{\partial \bar{W}_i}{\partial e_i}\right| \dot{e} \\ &\le \frac{c_i \lambda_i}{\lambda_{Wi}} \left|\bar{R}_i \bar{A}_{21}\delta + \bar{A}_{Ri} e\right| + \frac{c_i M_{ei}}{\underline{\alpha}_{Wi}} \frac{\lambda_i}{\lambda_{Wi}} \bar{W}_i \\ &\le c_i \left|\bar{R}_i \bar{A}_{21}\delta + \bar{A}_{Ri} e\right| + L_{1i} \bar{W}_i. \end{aligned}$$

Case 3: $r_i = 1$, $\frac{\lambda_i}{\lambda_{Wi}} W_i(\kappa_i, e_i) \le W_i\left(\kappa_i, e_i + \begin{bmatrix} 0_{n\times n} \\ G_i s_i \end{bmatrix}\right)$.

Similarly as in Case 2, we have

$$
\begin{aligned}
\left\langle \frac{\partial \bar{W}_i}{\partial e_i}, \dot{e}_i \right\rangle &\leq \left| \frac{\partial \bar{W}_i}{\partial e_i} \right| \dot{e} \\
&\leq c_i \left| \bar{R}_i \bar{A}_{21} \delta + \bar{A}_{Ri} e \right| + \frac{c_i M_{ei}}{\underline{\alpha}_{Wi}} \frac{\lambda_{Wi}}{\lambda_i} \frac{\lambda_i}{\lambda_{Wi}} W_i(\kappa_i, e_i) \\
&\leq c_i \left| \bar{R}_i \bar{A}_{21} \delta + \bar{A}_{Ri} e \right| + L_{1i} W_i \left(\kappa_i, e_i + \begin{bmatrix} 0_{n \times n} \\ G_i s_i \end{bmatrix} \right) \\
&= c_i \left| \bar{R}_i \bar{A}_{21} \delta + \bar{A}_{Ri} e \right| + L_{1i} \bar{W}_i .
\end{aligned}
$$

Hence, the inequality (7.24) holds.

References

1. Nowzari, C., Garcia, E., Corts, J.: Event-triggered communication and control of networked systems for multi-agent consensus, Automatica, **105**, 1-27, (2019)
2. Wang, A., Mu, B., Shi, Y.: Event-triggered consensus control for multiagent systems with time-varying communication and event-detecting delays," IEEE Trans. Control Syst. Technol., **27**(2), 507-515, (2019)
3. Xiao, F., Chen, T., Gao, H.: Synchronous hybrid event- and time-driven consensus in multiagent networks with time delays, IEEE Trans. Cybern., **46**(5), 1165-1174, (2016)
4. Zhu, W. Jiang, Z.: Event-based leader-following consensus of multi-agent systems with input time delay, IEEE Trans. Autom. Control, **60**(5), 1362-1367, (2015)
5. E. Garcia, Y. Cao, and D. W. Casbeer, Periodic event-triggered synchronization of linear multi-agent systems with communication delays, IEEE Trans. Autom. Control, **62**(1), 366-371, (2017)
6. Goebel, R., Teel, A. R.: Solutions to hybrid inclusions via set and graphical convergence with stability theory applications, Automatica, **42**(4), 573-587, (2006)
7. Goebel, R., Sanfelice, R. G., Teel, A. R.: Hybrid Dynamical Systems: Modeling, Stability, and Robustness. Princeton, NJ, USA: Princeton Univ. Press, (2012).
8. Dolk, V. S., Borgers, D. P., Heemels, W. P. M. H.: Output-based and decentralized dynamic event-triggered control with guaranteed $\mathscr{L}_p$-gain performance and Zeno-freeness, IEEE Trans. Autom. Control, **62**(1), 34-49, (2017)
9. Heemels, W. P. M. H., Teel, A. R., Wouw, N. van de, Nesic, D.: Networked control systems with communication constraints: tradeoffs between transmission intervals, felays and performance, IEEE Trans. Autom. Control, **55**(8), 1781-1796, (2010)
10. Carnevale, D., Teel, A. R., Nesic, D.: A Lyapunov proof of an improved maximum allowable transfer interval for networked control systems, IEEE Trans. Autom. Control, **52**(5), 892-897, (2007)
11. Ni, W. Cheng, D.: Leader-following consensus of multi-agent systems under fixed and switching topologies, Systems & Control Letters, **59**(3), 209-217, (2010)
12. Li, Z., Duan, Z., Chen, G., Huang, L.: Consensus of multiagent systems and synchronization of complex networks: a unified viewpoint, IEEE Trans. Circuits Syst. I, Reg. Papers, **57**(1), 213-224, (2010)

Chapter 8
Hybrid Event-Triggered Consensus of MAS with Packet Losses

8.1 Introduction

In Chaps. 6 and 7, event-triggered cooperative control problems of MAS are taken into consideration. The multiple agents with antagonistic relationship and communication network with time-delay are considered in Chaps. 6 and 7 respectively. Both these two chapters employ appropriate hybrid ETM to enhance the role of event-triggered control, that is, saving communication resources while ensuring exclusion of Zeno behavior. It is worth to be noted that though the communication network considered in these two chapters is not ideal (e.g., limited bandwidth or communication delay), the information transmission is reliable. That is, as long as the information is transmitted from a certain agent to its neighbors, it must be able to arrive the target agent, even if there exists communication delay. In the existing literature, due to the significance of cooperative control of MAS, [1–4], many research results regarding cooperative control of MAS with reliable communication have been obtained, such as [5–11]. However, in practical engineering applications, the information exchange between agents is usually performed via wireless network. It is well known that wireless network communication is more vulnerable to environment factors, the transmitted data sometimes loses and cannot arrive its target agent, in this case, we call the network communication as unreliable, i.e., the communication link between agents is unreliable. Since the communication link among agents is usually wireless, the network induced limited communication resources and packet losses issues should be considered, these issues are not fully exploited in the existing works. This chapter will consider the event-triggered cooperative control problem of MAS with packet losses.

To deal with the issue with regard to limited communication resources in the context of wireless network, distributed consensus control protocol should be designed to not only ensure desired closed-loop stability but also be able to reduce the usage of communication resources. Time-triggered control is one suitable method under wireless network communication and has been widely studied [10, 12–14], but it would result in excessive communications when the system tends to be stable. To

G. Zhao et al., *Cooperative Control of Multi-agent Systems*,
https://doi.org/10.1007/978-981-97-0968-7_8

overcome this limitation, event-triggered control was proposed in [15] and has been extensively used to consensus of MAS for the purpose of saving communication resources, such as some recent results in [11, 16] with respect to dynamic ETM, in [17, 18] with respect to communication delays, and in [19, 20] with respect to bipartite consensus. However, the communication links in these results are assumed to be reliable, i.e., no packet losses.

In the presence of packet losses, some results with respect to consensus of MAS are developed recently. In [21], the consensus problem of MAS was studied with Markovian packet loss communication channels. Bernoulli-distributed random packet losses and deterministic packet losses were considered in [22], and a switched system approach was proposed to analyze stability and consensus performance. Consensus tracking problem of MAS with Lipschitz nonlinearity and deterministic packet losses was studied in [23], where the sampled information is exchanged between agents. Based on parametric Riccati equation, a lower gain consensus protocol was proposed in [24] for MAS with packet losses and sampling transmission mechanism. In [25], the event-triggered consensus control strategy was developed for MAS with random packet losses that subject to Bernoulli distribution, and the maximum allowable sampling interval was calculated explicitly. An event-triggered active model predictive control approach was proposed in [9] for MAS with packet losses. In these existing results, there still exist some limitations that make the proposed consensus control strategies be not well suitable for network communication with limited resources and packet losses. For instance, sampled-data control was used in [22–24], which is easy to be implemented under network environment but often results in excessive communication resource utilization. The Markovian packet losses communication channels were assumed in [21], which may be not the case in practical applications. The packet losses in [9, 22, 25] are required to obey Bernoulli distribution and be *identical*.[1] It is well known that in practical MAS with wireless network, packet losses usually inevitably occur due to limited bandwidth, noise interference, cyber attack, and packet disorder. Thus, whether the packet transmitted in the communication channel from a certain agent to its neighbors loses is not always the same as other communication channels, i.e., *nonidentical packet losses*,[2] that is more meaningful and can cover identical packet losses as a special case. Due to the difficulty in the aspect of modelling and event-triggered consensus analysis, the event-triggered consensus problem of MAS with nonidentical random packet losses was rarely taken into account in the existing literature.

In addition, based on if there is a leader agent in the MAS, the consensus problems of MAS include types: leaderless consensus and leader-following consensus (consensus tracking). Most of the existing works regarding consensus of MAS only focus on leaderless consensus problem [21, 22, 25] or consensus tracking problem [23, 26, 27]. Notice that although the control objectives of leaderless and leader-following

[1] Identical packet losses means that the packet losses in all communication channels among agents occur at the same time.

[2] Nonidentical packet losses means that the packet losses in each communication channel from a certain agent to its neighbors can be different from other communication channels.

consensus problems are not exactly the same, in most cases, the control objective is achieved by driving properly defined error variables to the origin. Therefore, it is expected that by defining appropriate consensus error variables, a unified system model can be constructed for both leaderless and leader-following MAS, and the event-triggered leaderless and leader-following consensus problems can be tackled in a unified framework. Thus, as another motivation, we aim to propose a hybrid systems approach to tackle the event-triggered leaderless and leader-following consensus problems simultaneously.

In this chapter, based on the above discussions, we investigate event-triggered consensus problem of MAS with nonidentical packet losses. First, a novel hybrid event-triggering mechanism (ETM) with internal jump state is proposed, the state jumps to different values according to whether packet losses occur when sending information. A timer variable is introduced to enforce a lower bound on inter-event time such that the minimum inter-event time (MIET) is guaranteed to be strictly positive, and the explicit MIET can be calculated by means of an internal function. Second, the packet losses in communication channels are nonidentical and random, and do not need to obey Bernoulli distribution, it is only required that the number of maximum allowable successive packet losses (NMASPL) is upper bounded by a positive number. The nonidentical packet losses can cover the identical packet losses [22, 23, 23, 24] as a special case, and asymptotic stability rather than mean square stability [22, 25] is obtained. Third, by means of auxiliary variables, a unified hybrid model is constructed for leaderless and leader-following MAS with packet losses and hybrid ETM, and a hybrid systems framework is presented for hybrid ETM design and consensus analysis, and the relationship between MIET, NMASPL, and design parameters is given.

8.2 Networked MAS and Problem Formulation

This section formulates the leaderless and leader-following consensus problems for event-triggered MASs with nonidentical packet losses. The nonidentical assumption on packet losses is a less conservativeness condition, which actually allows different communication channels in a MAS network to have non-synchronized dropouts. And the event-triggered control is helpful for saving the limited communication resources.

Consider leaderless MAS with general linear dynamics, the agent's model is given by

$$\dot{x}_i(t) = Ax_i(t) + Bu_i(t), \tag{8.1}$$

where $x_i(t) \in \mathbb{R}^n$, $u_i(t) \in \mathbb{R}^p$ denotes the control input. A, B are matrices with appropriate dimensions and it is assumed that (A, B) is stabilizable. The control objective of leaderless consensus of MAS is to let all agents' states reach the same value in the sense that $\lim_{t\to\infty} \|x_i(t) - x_j(t)\| = 0, i, j \in [1, N]$. Toward this objective, an event-triggered control law is designed as follows

$$u_i(t) = K \sum_{j \in \mathcal{N}_i} (\hat{x}_{ji}(t) - x_i(t)). \tag{8.2}$$

If there is a leader in the MAS and the control objective is to let all follower agents' states track the leader agent's state in the sense that $\lim_{t\to\infty} \|x_i(t) - x_0(t)\| = 0$, $i \in [1, N]$, then a leader-following MAS is constituted. The leader is labeled as 0, and its dynamics is given by

$$\dot{x}_0(t) = Ax_0(t), \tag{8.3}$$

where $x_0(t) \in \mathbb{R}^n$ denotes the state. For the leader-following MAS, the event-triggered control law is designed as follows

$$u_i(t) = K \left(\sum_{j \in \mathcal{N}_i} (\hat{x}_{ji}(t) - x_i(t)) + b_i(x_0(t) - x_i(t)) \right) \tag{8.4}$$

where the gain matrix K is to be determined, and $\hat{x}_{ji}(t)$ is the estimation of agent j's state as seen by agent i. Its dynamics is as follows

$$\begin{aligned} \dot{\hat{x}}_{ji}(t) &= A\hat{x}_{ji}(t), t \in (t_k^j, t_{k+1}^j], \\ \hat{x}_{ji}(t^+) &= \begin{cases} x_j(t), t = t_k^j \ \wedge \ m_j^{loss} = 0 \\ \hat{x}_{ji}(t), t = t_k^j \ \wedge \ m_j^{loss} = 1 \end{cases} \end{aligned} \tag{8.5}$$

where t_k^j denotes the kth event-triggering instant of agent j, $\wedge$ denotes logical operator 'and'. The value of m_j^{loss} depends on whether there exist packet losses in the communication channel from agent j to its neighbors, in particular, $m_j^{loss} = 0$ means that there is no packet loss and the information transmission from agent j to its neighbors is successful, so, the estimation $\hat{x}_{ji}(t)$ is updated to be the newly received $x_j(t)$, and $m_j^{loss} = 1$ indicates that information transmission fails, so, $\hat{x}_{ji}(t)$ is not updated and remains unchanged. Under the wireless network environment, the information exchange among agents is usually performed in the manner of broadcasting. Therefore, in this chapter, we assume that the packet losses in communication channels from agent i to its neighbors are the same, which is reasonable in practice. Moreover, based on (8.5), it implies that no matter whether the state $x_j(t)$ is successfully transmitted from agent j to its neighbors or not, all the estimated states $\hat{x}_{ji}$, for $i \in \mathcal{N}_j$, are the same.

In addition, when $i = j$, we define a state $\hat{x}_{ii}(t)$ in agent i, which is useful to construct event-triggering conditions, the dynamics of $\hat{x}_{ii}(t)$ is given by

$$\begin{aligned} \dot{\hat{x}}_{ii}(t) &= A\hat{x}_{ii}(t), t \in (t_k^i, t_{k+1}^i], \\ \hat{x}_{ii}(t^+) &= x_i(t), \quad t = t_k^i, \end{aligned} \tag{8.6}$$

it is obvious that $\hat{x}_{ii}(t)$ is updated to be $x_i(t)$ at each triggering instant t_k^i, this is because $x_i(t)$ is not transmitted over network and it is always available for agent i itself.

The preliminaries about graph, which is used to describe the communication relationship between agents, have been introduced in Chap. 1. For proceeding this chapter, we make Assumption 8.1 to limit the communication connection among agents, and Assumption 8.2 to limit the number of maximum allowable successive packet losses (NMASPL).

Assumption 8.1 In the leaderless case, the communication graph $\mathcal{G}$ is connected and undirected. In leader-following case, at least one follower can receive information from the leader, i.e., there exists a spanning tree in the graph $\mathcal{G}$ with the leader agent 0 as the root.

Assumption 8.2 There exists a positive integer number $\rho_{\max}^i$ such that the number of successive packet losses in the communication from agent i to its neighbors is not larger than $\varrho_{\max}^i$.

The following lemma is useful for deriving the Lyapunov-based stability results.

Lemma 8.1 ([28]) *For connected undirected graph $\mathcal{G}$, if a vector $x \in \mathbb{R}^N$ satisfies $\mathbf{1}_N^T x = 0$, we have*

$$\min_{x \neq 0} \frac{x^T \hat{\mathscr{L}} x}{x^T x} = a(\hat{\mathscr{L}}),$$

where $\hat{\mathscr{L}} = \mathscr{L}$ in leaderless case and $\hat{\mathscr{L}} = \mathscr{H}$ in leader-following case, $a(\hat{\mathscr{L}})$ denotes the minimum positive eigenvalue of $\hat{\mathscr{L}}$.

As discussed in introduction, for sampled-data control (i.e., time-triggered control) of MAS, the information transmission among agents is performed at sampling instants, and it is required that the sampling periods (independent of system states) are upper bounded by $\bar{h}$, which is the maximum allowable sampling period. Due to the sampling periods are selected based on the worst-case situation of system [29, 30] rather than based on system states, it usually leads to excessive information transmission over network and further results in unnecessary waste of communication resources. It is well known that, in the context of wireless network, the communication resources are usually limited, thus, we consider event-triggered consensus control here. Based on event-triggered control, since the information transmission instants are determined based on system states, it is desired that larger average transmission intervals compared with time-triggered control can be achieved, while stability properties can still be ensured.

The event-triggered consensus control of MAS with packet losses brings new challenges, since the transmitted information at triggering instants may be lost. Thus, the effect of packet losses needs to be taken into account when constructing ETM. In addition, one important issue for constructing ETM is to exclude Zeno triggering behavior, which may lead to unpredicted system dynamics and is not expected in practice. To this end, time regularization technique is used in the constructed ETM, so that a strictly positive lower bound of the triggering intervals, which is the so-called minimum inter-event time (MIET), is always guaranteed in the sense that any two triggering instants (for a certain agent) is not less than the MIET. Specifically, we construct the ETM in the following form to determine the next triggering instant t_{k+1}^i for agent i,

$$t_{k+1}^i = \inf_{t>0}\{t \geq t_k^i + \tau_{miet}^i | \eta_i(t) \leq 0\}, \tag{8.7}$$

where $\tau_{miet}^i > 0$ denotes the MIET for agent i, which is strictly positive and can be pre-specified. $\eta_i(t)$ is an internal variable and $\eta_i(t) \leq 0$ is the triggering condition, which means that when $\eta_i(t) \leq 0$ is satisfied, an triggering event occurs and agent i broadcasts information to its neighbors. $\eta_i(t)$ evolves according to the following dynamics

$$\dot{\eta}_i(t) = \Psi_i^F(o_i(t), \eta_i(t)), t \in (t_k^i, t_{k+1}^i], \tag{8.8}$$
$$\eta_i(t^+) = \begin{cases} \Psi_i^{J1}(o_i(t)), t = t_k^i \wedge m_i^{ack} = 1 \\ \Psi_i^{J2}(o_i(t)), t = t_k^i \wedge m_i^{ack} = 0 \end{cases}$$

where $o_i(t)$ denotes locally available variables for agent i, e.g., $e_{ai}(t)$. $\Psi_i^F(o_i(t), \eta_i(t))$, $\Psi_i^{J1}(o_i(t))$ and $\Psi_i^{J2}(o_i(t))$ are flow map and jump map functions depending on $o_i(t)$ or $\eta_i(t)$, that are to be designed and determine how $\eta_i(t)$ evolves when triggering event occurs or not. The binary variable m_i^{ack} is a flag to denote whether the acknowledgement information is received by agent i or not. Specifically, $m_i^{ack} = 1$ means that after agent i transmit information to its neighbors, all neighbors successfully receive the information and send an acknowledgement frame to agent i, and at least one acknowledgement frame is received by agent i. $m_i^{ack} = 0$ corresponds to two cases: the first one is that the transmitted information at triggering instant t_k^i is lost, so, no neighbors send acknowledgement frame back to agent i; another case is that the transmitted information is successfully received by agent i's neighbors, but due to packet losses, the acknowledgement frames sent back by agent j ($j \in \mathcal{N}_i$) are lost, so, agent i does receive acknowledgement frame and thinks that packet loss occur.

Remark 8.1 For agent $j \in \mathcal{N}_i$, it is required that it sends an acknowledgement frame back to agent i when it receives information. Through the acknowledgement mechanism, agent i is able to know whether packet losses occur, and then to determine how $\eta_i(t)$ in the ETM changes at triggering instants. It is not conservative to assume that acknowledgement signals are available, because the well known transmission control protocol (TCP) also utilizes such mechanism, and this assumptions was also adopted in some recent works [2, 16, 31].

Notice that time regularization technique is utilized in the constructed ETM (8.7), the next triggering instant only occurs after at least τ_{miet}^i time units have elapsed, thus, Zeno behavior is naturally excluded. But we need to know that the variable τ_{miet}^i cannot be selected arbitrarily, it is firstly required to be upper bounded, and how to design τ_{miet}^i is also related to the number of successive packet losses, that will be given in the main results. Moreover, the proposed ETM (8.7) is dynamic rather than static, it has potential to achieve larger inter-event times in contrast with its counterpart of static ETM. In what follows, the main concern is to specify conditions for $\Psi_i^F(o_i(t), \eta_i(t))$, $\Psi_i^{J1}(o_i(t))$, $\Psi_i^{J2}(o_i(t))$, τ_{miet}^i and design the gain matrix K.

8.3 Hybrid Model of the MAS with Packet Losses

In this section, we aim to construct a unified hybrid model for leaderless and leader-following MAS. First, by defining appropriate consensus error variables and state estimation error variables, the closed-loop mathematical models including jump dynamics are derived for leaderless and leader-following MASs respectively. Then, to facilitate stability analysis, by means of several internal variables, a unified hybrid model in the hybrid systems framework as developed in [32] is constructed, such that both leaderless and leader-following consensus problems of MAS can be tackled using a hybrid systems approach.

8.3.1 *Leaderless MAS Model*

Define the consensus error variables $\delta_i(t)$ and state estimation error variables $e_{ii}(t)$, $e_{ij}(t)$, as follows

$$\begin{aligned}
\delta_i(t) &= \sum_{j\in\mathcal{N}_i}(x_j(t) - x_i(t)), \\
e_{ii}(t) &= \hat{x}_{ii}(t) - x_i(t), \\
e_{ij}(t) &= \hat{x}_{ij}(t) - x_i(t),
\end{aligned} \tag{8.9}$$

and define

$$\begin{aligned}
\delta(t) &= \mathrm{col}(\delta_1(t), \delta_2(t), \ldots, \delta_N(t)), \\
e_a(t) &= \mathrm{col}(e_{11}(t), e_{22}(t), \ldots, e_{NN}(t)), \\
&= \mathrm{col}(e_{a1}(t), e_{a2}(t), \ldots, e_{aN}(t)), \\
e_b(t) &= \mathrm{col}(e_{1j}(t), e_{2j}(t), \ldots, e_{Nj}(t)), \\
&= \mathrm{col}(e_{b1}(t), e_{b2}(t), \ldots, e_{bN}(t)),
\end{aligned} \tag{8.10}$$

where it is shown that $e_{ii}(t) = e_{ai}(t)$ for $i \in [1, N]$. Moreover, as discussed before, whether the packet losses occur or not in communication channels from agent i to its neighbors are the same, i.e, $\hat{x}_{ij_1}(t) = \hat{x}_{ij_2}(t)$ for any $j_1, j_2 \in \mathcal{N}_i$. Thus, we use a unified $e_{ij}(t)$ $(j \in \mathcal{N}_i)$ to denote the error variables in all agent i's neighbors, that is $e_{ij}(t) = e_{ij_1}(t) = e_{ij_2}(t)$ for any $j_1, j_2 \in \mathcal{N}_i$, and from (8.10), we also have $e_{ij}(t) = e_{bi}(t)$.

Based on (8.1), (8.5), (8.6), and (8.9), for $t \neq t_k^i$, we have

$$\begin{aligned}
\dot{x}_i(t) &= Ax_i(t) + BK(\delta_i(t) + (\mathscr{A}_r^i \otimes I_n)e_b(t)),\\
\dot{\delta}_i(t) &= A\delta_i(t) - BK(\mathscr{L}_r^i \otimes I_n)(\delta(t) + (\mathscr{A} \otimes I_n)e_b(t)),\\
\dot{e}_{ii}(t) &= Ae_{ii}(t) - BK(\delta_i(t) + (\mathscr{A}_r^i \otimes I_n)e_b(t)),\\
\dot{e}_{ij}(t) &= Ae_{ij}(t) - BK(\delta_i(t) + (\mathscr{A}_r^i \otimes I_n)e_b(t)),
\end{aligned}$$

and

$$\begin{aligned}
\dot{\delta}(t) &= (I_N \otimes A - \mathscr{L} \otimes BK)\delta(t) - (\mathscr{L}\mathscr{A} \otimes BK)e_b(t),\\
\dot{e}_a(t) &= (I_N \otimes A)e_a(t) - (I_N \otimes BK)\delta(t) - (\mathscr{A} \otimes BK)e_b(t),\\
\dot{e}_b(t) &= (I_N \otimes A - \mathscr{A} \otimes BK)e_b(t) - (I_N \otimes BK)\delta(t),
\end{aligned} \tag{8.11}$$

In addition, when $t = t_k^i$, from (8.5)–(8.9), we obtain

$$\begin{aligned}
&e_{ai}(t^+) = 0,\\
&e_{bi}(t^+) = 0, \text{ if } m_i^{loss} = 0,\\
&e_{bi}(t^+) = e_{bi}, \text{ if } m_i^{loss} = 1.
\end{aligned} \tag{8.12}$$

The flow dynamics (8.11) and jump dynamics (8.12) describe how the variables $\delta(t)$, $e_a(t)$, $e_b(t)$ evolve when the triggering event occurs or not.

8.3.2 *Leader-Following MAS Model*

In this case, define the consensus error variables

$$\delta_i(t) = x_i(t) - x_0(t), \tag{8.13}$$

and the state estimation error variables $e_{ii}(t)$, $e_{ij}(t)$ are defined in the same form as in (8.9). Moreover, the combined variables $\delta(t)$, $e_a(t)$, $e_b(t)$ have the same form as in (8.10).

Then, according to (8.1), (8.9), (8.10), and (8.3)–(8.13), take similar derivation process as in leaderless case, for $t \neq t_k^i$, we obtain

$$\begin{aligned}\dot{\delta}(t) &= (I_N \otimes A - \mathscr{H} \otimes BK)\delta(t) + (\mathscr{A} \otimes BK)e_b(t),\\ \dot{e}_a(t) &= (I_N \otimes A)e_a(t) - (\mathscr{A} \otimes BK)e_b(t) + (\mathscr{H} \otimes BK)\delta(t),\\ \dot{e}_b(t) &= (I_N \otimes A - \mathscr{A} \otimes BK)e_b(t) + (\mathscr{H} \otimes BK)\delta(t),\end{aligned} \tag{8.14}$$

Notice that in leader-following case, the information exchange among agents is performed in the same way as in leaderless case. Thus, the packet losses process is also the same as in leaderless case, and it implies that, at triggering instants t_k^i, the jump dynamics of $e_{ai}(t)$, $e_{bi}(t)$ can be also described in the form of (8.12). According to (8.11), (8.14), and (8.12), we can see that if take $\delta(t)$, $e_a(t)$, $e_b(t)$ as the closed-loop states, the dynamics of both leaderless and leader-following MAS with event-triggered control law and packet losses can be described in the same form, though the system matrices in flow dynamics are different. Therefore, it is expected that the event-triggered leaderless and leader-following consensus problems of MAS with packet losses can be addressed in a unified framework. However, due to there exist jump dynamics of $e_{ai}(t)$ and $e_{bi}(t)$, traditional consensus analysis results depending on continuous-time model of MAS are inapplicable for MAS as described by (8.11) (or (8.14)) and (8.12).

8.3.3 Hybrid Model Construction of MAS

In previous sections, though the dynamics of $\delta(t)$, $e_a(t)$, and $e_b(t)$ are described for leaderless and leader-following MAS respectively, the aforementioned form of dynamics is not appropriate for consensus analysis and ETM design, since the ETM is not integrated into the closed-loop system model. Based on (8.11), (8.14), and (8.12), we construct a unified model in the following hybrid systems form to describe the closed-loop dynamics of MAS with packet losses and dynamic ETM.

$$\mathbb{H} : \begin{cases} \dot{\xi} = \mathscr{F}(\xi), \ \text{if } \xi \in \mathbb{F} \\ \xi^+ \in \mathscr{J}(\xi), \ \text{if } \xi \in \mathbb{J} \end{cases} \tag{8.15}$$

where ξ is the closed-loop state that will be defined later, $\mathscr{F}(\xi)$ and $\mathscr{J}(\xi)$ denote flow dynamics and jump dynamics respectively, $\mathbb{F}$ and $\mathbb{J}$ are flow set and jump set. The hybrid systems form as given in (8.15) can be denoted as $\mathbb{H} = (\mathscr{F}, \mathscr{J}, \mathbb{F}, \mathbb{J})$ or $\mathbb{H}$ for short.

To construct the hybrid model, the integer variable $\kappa_i(t) \in \mathbb{N}_{\geq 0}$ is introduced to record the number of successive packet losses in communication channels from agent i to its neighbors since the last successful information transmission instant. $\kappa_i(t)$ evolves according to $\dot{\kappa}_i(t) = 0$ for $t \in (t_k^i, t_{k+1}^i]$, $\kappa_i(t_k^{i+}) = 0$ if agent i's information is received by its neighbors and the acknowledgement frame is successfully received at triggering instant t_k^i, and $\kappa_i(t_k^{i+}) = \kappa_i(t_k^i) + 1$ if the information transmitted from agent i is lost or it is not lost but no acknowledgement frame is received (i.e., the acknowledgement frames sent by agent i's neighbors are all lost). In addition, we introduce an internal timer variable $\tau_i(t) \in \mathbb{R}_{\geq 0}$ to keep track of the time elapsed since the last triggering instant for agent i, $\tau_i(t)$ evolves according to $\dot{\tau}_i(t) = 1$ for $t \in (t_k^i, t_{k+1}^i]$, and $\tau_i(t^+) = 0$ at triggering instant $t = t_k^i$.

Define

$$\begin{aligned} &\tau = \text{col}(\tau_1, \tau_2, \ldots, \tau_N),\ \kappa = \text{col}(\kappa_1, \kappa_2, \ldots, \kappa_N), \\ &\eta = \text{col}(\eta_1, \eta_2, \ldots, \eta_N),\ \xi = \text{col}(\delta, e_a, e_b, \tau, \kappa, \eta). \end{aligned} \tag{8.16}$$

Based on (8.11), (8.14), (8.12), (8.16), (8.7), (8.8), and the dynamics of κ_i and τ_i, we can write the closed-loop dynamics in hybrid systems form as shown in (8.15), where the flow map $\mathscr{F}(\xi)$ has the following form

$$\mathscr{F}(\xi) = (\bar{A}_{11}\delta + \bar{A}_{12}e_b, \bar{A}_{21}\delta + \bar{A}_{22}e_a + \bar{A}_{23}e_b, \bar{A}_{31}\delta + \bar{A}_{32}e_b, \mathbf{1}_N, \mathbf{0}_N, \Psi^F(o)), \tag{8.17}$$

where $\Psi^F(o) = \text{col}(\Psi_1^F(o_1, \eta_1), \ldots, \Psi_N^F(o_N, \eta_N))$, and the system matrices

$$\begin{aligned} &\bar{A}_{11} = I_N \otimes A - \mathscr{L} \otimes BK, \bar{A}_{12} = -\mathscr{L}\mathscr{A} \otimes BK, \\ &\bar{A}_{21} = -I_N \otimes BK, \bar{A}_{22} = I_N \otimes A, \bar{A}_{23} = -\mathscr{A} \otimes BK, \\ &\bar{A}_{31} = -I_N \otimes BK, \bar{A}_{32} = I_N \otimes A - \mathscr{A} \otimes BK, \end{aligned} \tag{8.18}$$

for leaderless MAS, and

$$\begin{aligned} &\bar{A}_{11} = I_N \otimes A - \mathscr{H} \otimes BK,\ \bar{A}_{12} = \mathscr{A} \otimes BK, \\ &\bar{A}_{21} = \mathscr{H} \otimes BK,\ \bar{A}_{22} = I_N \otimes A,\ \bar{A}_{23} = -\mathscr{A} \otimes BK, \\ &\bar{A}_{31} = \mathscr{H} \otimes BK,\ \bar{A}_{32} = I_N \otimes A - \mathscr{A} \otimes BK, \end{aligned} \tag{8.19}$$

for leader-following MAS.

The jump map $\mathscr{J}(\xi) = \bigcup_{i=1}^N \mathscr{J}_i(\xi)$, where $\mathscr{J}_i(\xi)$ is given by

$$\mathscr{J}_i(\xi) = \begin{cases} \mathscr{J}_{i1}(\xi), \text{if } \xi \in \mathbb{D}_i \wedge m_i^{loss} = 1, \\ \mathscr{J}_{i2}(\xi), \text{if } \xi \in \mathbb{D}_i \wedge m_i^{loss} = 0 \wedge m_i^{ack} = 0, \\ \mathscr{J}_{i3}(\xi), \text{if } \xi \in \mathbb{D}_i \wedge m_i^{loss} = 0 \wedge m_i^{ack} = 1, \end{cases} \tag{8.20}$$

with

$$\begin{aligned} &\mathscr{J}_{i1}(\xi) = \text{col}(\delta, \Lambda_i e_a, e_b, \Lambda_i \tau, \kappa + (I_N - \Lambda_i)\mathbf{1}_N, \Lambda_i \eta + (I_N - \Lambda_i)\Psi^{J2}(o)), \\ &\mathscr{J}_{i2}(\xi) = \text{col}(\delta, \Lambda_i e_a, \Lambda_i e_b, \Lambda_i \tau, \kappa + (I_N - \Lambda_i)\mathbf{1}_N, \Lambda_i \eta + (I_N - \Lambda_i)\Psi^{J2}(o)), \\ &\mathscr{J}_{i3}(\xi) = \text{col}(\delta, \Lambda_i e_a, \Lambda_i e_b, \Lambda_i \tau, \Lambda_i \kappa, \Lambda_i \eta + (I_N - \Lambda_i)\Psi^{J1}(o)), \end{aligned}$$

where $\Psi^{J1}(o) = \text{col}(\Psi_1^{J1}(o_1), \ldots, \Psi_N^{J1}(o_N))$, $\Psi^{J2}(o) = \text{col}(\Psi_1^{J2}(o_1), \ldots, \Psi_N^{J2}(o_N))$, $\Lambda_i \in \mathbb{R}^{N\times N}$ is a diagonal matrix with the iith entry is zero, the other diagonal entries are 1 and all the rest non-diagonal entries are zeros.

The jump set $\mathbb{J}_i$ corresponds to agent i is defined as

$$\mathbb{J}_i := \{\xi \in \mathbb{X} | \tau_i \geq \tau_{miet}^i \text{ and } \eta_i \leq 0\}, \tag{8.21}$$

where $\mathbb{X} := \mathbb{R}^{nN} \times \mathbb{R}^{nN} \times \mathbb{R}^{nN} \times \mathbb{R}^N \times \mathbb{N}_{\geq 0}^N \times \mathbb{R}_{\geq 0}^N$. The condition $\tau_i \geq \tau_{miet}^i$ is equivalent to the condition $t \geq t_k^i + \tau_{miet}^i$ in ETM (8.7), thus, the ETM (8.7) is embedded into the hybrid model.

The function $\mathscr{J}_{i1}(\xi)$ describes how ξ changes at triggering instants t_k^i when the packet losses occur. $\mathscr{J}_{i2}(\xi)$ describes how ξ changes when information transmitted by agent i is successfully received by its neighbors, but the acknowledgement frames sent back by the neighbors are lost. $\mathscr{J}_{i3}(\xi)$ describes the jump dynamics of ξ when the information is transmitted successfully and at least one acknowledgement frame is received by agent i.

Based on (8.21), the flow set $\mathbb{F}_i$ can be described as

$$\mathbb{F}_i := \{\xi \in \mathbb{X} | \tau_i < \tau_{miet}^i \text{ or } \eta_i > 0\}, \tag{8.22}$$

According to (8.17), (8.20), (8.22), and (8.21), the flow set $\mathbb{F}$ and jump set $\mathbb{J}$ can be represented as the following form

$$\mathbb{F} = \bigcap_{i=1}^{N} \mathbb{F}_i, \quad \mathbb{J} = \bigcup_{i=1}^{N} \mathbb{J}_i. \tag{8.23}$$

Thus, the leaderless and leader-following MAS with packet losses and dynamic ETM can be described by a unified hybrid systems form

$$\mathbb{H} := (\mathscr{F}(\xi), \mathscr{J}(\xi), \mathbb{F}, \mathbb{J}), \tag{8.24}$$

with $\mathscr{F}(\xi)$, $\mathscr{J}(\xi)$, $\mathbb{F}$, $\mathbb{J}$ in (8.17), (8.20), (8.22), (8.21).

The functions $\Psi_i^F(o_i, \eta_i)$, $\Psi_i^{J1}(o_i)$, and $\Psi_i^{J2}(o_i)$ are important components of the dynamic ETM (8.7) that need to be designed, and o_i is composed of locally available variables for agent i, such as e_{ai}, τ_i, κ_i, these variables will be used to construct the flow and jump map functions.

Based on the unified hybrid model, the event-triggered leaderless and leader-following consensus of MAS with packet losses are equivalent to stability of the hybrid system (8.24), which is defined as follows.

Definition 8.1 ([33]) Consider the hybrid system (8.24), the set $\{\xi \in \mathbb{X} | \delta = 0_{nN}, e_a = 0_{nN}, e_b = 0_{nN}\}$ is said to be uniformly globally asymptotically stable (UGAS) if there exists a function $\beta \in \mathscr{KL}$, such that for initial condition $\xi(0) \in \mathbb{X}$, the solution ξ satisfies

$$\|(\delta(t), e_a(t), e_b(t))\| \leq \beta(\|(\delta(0), e_a(0), e_b(0))\|, t). \tag{8.25}$$

8.4 Consensus Analysis and ETM Design Under Packet Losses

In this section, first, some conditions are supposed to be held, that are useful for stability analysis. Then, Lyapunov-based stability analysis and the design of flow/jump map functions $\Psi_i^F(o_i, \eta_i)/\Psi_i^{J1}(o_i)/\Psi_i^{J2}(o_i)$ are given, and how to determine the minimum inter-event time τ_{miet}^i and the number of maximum allowable successive packet losses (NMASPL) is also developed. Second, how to construct explicit form of Lyapunov function to satisfy the previous proposed stability conditions is given.

8.4.1 *ETM Design and Consensus Analysis*

For the purpose of stability analysis, inspired by some recent results [29, 31], the following conditions with regard to e_{ai}, e_{bi}, and consensus error variable δ are supposed to be satisfied.

Condition I. Consider $i \in [1, N]$, there exist locally Lipschitz functions $W_i : \mathbb{R}^n \to \mathbb{R}_{\geq 0}$, functions $H_{ai} : \mathbb{R}^{nN} \times \mathbb{R}^{nN} \times \mathbb{R}^{nN} \to \mathbb{R}_{\geq 0}$, $H_{bi} : \mathbb{R}^{nN} \times \mathbb{R}^{nN} \to \mathbb{R}_{\geq 0}$, and positive constants $\bar{\alpha}_W^{ai}$, $\bar{\alpha}_W^{bi}$, $\underline{\alpha}_W^{ai}$, $\underline{\alpha}_W^{bi}$, such that:

(1) For $e_{ai} \in \mathbb{R}^n$ and $e_{ai} \in \mathbb{R}^n$,

$$\begin{aligned} &\underline{\alpha}_W^{ai} \|e_{ai}\| \leq W_i(e_{ai}) \leq \bar{\alpha}_W^{ai} \|e_{ai}\|, \; W_i(e_{ai}^+) = 0, \\ &\underline{\alpha}_W^{bi} \|e_{bi}\| \leq W_i(e_{bi}) \leq \bar{\alpha}_W^{bi} \|e_{bi}\|, \; W_i(e_{bi}^+) \leq W_i(e_{bi}). \end{aligned} \tag{8.26}$$

(2) For $\delta \in \mathbb{R}^{nN}$, and almost all $e_a \in \mathbb{R}^{nN}$, $e_b \in \mathbb{R}^{nN}$,

$$\begin{aligned} &\left\langle \frac{\partial W_i(e_{ai})}{\partial e_{ai}}, \dot{e}_{ai} \right\rangle \leq H_{ai}(\delta, e_a, e_b), \\ &\left\langle \frac{\partial W_i(e_{bi})}{\partial e_{bi}}, \dot{e}_{bi} \right\rangle \leq H_{bi}(\delta, e_b). \end{aligned} \tag{8.27}$$

Condition II. There exist locally Lipschitz function $V : \mathbb{R}^{nN} \to \mathbb{R}_{\geq 0}$, positive constants $\underline{\alpha}_V$, $\bar{\alpha}_V$, $\gamma_i > 0$, $0 < \alpha_i < 1$, and class $\mathscr{K}_\infty$ functions $\rho, \sigma_a, \sigma_b : \mathbb{R} \to \mathbb{R}_{\geq 0}$, such that:

(1) For all $\delta \in \mathbb{R}^{nN}$,

$$\underline{\alpha}_V \|\delta\|^2 \leq V(\delta) \leq \bar{\alpha}_V \|\delta\|^2. \tag{8.28}$$

(2) For all $e_a \in \mathbb{R}^{nN}$, $e_b \in \mathbb{R}^{nN}$, and almost all $\delta \in \mathbb{R}^{nN}$,

$$\begin{aligned}&\langle \nabla V(\delta), \bar{A}_{11}\delta + \bar{A}_{12}e_a + \bar{A}_{13}e_b \rangle \\ &\leq -\rho(\|\delta\|) - \sum_{i=1}^{N}(H_{ai}^2(\delta, e_a, e_b) + H_{bi}^2(\delta, e_b)) - \sum_{i=1}^{N}(\sigma_a(W_i(e_{ai})) + \sigma_b(W_i(e_{bi}))) \\ &+ \sum_{i=1}^{N}\alpha_i\gamma_i^2(W_i^2(e_{ai}) + W_i^2(e_{bi})).\end{aligned} \tag{8.29}$$

The aforementioned conditions are useful for stability analysis of the hybrid system (8.24), and the explicit form of functions $W_i(e_{ai})$, $W_i(e_{bi})$, and $V(\delta)$ will be constructed in next section. In addition, we note that the variable τ is a closed-loop state, and the flow/jump set $\mathbb{F}/\mathbb{J}$ is embedded into the hybrid system model via defining appropriate sets $\mathbb{F}_i$ and $\mathbb{J}_i$, in which the condition with respect to τ_i is an important component. For the purpose of stability analysis and selecting the MIET τ_{miet}^i for agent i, we introduce an auxiliary function $\phi_i(\tau_i) : \mathbb{R}_{\geq 0} \to \mathbb{R}_{>0}$ with the following dynamics

$$\dot{\phi}_i(\tau_i) = -\gamma_i(\phi_i^2(\tau_i) + 1), \tag{8.30}$$

where $\gamma_i > 0$ is the same parameter as used in condition (8.29), and the initial condition $\phi_i(0)$ needs to be pre-specified. The functions $\phi_i(\tau_i)$, will be used to construct Lypunov function for closed-loop system stability analysis, thus, the positiveness of $\phi_i(\tau_i)$ is generally required. On the basis of (8.30), it is clear that, by selecting positive $\phi_i(0) > 0$, there exists an upper bound τ_M^i for τ_i such that if $\tau_i < \tau_M^i$, $\phi_i(\tau_i) > 0$ is ensured. The value of τ_M^i can be chosen according to the following lemma.

Lemma 8.2 *Consider the functions $\phi_i(\tau_i)$ ($i \in [1, N]$) with dynamics described by (8.30), if the initial conditions $\phi_i(0)$ are strictly positive constants, $\gamma_i > 0$, and τ_M^i is chosen to be*

$$\tau_M^i = \frac{\arctan \phi_i(0)}{\gamma_i}. \tag{8.31}$$

Then, for any $\tau_i < \tau_M^i$, the condition $\phi_i(\tau_i) > 0$ is satisfied.

Proof Based on (8.30), the function $\phi_i(\tau_i)$ is strictly decreasing. Assume that τ_M^i is a constant that satisfies $\phi_i(\tau_M^i) = 0$, then, for $\tau_i < \tau_M^i$, the condition $\phi_i(\tau_i) > 0$ holds. So, we need to compute a τ_M^i to satisfy $\phi_i(\tau_M^i) = 0$. From (8.30), we have

$$d\tau_i = -\frac{1}{\gamma_i}\frac{d\phi_i(\tau_i)}{\phi_i^2(\tau_i) + 1} \tag{8.32}$$

Integrate both sides of the above equation from $\tau_i = 0$ to $\tau_i = \tau_M^i$, we obtain

$$\tau_M^i - 0 = -\frac{1}{\gamma_i}(\arctan(\phi_i(\tau_M^i)) - \arctan(\phi_i(0))) \tag{8.33}$$

Note that $\phi_i(\tau_M^i) = 0$, thus, $\tau_M^i = \frac{\arctan \phi_i(0)}{\gamma_i}$, and it implies that for any $\tau_i < \tau_M^i$, the condition $\phi_i(\tau_i) > 0$ is satisfied. □

Now, we are in the position to present the consensus analysis and ETM design results, that are stated as follows.

Theorem 8.1 *Consider the hybrid system $\mathbb{H}$ as shown in (8.24). Suppose that the conditions in Condition I and Condition II hold, the flow map function $\Psi_i^F(o_i, \eta_i)$ in the ETM (8.7) has the following form*

$$\Psi_i^F(o_i, \eta_i) = \omega_i(\alpha_i, \kappa_i, \tau_{miet}^i)(W_i^2(e_{ai}) + W_i^2(e_{bi})) - \chi_i(\eta_i) \tag{8.34}$$

where

$$\omega_i(\alpha_i, \kappa_i, \tau_{miet}^i) = \begin{cases} 1 - \alpha_i, \ \textit{if}\ 0 \le \tau_i \le \tau_{miet}^i \\ -(\alpha_i + \phi_i^2((\kappa_i + 1)\tau_{miet}^i), \ \textit{if}\ \tau_i > \tau_{miet}^i \end{cases}$$

where $0 < \alpha_i < 1$, the function $\chi_i(\eta_i) : \mathbb{R}_{\ge 0} \to \mathbb{R}_{\ge 0}$, τ_{miet}^i satisfies the following condition

$$\tau_{miet}^i < \frac{\arctan \phi_i(0)}{\gamma_i(\varrho_{\max}^i + 1)} \tag{8.35}$$

where $\gamma_i > 0$, $\varrho_{\max}^i$ denotes the number of maximum allowable successive packet losses for agent i.

The jump map functions $\Psi_i^{J1}(o_i)$ and $\Psi_i^{J2}(o_i)$ are designed as the following form

$$\begin{aligned} \Psi_i^{J1}(o_i) &= \gamma_i\phi_i((\kappa_i + 1)\tau_{miet}^i)W_i^2(e_{ai}), \\ \Psi_i^{J2}(o_i) &= \gamma_i\phi_i((\kappa_i + 1)\tau_{miet}^i)(W_i^2(e_{ai}) + W_i^2(e_{bi})), \end{aligned} \tag{8.36}$$

Then, the hybrid system $\mathbb{H}$ is UGAS, which indicates that the leaderless and leader-following consensus of MAS with packet losses is achieved. Moreover, $\tau_{miet}^i > 0$ implies that $t_{k+1}^i - t_k^i \ge \tau_{miet}^i$, i.e., Zeno behavior is ensured to be excluded.

Proof Consider the following Lypaunov function

$$U(\xi) = V(\delta) + \sum_{i=1}^{N} \gamma_i\bar{\phi}_i(\kappa_i, \tau_i)(W_i^2(e_{ai}) + W_i^2(e_{bi})) + \sum_{i=1}^{N} \eta_i, \tag{8.37}$$

where $V(\delta)$, $W_i(e_{ai})$, $W_i(e_{bi})$ satisfy the given conditions in Condition I and Condition II. $\bar{\phi}_i(\kappa_i, \tau_i) = \phi_i(\kappa_i \tau^i_{miet} + \min(\tau_i, \tau^i_{miet}))$. From Condition I and II, it is known that $V(\delta)$, $W_i(e_{ai})$, and $W_i(e_{bi})$ are radially unbounded. Based on (8.7), (8.8), (8.34), and (8.36), $\eta \geq 0$ can be guaranteed. Besides, based on the form of $\bar{\phi}_i(\kappa_i, \tau_i)$, $\kappa_i \leq \varrho^i_{\max}$, the condition (8.35), and Lemma 8.2, it implies that $\bar{\phi}_i(\kappa_i, \tau_i) > 0$ is guaranteed. Thus, $U(\xi)$ is a suitable Lyapunov function for stability analysis.

First, consider the jump dynamics of $U(\xi)$, according to whether the transmitted information from a certain agent i is lost and whether the acknowledgement frame sent back by agent i's neighbors is received, there are three cases of jump dynamics (see (8.20)):

① $\xi \in \mathbb{J}_i$ and $m_i^{loss} = 1$

This case means that packet loss occurs when agent i sending information. According to the definition of $\mathscr{J}_{i1}(\xi)$ and (8.36), we have $e^+_{ai} = 0$, $e^+_{bi} = e_{bi}$, $\kappa_i^+ = \kappa_i + 1$, $\tau_i^+ = 0$, $\eta_i^+ = \Psi_i^{J1}(o_i)$, thus, we have

$$\begin{aligned}
&U(\xi^+) - U(\xi)\\
=&\gamma_i \bar{\phi}_i(\kappa_i^+, \tau_i^+)(W_i^2(e^+_{ai}) + W_i^2(e^+_{bi})) + \eta_i^+ - \gamma_i \bar{\phi}_i(\kappa_i, \tau_i)(W_i^2(e_{ai}) + W_i^2(e_{bi})) - \eta_i\\
=&\gamma_i \phi_i((\kappa_i + 1)\tau^i_{miet} + \min\{0, \tau^i_{miet}\}) W_i^2(e_{bi}) + \gamma_i \phi_i((\kappa_i + 1)\tau^i_{miet}) W_i^2(e_{ai})\\
&- \gamma_i \bar{\phi}_i(\kappa_i, \tau_i)(W_i^2(e_{ai}) + W_i^2(e_{bi}))\\
\leq& 0,
\end{aligned} \tag{8.38}$$

where the fact that $\tau_i \geq \tau^i_{miet}$ and $\phi_i((\kappa_i + 1)\tau^i_{miet}) = \phi_i(\kappa_i \tau^i_{miet} + \min\{\tau_i, \tau^i_{miet}\}) = \bar{\phi}_i(\kappa_i, \tau_i)$ is used in the last inequality.

② $\xi \in \mathbb{J}_i$ and $m_i^{loss} = 0$ and $m_i^{ack} = 0$

This case means that no packet losses occur when agent i sending information, but no acknowledgement frame is received. According to the definition of $\mathscr{J}_{i2}(\xi)$ and (8.36), we have $e^+_{ai} = 0$, $e^+_{bi} = 0$, $\kappa_i^+ = \kappa_i + 1$, $\tau_i^+ = 0$, $\eta_i^+ = \Psi_i^{J1}(o_i)$, and

$$\begin{aligned}
&U(\xi^+) - U(\xi)\\
=&\gamma_i \bar{\phi}_i(\kappa_i^+, \tau_i^+)(W_i^2(e^+_{ai}) + W_i^2(e^+_{bi})) + \eta_i^+ - \gamma_i \bar{\phi}_i(\kappa_i, \tau_i)(W_i^2(e_{ai}) + W_i^2(e_{bi})) - \eta_i\\
=&0 + \gamma_i \phi_i((\kappa_i + 1)\tau^i_{miet}) W_i^2(e_{ai}) - \gamma_i \bar{\phi}_i(\kappa_i, \tau_i)(W_i^2(e_{ai}) + W_i^2(e_{bi}))\\
\leq& - \gamma_i \bar{\phi}_i(\kappa_i, \tau_i) W_i^2(e_{bi}) \leq 0.
\end{aligned} \tag{8.39}$$

③ $\xi \in \mathbb{J}_i$ and $m_i^{loss} = 0$ and $m_i^{ack} = 1$

This case means that no packet losses occur when agent i sending information, and acknowledgement frame is received. According to the definition of $\mathscr{J}_{i2}(\xi)$ and (8.36), we have $e^+_{ai} = 0$, $e^+_{bi} = 0$, $\kappa_i^+ = 0$, $\tau_i^+ = 0$, $\eta_i^+ = \Psi_i^{J2}(o_i)$, thus, it has

$$
\begin{aligned}
&U(\xi^{+}) - U(\xi) \\
=&\gamma_i \bar{\phi}_i(\kappa_i^{+}, \tau_i^{+})(W_i^2(e_{ai}^{+}) + W_i^2(e_{bi}^{+})) + \eta_i^{+} - \gamma_i \bar{\phi}_i(\kappa_i, \tau_i)(W_i^2(e_{ai}) + W_i^2(e_{bi})) - \eta_i \\
=&0 + \gamma_i \phi_i((\kappa_i + 1)\tau_{miet}^{i})(W_i^2(e_{ai}) + W_i^2(e_{bi})) - \gamma_i \bar{\phi}_i(\kappa_i, \tau_i)(W_i^2(e_{ai}) + W_i^2(e_{bi})) \\
\leq&0.
\end{aligned} \tag{8.40}
$$

Thus, with the proposed ETM design and the given conditions, it guarantees that the Lyapunov function $U(\xi)$ is non-increasing at ξ's jump dynamics.

Then, consider the evolution of $U(\xi)$ along ξ's flow dynamics. Due to $U(\xi)$ is not differentiable everywhere, we use notation $\langle \nabla U(\xi), \mathscr{F}(\xi) \rangle$ to denote the derivative of $U(\xi)$, with $\mathscr{F}(\xi)$ given by (8.17). Based on Condition I and Condition II, for $0 \leq \tau_i < \tau_{miet}^{i}$ and almost all (δ, e_a, e_b), we have

$$
\begin{aligned}
&\langle \nabla U(\xi), \mathscr{F}(\xi) \rangle \\
=&-\rho(\|\delta\|) - \sum_{i=1}^{N}(H_{ai}^2(\delta, e_a, e_b) + H_{bi}^2(\delta, e_b)) - \sum_{i=1}^{N}(\sigma_a(W_i(e_{ai})) + \sigma_b(W_i(e_{bi}))) \\
&+\sum_{i=1}^{N}\alpha_i \gamma_i^2 (W_i^2(e_{ai}) + W_i^2(e_{bi})) + \sum_{i=1}^{N} 2\gamma_i \bar{\phi}_i (W_i(e_{ai}) H_{ai}(\delta, e_a, e_b) + W_i(e_{bi}) H_{bi}(\delta, e_b)) \\
&-\sum_{i=1}^{N}\gamma_i^2(\bar{\phi}_i^2 + 1)(W_i^2(e_{ai}) + W_i^2(e_{bi})) + \sum_{i=1}^{N}\Psi_i^F(o_i, \eta_i)
\end{aligned} \tag{8.41}
$$

According to the expression of $\Psi_i^F(o_i, \eta_i)$ given by (8.34), it results that

$$
\begin{aligned}
&\langle \nabla U(\xi), \mathscr{F}(\xi) \rangle \\
\leq&-\rho(\|\delta\|) - \sum_{i=1}^{N}(\sigma_a(W_i(e_{ai})) + \sigma_b(W_i(e_{bi}))) - \sum_{i=1}^{N}(H_{ai}(\delta, e_a, e_b) - \gamma_i \bar{\phi}_i W_i(e_{ai}))^2 \\
&-\sum_{i=1}^{N}(H_{bi}^2(\delta, e_b) - \gamma_i \bar{\phi}_i W_i(e_{bi}))^2 + \sum_{i=1}^{N}(\alpha_i - 1)\gamma_i^2 \bar{\phi}_i^2 W_i^2(e_{ai}) + \sum_{i=1}^{N}\Psi_i^F(o_i, \eta_i) \\
\leq&-\rho(\|\delta\|) - \sum_{i=1}^{N}(\sigma_a(W_i(e_{ai})) + \sigma_b(W_i(e_{bi})) + \chi_i(\eta_i))
\end{aligned} \tag{8.42}
$$

Next, consider the case $\tau_i \geq \tau_{miet}^{i}$, in this case, $\dot{\bar{\phi}}_i = 0$, this is because $\bar{\phi}_i(\kappa_i, \tau_i) = \phi_i((\kappa_i + 1)\tau_{miet}^{i})$ is a piecewise constant for $\tau_i > \tau_{miet}^{i}$. Thus, for $\tau_i > \tau_{miet}^{i}$ and almost all (δ, e_a, e_b), we have

$$
\begin{aligned}
&\langle \nabla U(\xi), \mathscr{F}(\xi)\rangle \\
&= - \rho(\|\delta\|) - \sum_{i=1}^{N}(H_{ai}^2(\delta, e_a, e_b) + H_{bi}^2(\delta, e_b)) - \sum_{i=1}^{N}(\sigma_a(W_i(e_{ai})) + \sigma_b(W_i(e_{bi}))) \\
&\quad + \sum_{i=1}^{N}\alpha_i\gamma_i^2(W_i^2(e_{ai}) + W_i^2(e_{bi})) + \sum_{i=1}^{N}\Psi_i^F(o_i, \eta_i) \\
&\quad + \sum_{i=1}^{N}2\gamma_i\bar{\phi}_i(W_i(e_{ai})H_{ai}(\delta, e_a, e_b) + W_i(e_{bi})H_{bi}(\delta, e_b))
\end{aligned}
\tag{8.43}
$$

According to the expression of $\Psi_i^F(o_i, \eta_i)$ given by (8.34), and notice that $\bar{\phi}_i(\kappa_i, \tau_i) = \phi_i((\kappa_i + 1)\tau_{miet}^i)$ when $\tau_i > \tau_{miet}^i$. Hence, based on (8.34) and (8.43), we can derive that

$$
\begin{aligned}
&\langle \nabla U(\xi), \mathscr{F}(\xi)\rangle \\
&\leq - \rho(\|\delta\|) - \sum_{i=1}^{N}(\sigma_a(W_i(e_{ai})) + \sigma_b(W_i(e_{bi}))) - \sum_{i=1}^{N}(H_{ai}(\delta, e_a, e_b) - \gamma_i\bar{\phi}_i W_i(e_{ai}))^2 \\
&\quad - \sum_{i=1}^{N}(H_{bi}^2(\delta, e_b) - \gamma_i\bar{\phi}_i W_i(e_{bi}))^2 - \sum_{i=1}^{N}\chi_i(\eta_i) \\
&\leq - \rho(\|\delta\|) - \sum_{i=1}^{N}(\sigma_a(W_i(e_{ai})) + \sigma_b(W_i(e_{bi})) + \chi_i(\eta_i))
\end{aligned}
\tag{8.44}
$$

According to (8.38), (8.39), (8.40), (8.42), (8.44), and based on standard Lyapunov stability theory, we can conclude that the (δ, e_a, e_b) dynamics is UGAS, and δ is asymptotically stable implies the leaderless/leader-following consensus of MAS with packet losses. Moreover, $\tau_{miet}^i > 0$ imposes that the triggering intervals are strictly positive and Zeno behavior is avoided. □

Remark 8.2 The event-triggered consensus of MAS has been widely studied in the literature [11, 17, 25, 28, 34], but the packet dropouts was rarely considered, this is mainly due to packet losses introduce difficulties for system modelling and ETM design. In the presence of packet losses, we propose a hybrid model to describe the MAS dynamics, and the proposed ETM design is explicitly related to the number of successive packet losses, which is novel compared with the existing ETM design results. Moreover, a positive τ_{miet}^i is embedded into the triggering conditions to enforce that the minimum triggering interval is not less than τ_{miet}^i, such that Zeno behavior is absolutely excluded.

Remark 8.3 Based on (8.35), it is shown that the MIET τ_{miet}^i and NMASPL $\varrho_{\max}^i$ depend on $\phi_i(0)$ and γ_i, that can be selected arbitrarily. Once $\phi_i(0)$ and γ_i are given, the MIET τ_{miet}^i is inverse proportional to the NMASPL $\varrho_{\max}^i$, that is, there exists tradeoff between τ_{miet}^i and $\varrho_{\max}^i$. Besides, with the proposed approach, each agent can select its own MIET τ_{miet}^i, thus, the NMASPL $\varrho_{\max}^i$ for a certain agent can be different from each other.

8.4.2 Construction of Storage Functions $V(\delta)$, $W_i(e_{ai})$, $W_i(e_{bi})$

In previous section, we note that the conditions in Condition I and Condition II are important for stability analysis and ETM design. However, these conditions are assumed to be satisfied, and the explicit form of $V(\delta)$, $W_i(e_{ai})$, and $W_i(e_{bi})$ are not given. In this section, in order to verify stability and implement the proposed ETM by using the results in Theorem 8.1, the explicit form of storage functions $V(\delta)$, $W_i(e_{ai})$, $W_i(e_{bi})$ will be constructed, such that the conditions in Condition I and Condition II are satisfied.

First, consider the functions $W_i(e_{ai})$ and $W_i(e_{bi})$. To satisfy the conditions (8.26) and (8.27), we construct these functions in the form of $W_i(e_{ai}) = \|e_{ai}\|$ and $W_i(e_{bi}) = \|e_{bi}\|$. According to the jump map $G_i(\xi)$ in (8.20), it is known that $e_{ai}^+ = 0$ and $e_{bi}^+ = 0$ or $e_{bi}^+ = e_{bi}$, thus, the conditions $W_i(e_{ai}^+) = 0$ and $W_i(e_{bi}^+) \leq W_i(e_{bi})$ hold. Besides, let $\bar{\alpha}_W^{ai} = \bar{\alpha}_W^{bi} = 1$, $\underline{\alpha}_W^{ai} = \underline{\alpha}_W^{bi} = 1$, all the conditions in (8.26) are satisfied with $W_i(e_{ai}) = \|e_{ai}\|$ and $W_i(e_{bi}) = \|e_{bi}\|$. Then, we compute the values of $\left\langle \frac{\partial W_i(e_{ai})}{\partial e_{ai}}, \dot{e}_{ai} \right\rangle$ and $\left\langle \frac{\partial W_i(e_{bi})}{\partial e_{bi}}, \dot{e}_{bi} \right\rangle$, and derive explicit form of $H_{ai}(\delta, e_a, e_b)$ and $H_{bi}(\delta, e_b)$. Based on (8.27), we have

$$\left\langle \frac{\partial W_i(e_{ai})}{\partial e_{ai}}, \dot{e}_{ai} \right\rangle \leq \left\| \frac{\partial (e_{ai}^T e_{ai})^{\frac{1}{2}}}{\partial e_{ai}} \right\| \|\dot{e}_{ai}\| = \|\dot{e}_{ai}\| . \tag{8.45}$$

Similarly, $\left\langle \frac{\partial W_i(e_{bi})}{\partial e_{bi}}, \dot{e}_{bi} \right\rangle \leq \|\dot{e}_{bi}\|$. Denote $R_i \in \mathbb{R}^{1\times N}$ as a row matrix with the ith element is 1 and all the rest elements are zeros. According to (8.17), we obtain

$$\begin{aligned}
\|\dot{e}_{ai}\| &= \left\|(R_i \otimes I_n)(\bar{A}_{21}\delta + \bar{A}_{22}e_a + \bar{A}_{23}e_b)\right\| \\
&= \underbrace{\left\|\bar{R}_{\delta i}^a \delta + \bar{R}_{ai}^a e_a + \bar{R}_{bi}^a e_b\right\|}_{H_{ai}(\delta, e_a, e_b)}, \\
\|\dot{e}_{bi}\| &= \left\|(R_i \otimes I_n)(\bar{A}_{31}\delta + \bar{A}_{32}e_b)\right\| \\
&= \underbrace{\left\|\bar{R}_{\delta i}^b \delta + \bar{R}_{bi}^b e_b\right\|}_{H_{bi}(\delta, e_b)} .
\end{aligned} \tag{8.46}$$

Thus, if the functions $H_{ai}(\delta, e_a, e_b)$ and $H_{bi}(\delta, e_b)$ are defined as above form, the conditions in (8.27) are satisfied.

Then, consider the function $V(\delta)$ and conditions (8.28), (8.29). It is required to provide an explicit form of $V(\delta)$ such that the conditions in Condition II are satisfied. The following results present how to construct $V(\delta)$ and give linear matrix inequality conditions to guarantee that Condition II holds.

Proposition 8.1 *Choose the function $V(\delta) = \delta^T (I_N \otimes P)\delta$ and let the gain matrix $K = \mu B^T P$, where the positive definite matrix P is a solution to the following Riccati equation*

$$PA + A^T P - 2\mu a(\hat{\mathcal{L}}) PBB^T P + Q = 0, \tag{8.47}$$

where $Q > 0$ is a matrix with appropriate dimensions, $\mu > 0$ is a constant, $\hat{\mathcal{L}}$ is defined in Lemma 8.1. If there exist constants $0 < \sigma < 1$, $\gamma_i > 0$, $0 < \alpha_i < 1$, and $\epsilon > 0$, such that the following inequality condition is satisfied

$$\begin{bmatrix} \Xi_{11} & \sum_{i=1}^{N} \bar{R}_{\delta i}^{aT} \bar{R}_{ai}^{a} & \Xi_{13} \\ * & \Xi_{22} & \sum_{i=1}^{N} \bar{R}_{ai}^{aT} \bar{R}_{bi}^{a} \\ * & * & \Xi_{33} \end{bmatrix} \leq 0, \tag{8.48}$$

where

$$\begin{aligned} \Xi_{11} &= -I_N \otimes (1-\sigma) Q + \sum_{i=1}^{N} (\bar{R}_{\delta i}^{aT} \bar{R}_{\delta i}^{a} + \bar{R}_{\delta i}^{bT} \bar{R}_{\delta i}^{b}), \\ \Xi_{13} &= (I_N \otimes P) \bar{A}_{12} + \sum_{i=1}^{N} (\bar{R}_{\delta i}^{aT} \bar{R}_{bi}^{a} + \bar{R}_{\delta i}^{bT} \bar{R}_{bi}^{b}), \\ \Xi_{22} &= -\Gamma \otimes I_n + \epsilon I_{nN} + \sum_{i=1}^{N} \bar{R}_{ai}^{aT} \bar{R}_{ai}^{a}, \\ \Xi_{33} &= -\Gamma \otimes I_n + \epsilon I_{nN} + \sum_{i=1}^{N} (\bar{R}_{bi}^{aT} \bar{R}_{bi}^{a} + \bar{R}_{bi}^{bT} \bar{R}_{bi}^{b}), \end{aligned} \tag{8.49}$$

where $\Gamma = diag\{\alpha_1 \gamma_1^2, \ldots, \alpha_N \gamma_N^2\}$, $\bar{R}_{\delta i}^{a}$, $\bar{R}_{ai}^{a}$, $\bar{R}_{bi}^{a}$, $\bar{R}_{\delta i}^{b}$, $\bar{R}_{bi}^{b}$ are defined in (8.46). Then, the conditions (8.28) are satisfied with the given $V(\delta)$, and the conditions (8.29) are satisfied with $H_{ai}(\delta, e_a, e_b)$ and $H_{bi}(\delta, e_b)$ given in (8.46).

Proof Since the matrix P is positive definite, it is obvious that by selecting $\underline{\alpha}_V = \lambda_{\min}(I_N \otimes P)$ and $\bar{\alpha}_V = \lambda_{\max}(I_N \otimes P)$, the conditions (8.28), for $i \in [1, N]$, are satisfied. Then, consider the evolution of $V(\delta)$ along δ's flow dynamics, and note that δ is a continuous variable and there exists no jump dynamics of δ, we have

$$\begin{aligned} &\langle \nabla V(\delta), \bar{A}_{11}\delta + \bar{A}_{12} e_b \rangle = \dot{V}(\delta) \\ =&\delta^T ((I_N \otimes P)\bar{A}_{11} + \bar{A}_{11}^T (I_N \otimes P))\delta + 2\delta^T (I_N \otimes P)\bar{A}_{12} e_b. \end{aligned}$$

From (8.18) and (8.19), $K = \mu B^T P$ and note that $\hat{\mathcal{L}} = \mathcal{L}$ in leaderless case and $\hat{\mathcal{L}} = \mathcal{H}$ in leader-following case, we obtain

$$\begin{aligned} &\langle \nabla V(\delta), \bar{A}_{11}\delta + \bar{A}_{12} e_b \rangle \\ =&\delta^T (I_N \otimes (PA + A^T P) - 2\mu \hat{\mathcal{L}} \otimes PBB^T P)\delta + 2\delta^T (I_N \otimes P)\bar{A}_{12} e_b \\ \leq& -\delta^T (I_N \otimes Q)\delta + 2\delta^T (I_N \otimes P)\bar{A}_{12} e_b, \end{aligned} \tag{8.50}$$

where the last inequality is obtained based on (8.47).

Then, let the matrix in (8.48) right multiplied by vector $\mathrm{col}(\delta, e_a, e_b)$ and left multiplied by its transpose, it yields that

$$\begin{aligned}
& -\delta^T(I_N \otimes Q)\delta + 2\delta^T(I_N \otimes P)\bar{A}_{12}e_b \\
\leq & -\delta^T(I_N \otimes \sigma Q)\delta + \epsilon(\|e_a\|^2 + \|e_b\|^2) - \sum_{i=1}^{N}(\bar{R}^a_{\delta i}\delta + \bar{R}^a_{ai}e_a + \bar{R}^a_{bi})^T(\bar{R}^a_{\delta i}\delta + \bar{R}^a_{ai}e_a + \bar{R}^a_{bi}) \\
& - \sum_{i=1}^{N}(\bar{R}^b_{\delta i}\delta + \bar{R}^b_{bi}e_b)^T(\bar{R}^b_{\delta i}\delta + \bar{R}^b_{bi}e_b) + \sum_{i=1}^{N}(e_a^T(\Gamma \otimes I_n)e_a + e_b^T(\Gamma \otimes I_n)e_b) \\
= & -\delta^T(I_N \otimes \sigma Q)\delta + \sum_{i=1}^{N}(\epsilon\|e_a\|^2 + \epsilon\|e_b\|^2) \\
& - \sum_{i=1}^{N}(H^2_{ai}(\delta, e_a, e_b) + H^2_{bi}(\delta, e_b)) + \sum_{i=1}^{N}\alpha_i\gamma_i^2(\|e_a\|^2 + \|e_b\|^2).
\end{aligned} \tag{8.51}$$

According to (8.50) and (8.51), and based on $W_i(e_{ai}) = \|e_{ai}\|$, $W_i(e_{ai}) = \|e_{ai}\|$, and $-\delta^T(I_N \otimes Q)\delta \leq -\lambda_{\min}(\sigma Q)\|\delta\|^2$, it means that the condition (8.29) in Condition II is satisfied. □

The explicit form of function $V(\delta)$, $W_i(e_{ai})$, $W_i(e_{bi})$ are constructed in this section, based on the obtained results, the stability of hybrid system (8.24) can be verified via Proposition 8.1, which also implies the leaderless and leader-following consensus of MAS with dynamic ETM and packet losses.

Remark 8.4 The obtained results in this chapter have some advantages compared with [25], where event-triggered consensus problem of MAS with packet losses was also considered. To be specific, first, the packet losses do not need to obey the Bernoulli distribution rule, alternatively, it is only required that the NMASPL is upper bounded; second, the packet losses are allowed to be non-identical; third, the proposed hybrid systems approach can tackle both leaderless and leader-following consensus of MAS in a unified framework.

8.5 Simulation Study

In this section, numerical examples are given to show the efficiency of the proposed hybrid systems approach.

8.5.1 Leaderless MAS

Consider leaderless MAS with six agents, the dynamics of each agent has the same form as (8.1), with systems matrices $A = [01; -10]$, $B = [11]^T$. The agents are

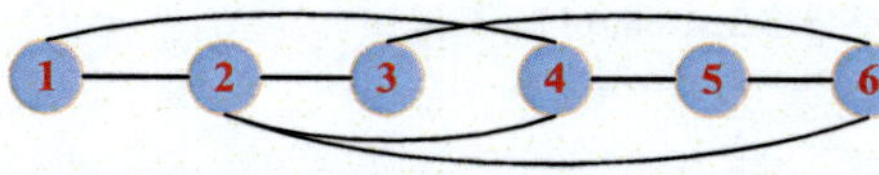

Fig. 8.1 Communication topology (leaderless MAS)

connected by an undirected graph as shown in Fig. 8.1, the corresponding Laplacian matrix $\mathscr{L}$ can be easily obtained from the communication graph. Based on Lemma 8.1, it can be computed that $a(\hat{\mathscr{L}}) = 1.382$, choose the parameters $\mu = 0.1$, $Q = 20I$, the gain matrix K can be obtained by solving the Riccati equation (8.47) as $K = [1.1708\ 0.2764]$. The parameters α_i and γ_i are chosen as $\alpha_i = 0.1$ and $\gamma_i = 8$, $\chi_i(\eta_i) = \pi_i \eta_i$ with $\pi_{1,2} = 2, \pi_{3,4} = 4, \pi_{5,6} = 5$, the initial conditions $\phi_i(0)$ are given as $\phi_i(0) = i + 4$ for $i \in [1, N]$. With the above given conditions, the parameters τ_M^i (see (8.31)) can be calculated as $\tau_M^1 = 0.1717$, $\tau_M^2 = 0.1757$, $\tau_M^3 = 0.1786$, $\tau_M^4 = 0.1808$, $\tau_M^5 = 0.1825$, $\tau_M^6 = 0.1839$, such that $\phi_i(\tau_i) > 0$ can be guaranteed with $\tau_i < \tau_M^i$. In order to ensure Zeno-freeness of triggering behavior, the lower bound of triggering intervals τ_{miet}^i in ETM (8.7) is chosen as $\tau_{miet}^{1,4} = 0.02$, $\tau_{miet}^{2,5} = 0.03$, $\tau_{miet}^{3,6} = 0.05$ in the simulation. Then, according to (8.35), the number of maximum allowable successive packet losses $\varrho_{\max}^i$ is $\varrho_{\max}^1 = 7$, $\varrho_{\max}^2 = 4$, $\varrho_{\max}^3 = 2$, $\varrho_{\max}^4 = 8$, $\varrho_{\max}^5 = 5$, $\varrho_{\max}^6 = 2$.

To do the simulation, the initial conditions are chosen as $x_i(0) = (-1)^i \times [1\ 2]^T$, $\eta_i(0) = 5$, $\tau_i(0) = \kappa_i(0) = 0$. Based on the given parameters and conditions, it can be verified that the obtained stability conditions in Proposition 8.1 and Theorem 8.1 are satisfied. The state trajectories of the leaderless MAS are shown in Fig. 8.2, which clearly demonstrates that the leaderless consensus is achieved. The corresponding control signals u_i are shown in Fig. 8.3. To show that whether packet losses occur when agents transmitting information via communication channels, Fig. 8.4 presents the event-triggering instants and packet losses for each agent. Both the small red and blue boxes denote triggering instants, the difference is that a red box denotes successful data transmission at triggering instant, and a blue box denotes that the transmitted data is lost. From Fig. 8.4, the number of successive packet losses is clearly shown and it can be seen that the minimum triggering interval is not less

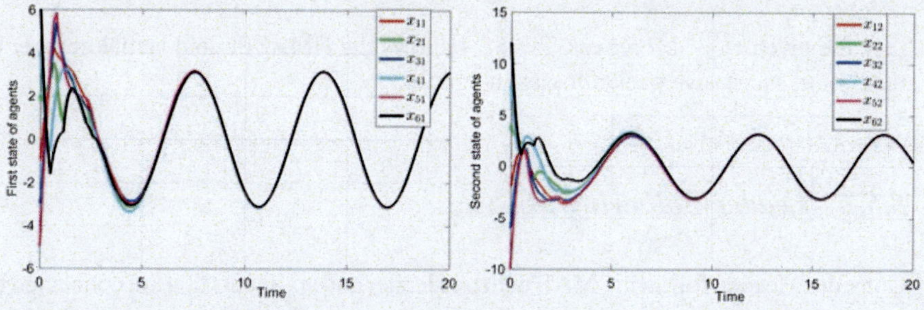

Fig. 8.2 The first states and second states of leaderless MAS

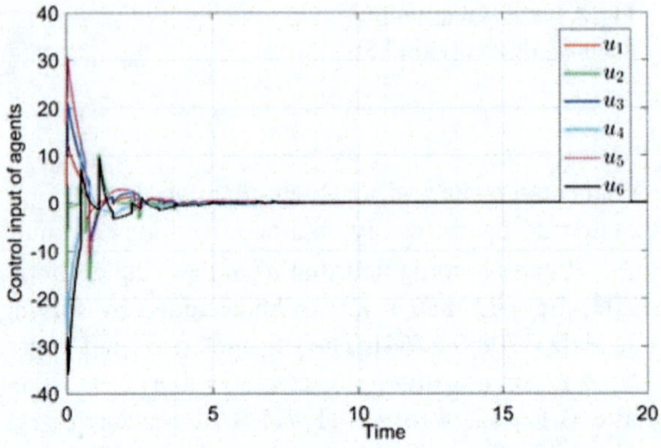

Fig. 8.3 Control input u_i of leaderless MAS

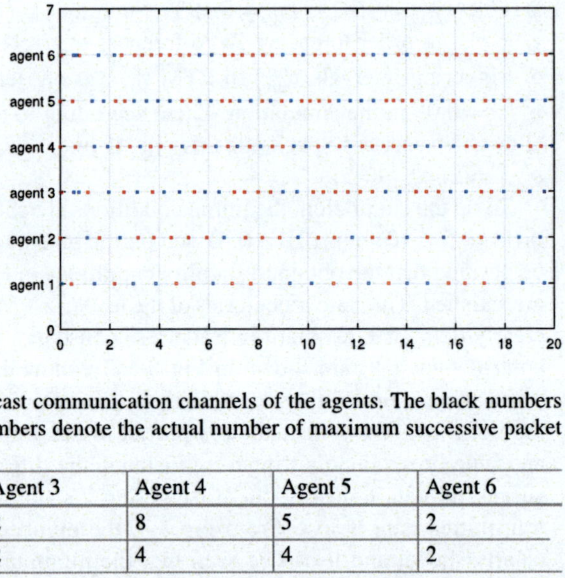

Fig. 8.4 Event-triggering instants and packet losses (leaderless MAS). The red box denotes successful data transmission at triggering instant, and blue box denotes that data transmission fails (i.e., packet loss occurs at triggering instant)

Table 8.1 Packet losses in broadcast communication channels of the agents. The black numbers denote the NMASPL, the blue numbers denote the actual number of maximum successive packet losses in simulation

Agent 1	Agent 2	Agent 3	Agent 4	Agent 5	Agent 6
7	4	2	8	5	2
3	3	2	4	4	2

than the given τ^i_{miet}. Moreover, Table 8.1 shows the NMASPL and actual number of maximum successive packet losses in simulation.

8.5.2 Leader-Following MAS

Consider a leader-following MAS with one leader and six agents that are connected by a communication graph as shown in Fig. 8.5. Based on Fig. 8.5, it can be obtained that $a(\hat{\mathscr{L}}) = 0.382$. Select $\mu = 0.1$ and $Q = 20I$, the gain matrix K can be calculated as $K = [2.1392\ 0.8120]$. Choose the functions $\chi_i(\eta_i) = \pi_i \eta_i$ with $\pi_{1,2} = 6$, $\pi_{3,4} =$

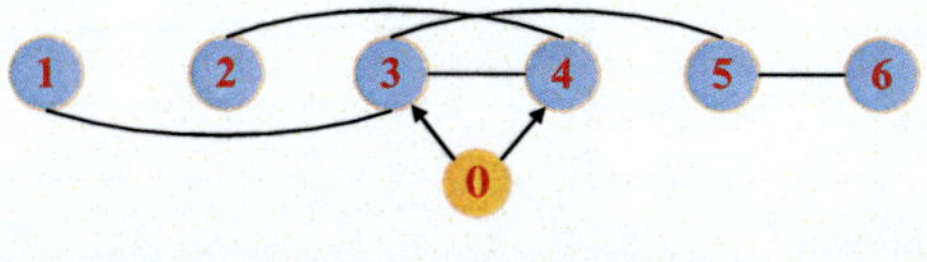

Fig. 8.5 Communication topology (leader-following MAS)

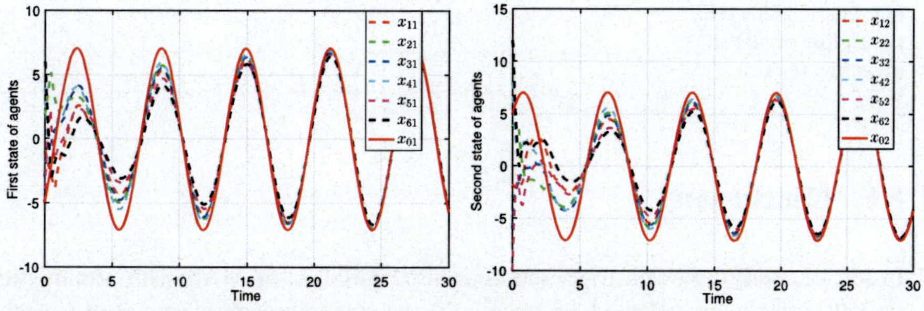

Fig. 8.6 The first states and second states of leader-following MAS

Fig. 8.7 Control input u_i of leader-following MAS

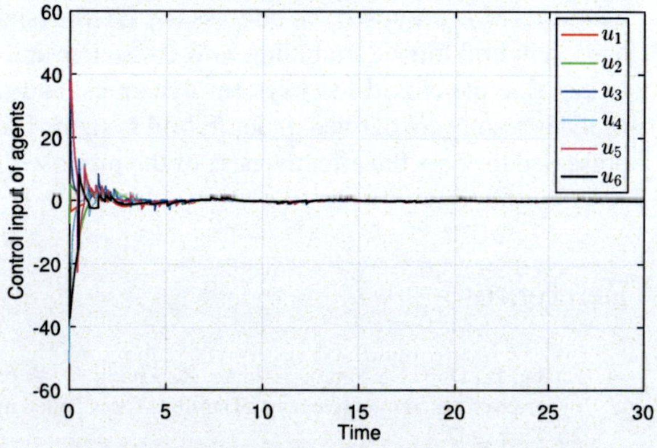

7, $\pi_{5,6} = 8$, for the sake of simplicity, the other parameters are selected the same as in the leaderless case, and the minimum inter-event time τ^i_{miet} and initial conditions are also chosen to be the same as in leaderless case. With these conditions, the simulation results are shown in Figs. 8.6, 8.7 and 8.8. From Fig. 8.6, it is shown that, the state trajectories of all follower agents can track the leader's states asymptotically. The control signals u_i of the agents are given in Fig. 8.7. To show the event-triggering instants and whether there exist packet losses at triggering instants, Fig. 8.8 presents the triggering instants and packet losses of the follower agents via small boxes, where red and blue boxes denotes successful and unsuccessful data transmission at triggering instants respectively. It can be seen that as long as the successive packet losses are not larger than the given NMASPL, the consensus tracking tracking under leader-following case can be achieved.

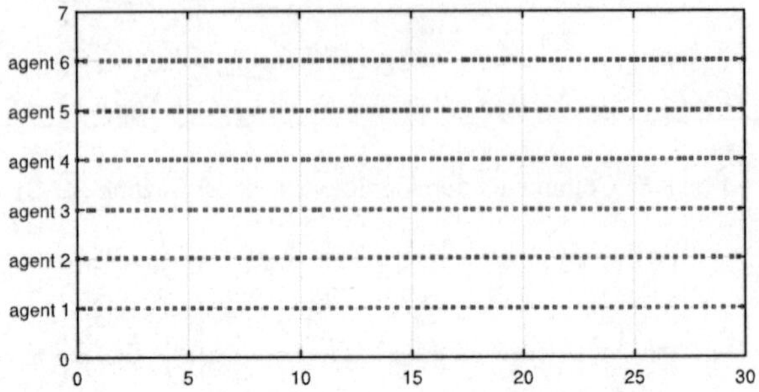

Fig. 8.8 Event-triggering instants and packet losses (leader-following MAS). The red box denotes successful data transmission at triggering instant, and blue box denotes that data transmission fails (i.e., packet loss occurs at triggering instant)

8.6 Conclusion

In this chapter, the event-triggered consensus problem of MAS with nonidentical packet losses is investigated, and a hybrid systems approach is proposed to tackle the leaderless and leader-following consensus problems in a unified framework. The distributed consensus control protocol is proposed based on an estimator, and a novel hybrid ETM is proposed. To address the ETM design and consensus analysis problems, a hybrid model, including flow dynamics and jump dynamics, is constructed to describe the closed-loop system dynamics, and the ETM design and consensus analysis results are derived under hybrid systems framework. Simulation results are presented to show the effectiveness of the proposed approach.

References

1. Zhang, B., Dou, C., Yue, D., Zhang, Z., Zhang, T.: A packet loss dependent event-triggered cyber-physical cooperative control strategy for islanded microgrid. IEEE Trans. Cybern. **51**(1), 267-282 (2021)
2. Zhang, D., Tang, Y., Ding, Z., Qian, F.: Event-based resilient formation control of multiagent systems. IEEE Trans. Cybern. **51**(5), 2490-2503 (2021)
3. Sun, Q., Han, R., Zhang, H., Zhou, J., Guerrero. J.: A multiagent-based consensus algorithm for distributed coordinated control of distributed generators in the energy internet. IEEE Trans. Smart Grid. **6**(6), 3006-3019 (2015)
4. Du, H., Wen, G., Cheng, Y., He, Y., Jia, R.: Distributed finite-time cooperative control of multiple high-order nonholonomic mobile robots. IEEE Trans. Neural Netw. Learn. Syst. **28**(12), 2998-3006 (2017)
5. Dimarogonas, D.V., Frazzoli, E., Johansson, K.H.: Distributed event-triggered control for multiagent systems. IEEE Trans. Autom. Control **57**(5), 1291-1297 (2012)
6. Garcia, E., Cao, Y., Casbeer, D.W.: Periodic event-triggered synchronization of linear multiagent systems with communication delays. IEEE Trans. Autom. Control **62**(1), 366-371 (2017)
7. Li, H., Liao, X., Huang, T., Zhu, W.: Event-triggering sampling based leader-following consensus in second-order multi-agent systems. IEEE Trans. Autom. Control. **60**(7), 1998-2003 (2015)
8. Fan, Y., Feng, G., Wang, Y., Song, C.: Distributed event-triggered control of multi-agent systems with combinational measurements. Automatica **49**(2), 671-675 (2013)

9. Yang, H., Zhao, H., Xia, Y., Zhang, J.: Event-triggered active MPC for nonlinear multiagent systems with packet losses. IEEE Trans. Cybern. **51**(6), 3093-3102 (2021)
10. Wen, G., Yu, W., Chen, M., Yu, X., Cheng, G.: H1 pinning synchronization of directed networks with aperiodic sampled-data communications. IEEE Trans. Circuits Syst. I-Regul. Pap. **61**(11), 3245-3255 (2014)
11. Hu, W., Yang, C., Huang, T., Gui, W.: A distributed dynamic event-triggered control approach to consensus of linear multiagent systems with directed networks. IEEE Trans. Cybern. **50**(2), 2996-3008 (2020)
12. Zhang, W., Tang, Y., Han, Q.L., Liu, Y.: Sampled-data consensus of linear time-varying multiagent networks with time-varying topologies. IEEE Trans. Cybern. **52**(1), 128-137 (2022)
13. Ge, X., Han, Q.L., Ding, D., Zhang, X.M., Ning, B.: A survey on recent advances in distributed sampled-data cooperative control of multiagent systems. Neurocomputing **275**(5), 1684-1701 (2018)
14. Liu, W., Huang, J.: Leader-following consensus for linear multi-agent systems via asynchronous sampled-data control. IEEE Trans. Autom. Control **65**(7), 3215-3222 (2019)
15. Tabuada, P.: Event-triggered real-time scheduling of stabilizing control tasks. IEEE Trans. Autom. Control **52**(9), 1680-1685 (2007)
16. Xu, W., Hu, G., Ho. D.W., Feng, Z.: Distributed secure cooperative control under denial-of-service attacks from multiple adversaries. IEEE Trans. Cybern. **50**(8), 3458-3467 (2020)
17. Deng, C., Er, M.J., Yang, G., Wang, N.: Event-triggered consensus of linear multiagent systems with time-varying communication delays. IEEE Trans. Cybern. **50**(7), 2916-2925 (2020)
18. Wang, A., Mu, B., Shi, Y.: Event-triggered consensus control for multiagent systems with time-varying communication and event detecting delays. IEEE Trans. Control Syst. Technol. **27**(2), 507-515 (2019)
19. Cai, Y., Zhang, H., Yu, R., He, Q.: Fully distributed bipartite output consensus of heterogeneous linear multiagent systems based on event-triggered transmission mechanism. Int J Robust Nonlinear Control **30**(8), 3382-3410 (2020)
20. Yu, H., Chen, X., Chen, T., Hao, F.: Event-triggered bipartite consensus for multi-agent systems: A Zeno-free analysis. IEEE Trans. Autom. Control **65**(11), 4866-4873 (2020)
21. Xu, L., Mo, Y., Xie, L.: Distributed consensus over markovian packet loss channels. IEEE Trans. Autom. Control **65**(1), 279-286 (2020)
22. Zhang, W., Tang, Y., Huang, T., Kurths, J.: Sampled-data consensus of linear multi-agent systems with packet losses. IEEE Trans. Neural Netw. Learn. Syst. **28**(11), 2516-2527 (2017)
23. Jiang, X., Xia, G., Feng, Z., Jiang, Z.: Consensus tracking of data-sampled nonlinear multi-agent systems with packet loss and communication delay. IEEE Trans. Netw. Sci. Eng. **8**(1), 126-137 (2021)
24. Xing, M., Deng, F., Hu, Z.: Sampled-data consensus for multiagent systems with time delays and packet losses. IEEE Trans. Syst., Man, Cybern.: Syst. **50**(1), 203-210 (2020)
25. Wang, F., Wen, G., Peng, Z., Huang, T., Yu, Y.: Event-triggered consensus of general linear multiagent systems with data sampling and random packet losses. IEEE Trans. Syst., Man, Cybern.: Syst. **51**(2), 1313-1321 (2021)
26. Jiang, X., Xia, G., Feng, Z., Li, T.: Non-fragile H1 consensus tracking of nonlinear multi-agent systems with switching topologies and transmission delay via sampled-data control. Inf. Sci. **509**(1), 210-226 (2020)
27. Ding, L., Han, Q.L., Guo, G.: Network-based leader-following consensus for distributed multi-agent systems. Inf. Sci. **49**(7), 2281-2286 (2013)
28. Zhao, G., Hua, C.: A hybrid dynamic event-triggered approach to consensus of multiagent systems with external disturbances. IEEE Trans. Autom. Control **66**(7), 3213-3220 (2021)
29. Heemels, W., Teel, A.R., Wouw, N., Nesic, D.: Networked control systems with communication constraints: Tradeoffs between transmission intervals, delays and performance. IEEE Trans. Autom. Control **55**(8), 1781-1796 (2010)
30. Carnevale, D., Teel, A.R., Nesic, D.: A Lyapunov proof of an improved maximum allowable transfer interval for networked control systems. IEEE Trans. Autom. Control **52**(5), 892-897 (2007)

31. Dolk, V., Heemels, W.: Event-triggered control systems under packet losses. Automatica **80**, 143-155 (2017)
32. Goebel, R., Sanfelice, R.G., Teel, A.R..: Hybrid Dynamical Systems: Modeling, Stability, Robustness. Princeton, NJ, USA: Princeton Univ. Press, (2012).
33. Heijmans, S., Postoyan, R., Nesic, D., Heemels, W.: Computing minimal and maximal allowable transmission intervals for networked control systems using the hybrid systems approach. IEEE Control Syst. Lett. **1**(1), 56-61 (2017)
34. Ruan, X., Feng, J., Xu, C., Wang, J.: Observer-based dynamic event-triggered strategies for leader-following consensus of multi-agent systems with disturbances. IEEE Trans. Netw. Sci. Eng. **7**(4), 3148-3158 (2020)

Part III
Reset Control of Multi-agent Systems

Chapter 9
Sampled-Data Based Reset Control of MAS

9.1 Introduction

In previous chapters, the topics with respect to sampled-data based and event-triggered cooperative control of MAS have been introduced. Since the cooperative control method of MAS has potential to be applied to many engineering applications [2–6], it has attracted attention of a large number of researchers in the field of automatic control, such as [7–14]. However, in these literature, the authors mainly focus on asymptotic stability behavior of system dynamics, by contrast, the transient performance is rarely studied except for several recent results [15–17]. Here, asymptotic stability behavior means that the consensus error variables of MAS converge to the origin eventually, the common transient performance includes overshoot and settling time. Based on practical application requirements, the fundamental control theory tells that the basic design criteria for a control system are stability, accuracy and rapidity, that corresponds to asymptotic behavior and transient performance. So, from the point of view of engineering applications, transient performance has the same significance as asymptotic stability. But how to improve transient performance while guaranteeing asymptotic consensus property is not easy.

Fortunately, with respect to the topic of how to improve transient performance, reset control method provides us inspirations to address this topic. Reset control is a control method that can reset the controller state when the system state meets certain reset conditions. The reset controller consists of a baseline controller and a state reset law. The baseline controller is a linear continuous/discrete time controller, and the state reset law includes two parts: reset trigger condition and reset mapping. Reset control has opened up a new field to break through the performance limitations of traditional linear control methods. Due to the constraint relationship between Bode amplitude and phase, traditional linear feedback control methods will generate significant phase lag, which greatly restricts the control performance of linear feedback control methods. Compared with the linear control method, reset control can reduce the phase lag without changing the amplitude frequency characteristics of the system, thus increasing the phase margin of the system, which brings many advantages

G. Zhao et al., *Cooperative Control of Multi-agent Systems*,
https://doi.org/10.1007/978-981-97-0968-7_9

to reset control: (1) Breaking through the performance limitations of linear control, such as the performance limitations of phase margin caused by linear control in the aircraft system [18]; (2) Improve the dynamic performance of the baseline system, such as improving position tracking performance in remote operation systems [19]; (3) Compensation for time delay, such as time delay compensation for temperature control systems of heat exchangers in chemical processes [20]; (4) Interference suppression, such as intermediate frequency interference suppression in hard drive systems [21]. Therefore, research on reset control not only has important theoretical significance, but also has broad prospects in practical applications.

The idea of reset control was originally proposed in [22], where the so-called Clegg integrator was introduced, and the state of the integrator is reset to zero when its input and output have opposite signs. Clegg integrator was extended to more general cases with zero-crossing reset conditions [23] and sector reset conditions [24]. The reset property, i.e., the controller states have jump dynamics when appropriate specified reset conditions are satisfied, provides reset control the ability to improve transient performance of linear systems, and this has been confirmed via examples [25–27] and theoretically [28]. Motivated by the advantages of reset control, in this chapter, we introduce how to apply reset control method to consensus of MAS with the objective of improving transient performance (e.g. settling time), since transient performance is important in some MASs such as multi-robot coordination and formation control of air vehicles. However, due to the controller state has jump dynamics when certain reset conditions are satisfied, when reset controller state is added into the consensus strategy, it leads to inapplicability of traditional Lyapunov-based consensus analysis method. For example, a quasi-resetting method was proposed in [17], and backward Euler method was used to implement the integral behavior. In [16], quasi-periodic reset mechanism was applied to leaders of MAS in centralized manner. In [15], the superiority of reset based consensus control approach is illustrated compared with traditional consensus algorithm, but the centralized resets in [15] could be further relaxed to be distributed resets such that it is more general.

In this chapter, compared with [15], a novel reset element is proposed, which is able to handle nonzero initial conditions and is implemented in a distributed way, and time regularization technique is utilized to avoid Zeno behavior. Furthermore, consider the network communication environment for multiple agents, we consider how to relax the requirement of continuous communication between neighboring agents. In particular, we design asynchronous sampled-data based reset consensus strategy to to avoid continuous communication and clock synchronization is not needed. Based on the proposed reset element, proportional integral + reset (PI+R) type control law is applied to consensus of MAS in both cases with continuous communication and with asynchronous sampling. The main results of this chapter include the following aspects: (1) reset control is utilized to consensus of MAS such that the transient consensus performance can be improved compared with traditional consensus protocols; (2) a hybrid systems framework is proposed to deal with the reset-induced state jump dynamics and sampling-induced error jump dynamics;

(3) compared with the existing work [15], centralized coordination of reset is not needed, asynchronous sampling based consensus protocol avoids continuous communication between neighboring agents and clock synchronization is not required.

9.2 System Model and Reset Control Law

In this section, we formulate the consensus problem of MAS with reset based control strategy. Since reset control of multi-input multi-output systems is still an open research topic, in this part, we only consider reset control design for single-input single-output systems. Thus, the MAS with first-order dynamics is considered in this chapter. Next, we first give the MAS model with first-order dynamics. Then, different from the existing results, we develop a consensus control law in PI+R type, such that the extra reset element in consensus control law can be used to improve transient performance.

Consider MAS with N agents that have the following single integrator dynamics

$$\dot{x}_i(t) = u_i(t), \quad i \in [1, N], \tag{9.1}$$

where $x_i(t) \in \mathbb{R}$ is the state and $u_i(t) \in \mathbb{R}$ is the control input. To be convenient, define the combined measurement variable $\delta_i(t)$ as follows

$$\delta_i(t) = \sum_{j \in \mathcal{N}_i} (x_j(t) - x_i(t)), \quad \text{for } i \in [1, N]. \tag{9.2}$$

For the purpose of achieving consensus of MAS and improving transient control performance, design a type of PI+R (Proportional Integral + Reset) control law as follows

$$u_i(t) = K_P \delta_i(t) + K_I x_{ri}(t), \tag{9.3}$$

where $x_{ri}(t)$ is the state of the reset element, and the dynamics of the reset element is given by

$$\left.\begin{aligned} \dot{\tau}_i(t) &= 1 \\ \dot{x}_{ri}(t) &= \delta_i(t) \end{aligned}\right\} \text{ If } (x_{ri}(t), \delta_i(t), \tau_i(t)) \in \mathbb{F}_i,$$
$$\left.\begin{aligned} \tau_i(t^+) &= 0 \\ x_{ri}(t^+) &= a_{ri} x_{ri} \end{aligned}\right\} \text{ If } (x_{ri}(t), \delta_i(t), \tau_i(t)) \in \mathbb{J}_i, \tag{9.4}$$

where $a_{ri} = 0$ if $x_{ri}(t)\delta_i(t) \le 0$, otherwise, $a_{ri} = p_i$ with $0 < p_i < 1$. With slightly abuse of notation, we use $x_{ri}(t^+)$ to denote the state $x_{ri}(t)$ right after reset, i.e., $x_{ri}(t^+) = \lim_{\epsilon \to 0^+} x_{ri}(t + \epsilon)$. The notation $\tau_i(t^+)$ is similar. Note that $x_{ri}(t)$ has jump dynamics at reset instants, so it is not differentiable everywhere, the derivative $\dot{x}_{ri}(t)$ denotes the generalized directional derivative. The flow set $\mathbb{F}_i$ and jump set $\mathbb{J}_i$ are defined in the following

$$\begin{aligned}\mathbb{F}_i =&\{(x_{ri}(t), \delta_i(t), \tau_i(t)) \in \mathbb{R}\times\mathbb{R}\times\mathbb{R} | \tau_i(t) \le \rho_i \vee \\ &(\tau_i(t) \ge \rho_i \wedge x_{ri}(t)\delta_i(t) \ge 0 \wedge \Psi_i(x_{ri}(t), \delta_i(t)) \ge 0)\} \\ \mathbb{J}_i =&\{(x_{ri}(t), \delta_i(t), \tau_i(t)) \in \mathbb{R}\times\mathbb{R}\times\mathbb{R} | \tau_i(t) \ge \rho_i \wedge (x_{ri}(t)\delta_i(t) \le 0 \vee \Psi_i(x_{ri}(t), \delta_i(t)) \le 0)\}\end{aligned} \tag{9.5}$$

where the notations $\wedge$ and $\vee$ denote 'and' and 'or' respectively. $\tau_i(t)$ is a timer variable that is used to record the time elapsed since the last reset instant, the role of the variable $\tau_i(t)$ is to achieve time regularization in order to avoid Zeno behavior, and the time interval between reset instants is enforced to be lower bounded by $\rho_i > 0$. $0 < p_i < 1$ is a proportion coefficient. $\Psi_i(x_{ri}(t), \delta_i(t))$ is the reset function (RF). How to select these parameters and design the RF will be given later.

Remark 9.1 Compared with [15], where the zero-crossing reset conditions were used to determine when reset event occurs, sector reset conditions with time regularization in this section are developed, which can overcome the possible ill-posedness problem of the results in [15]. It is worth to be noted that time regularization technique was also used to exclude Zeno reset behavior in [15], however, the centralized coordination of reset was required in [15], which is a restricted requirement that can be relaxed by using the proposed (9.4).

The preliminaries about graph, which is used to describe the communication relationship between agents, have been introduced in Chap. 1. For proceeding this chapter, we make Assumption 3.1 to limit the communication connection among agents. Then, according to the graph theory, it can obtain the following Lemma 9.1, which is crucial for stability analysis.

Assumption 9.1 The communication graph $\mathscr{G}$ is undirected and connected.

Lemma 9.1 ([1]) *For the directed and strongly connected graph $\mathscr{G}$ with Laplacian matrix $\mathscr{L}$, there always exists a vector $\zeta = [\zeta_1\ \zeta_2\ \cdots\ \zeta_N]^T$ with all the elements being positive such that $\zeta^T\mathscr{L} = 0$ and $\sum_{i=1}^N \zeta_i = 1$. Define $Z = diag\{\zeta_1, \zeta_1, \ldots, \zeta_N\}$, then, it has that $\hat{\mathscr{L}} = \frac{1}{2}(Z\mathscr{L} + \mathscr{L}^T Z)$ is symmetric and $\sum_{j=1}^N \hat{\mathscr{L}}_{ij} = \sum_{j=1}^N \hat{\mathscr{L}}_{ji} = 0$ for $i \in [1, N]$. Moreover, the following condition holds*

$$\lambda_2(\mathscr{L}) = \min_{x^T\zeta=0, x\neq 0} \frac{x^T\hat{\mathscr{L}}x}{x^T Z x} > 0 \tag{9.6}$$

with $\lambda_2(\mathscr{L})$ is the minimum positive eigenvalue of $\mathscr{L}$.

In this chapter, the control objective is to design PI+R control strategy that not only realizes consensus of the MAS [9], that is, the following conditions are satisfied,

$$\lim_{t\to\infty} \|x_i(t) - x_j(t)\| = 0,\ i, j \in [1, N], \tag{9.7}$$

but also achieves better transient performance (e.g., settling time) compared with static/dynamic state feedback and the existing reset control methods.

9.3 Stability Analysis and Reset Control Design

This section includes two parts: in the first part, we consider consensus problem of MAS with continuous communication, which means that the agent i's neighbors' information is available for agent i in real time. In this case, the stability analysis and reset control design is relatively easy in contrast with sampling communication, and it can be viewed as the basis for stability analysis and reset control design in the case with sampling communication. In the second part, we consider the consensus problem of MAS with asynchronous sampling communication, it is assumed that only sampled information of neighbors at their sampling instants is available for a certain agent. In this case, the sampling communication will induce extra error signals (the error between real-time inforamtion and sampled information) compared with the case with continuous communication. To deal with the reset control based consensus problem in both cases, we construct hybrid model to completely describe the closed-loop dynamics, the hybrid model contains the reset function, reset map, sampling period. Then, the stability analysis and reset control design are given under hybrid systems framework.

9.3.1 *Consensus Based on Continuous Communication*

Due to reset element state $x_{ri}(t)$ will be reset to $a_{ri}x_{ri}(t)$ when $(x_{ri}(t), \delta_i(t), \tau_i(t)) \in \mathbb{D}_i$, and continuously evolves at the rest time. It means that $x_{ri}(t)$ has both flow dynamics and jump dynamics, thus, traditional analysis and synthesis method based on continuous-time system model is inapplicable in this case. For the purpose of stability analysis and reset control design, we next construct a hybrid dynamical system model for the closed-loop MAS with reset control law, which can appropriately describe the system dynamics. The aim is to construct a hybrid model in the following form

$$\mathbb{H}: \begin{cases} \dot{\xi} = \mathscr{F}(\xi), & \text{If } \xi \in \mathbb{F}, \\ \xi^+ \in \mathscr{J}(\xi), & \text{If } \xi \in \mathbb{J}, \end{cases} \tag{9.8}$$

where $\mathscr{F}(\xi)$ describes the flow dynamics, $\mathscr{J}(\xi)$ describes the jump dynamics, $\mathbb{F}$ and $\mathbb{J}$ denote the flow set and jump set respectively. ξ^+ is short for $\xi(t^+)$, which is defined as $\xi(t^+) = \lim\limits_{\epsilon \to 0^+} \xi(t + \epsilon)$.

In order to describe the dynamics of closed-loop MAS in the form of (9.8), define combinational variables

$$\delta = \left[\delta_1\ \delta_2\ \cdots\ \delta_N\right]^T,\ x_r = \left[x_{r1}\ x_{r2}\ \cdots\ x_{rN}\right]^T,\ \tau = \left[\tau_1\ \tau_2\ \cdots\ \tau_N\right]^T,\ \xi = \left[\delta^T\ x_r^T\ \tau^T\right]^T. \quad (9.9)$$

Then, based on (9.1), (9.3), (9.4) and (9.9), the flow dynamics of the closed-loop MAS can be described as the form of

$$\dot{\xi} = \mathscr{F}(\xi), \text{ If } \xi \in \mathbb{F}, \quad (9.10)$$

where

$$\mathscr{F}(\xi) = \begin{bmatrix} -K_P\mathscr{L}\delta - K_I\mathscr{L}x_r \\ \delta \\ 1_N \end{bmatrix}.$$

The flow set $\mathbb{F} := \bigcap_{i=1}^N \mathbb{F}_i$, with $\mathbb{F}_i := \{\xi \in \mathbb{X} | (x_{ri}, \delta_i, \tau_i) \in \mathbb{C}_i\}$, where $\mathbb{X} := \{(\delta, x_r, \tau) \in \mathbb{R}^N \times \mathbb{R}^N \times \mathbb{R}^N_{\geq 0}\}$.

Similarly, the jump dynamics of the closed-loop MAS can be described by

$$\xi^+ \in \bigcup_{i=1}^N \mathscr{J}_i(\xi), \text{ If } \xi \in \mathbb{J}, \quad (9.11)$$

where $\mathbb{J} := \bigcup_{i=1}^N \mathbb{J}_i$ with $\mathbb{J}_i := \{\xi \in \mathbb{X} | (x_{ri}, \delta_i, \tau_i) \in \mathbb{D}_i\}$, and

$$\mathscr{J}_i(\xi) := \begin{cases} \mathscr{J}_{1i}(\xi), & \text{If } \xi \in \mathbb{J}_i \wedge x_{ri}\delta_i \leq 0, \\ \mathscr{J}_{2i}(\xi), & \text{If } \xi \in \mathbb{J}_i \wedge \Psi_i(x_{ri}, \delta_i) \leq 0, \end{cases} \quad (9.12)$$

where

$$\mathscr{J}_{1i}(\xi) = \begin{bmatrix} \delta \\ (I_N - \Gamma_i)x_r \\ (I_N - \Gamma_i)\tau \end{bmatrix},\ \mathscr{J}_{2i}(\xi) = \begin{bmatrix} \delta \\ (I_N - (1 - p_i)\Gamma_i)x_r \\ (I_N - \Gamma_i)\tau \end{bmatrix}.$$

The function $\mathscr{J}_i(\xi)$ denotes how ξ jumps at reset instants. From (9.5) and (9.12), we can see that there exist two types of resets, in particular, the reset element state x_{ri} is reset to zero when $x_{ri}\delta_i \leq 0$ holds, x_{ri} is reset to $p_i x_{ri}$ when $\Psi_i(x_{ri}, \delta_i) \leq 0$ holds. If both $x_{ri}\delta_i \leq 0$ and $\Psi_i(x_{ri}, \delta_i) \leq 0$ hold, $x_{ri}\delta_i \leq 0$ has the priority and x_{ri} is reset to zero. Γ_i is a diagonal matrix with ii-th entry is equal to one and all the rest entries are equal to zeros.

Now, the closed-loop MAS have been constructed as a hybrid system model as shown in (9.10) and (9.11), and it indicates that the consensus of agents' states is achieved if the constructed hybrid system is asymptotically stable. The stability property of the hybrid system consisting of (9.10) and (9.11) is defined as follows.

Definition 9.1 [29] For the constructed hybrid model as described by (9.10) and (9.11). The set given by $\{\xi \in \mathbb{X} | \delta = 0_N, x_r = 0_N\}$ is said to be uniformly globally asymptotically stable (UGAS) if there exists a function $\beta \in \mathscr{K}\mathscr{L}$ such that, for any initial condition $\xi(0) \in \mathbb{X}$, the following condition holds

$$\|(\delta(t), x_r(t))\| \leq \beta(\|(\delta(0), x_r(0))\|, t). \tag{9.13}$$

For the purpose of stability analysis, we introduce an internal function $\phi_i(\tau_i) : \mathbb{R}_{\geq 0} \to \mathbb{R}_{\geq 0}$ for $i \in [1, N]$, with respect to the timer variable τ_i, which is helpful for stability analysis

$$\begin{aligned} \dot{\phi}_i(\tau_i) &= -s(\tau_i)\theta_i(\phi_i^2(\tau_i) + 1), \\ s(\tau_i) &= \max\{\text{sign}(\rho_i - \tau_i), 0\}, \end{aligned} \tag{9.14}$$

where $\theta_i > 0$ is a constant and the initial condition $\phi_i(0)$ is given to be positive, sign$(\cdot)$ denotes the sign function. The function $\phi_i(\tau_i)$ is useful for consensus analysis of the MAS described by hybrid model, and how to determine parameters ρ_i depends on $\phi_i(\tau_i)$, specifically, the value of ρ_i can be chosen based on the result in the following lemma.

Lemma 9.2 *For given initial conditions $\phi_i(0) > 0$, $\theta_i > 0$, $i \in [1, N]$, if ρ_i satisfies the following condition*

$$0 < \rho_i < \rho_M^i \tag{9.15}$$

with $\rho_M^i = \frac{1}{\theta_i}\arctan(\phi_i(0))$. Then, $\phi_i(\tau_i) > 0$ is ensured for $\tau_i \leq \rho_i$.

Proof According to the differential equation in (9.14), the upper bound ρ_M^i can be easily calculated with initial condition $\phi_i(0) > 0$, so, the detailed calculation process is omitted here. □

Now, we are ready to state the main results.

Theorem 9.1 *Consider the MAS with hybrid model (9.10) and (9.11). Let $K_P > 0$ and $K_I = \alpha K_P$ with $\alpha > 0$, if there exist constants $0 < \sigma < 1$, $0 < \epsilon < \theta_i^2$, $P > 0$, $0 < p_i < 1$ such that $PK_I \leq \frac{1}{2}\theta_i^2$ and the following inequality conditions hold*

$$\begin{aligned} \begin{bmatrix} -(1-\sigma)(Z \otimes Q) + I & 0 \\ 0 & -\frac{1}{2}\Theta + \epsilon I \end{bmatrix} &\leq 0, \\ 2\lambda_2(\mathscr{L}) - \alpha \left\|\mathscr{L}^2 Z\right\| &> 0, \\ p_i^2\phi_i(0) &\leq \phi_i(\rho_i), \end{aligned} \tag{9.16}$$

where $Q = (2\lambda_2(\mathscr{L}) - \alpha \left\|\mathscr{L}^2 Z\right\|)PK_P$, Θ is a diagonal matrix with $\Theta = diag\{\theta_1^2, \theta_2^2, \ldots, \theta_N^2\}$. Design reset function $\Psi_i(x_{r_i}, \delta_i)$ as the following form

$$\Psi_i(x_{r_i}, \delta_i) = \sigma_\delta\zeta_i\delta_i^T Q\delta_i - \theta_i^2(\phi_i^2(\tau_i) + 1)\|x_{ri}\|^2 \tag{9.17}$$

where $0 < \sigma_\delta < \sigma$, ζ_i *is the diagonal entry of* Z. *Then, the* (δ, x_r) *dynamics is UGAS, which implies that the consensus of the MAS is achieved.*

Proof For stability analysis and reset control design, choose the Lyapunov function

$$V(\xi) = \delta^T(Z \otimes P)\delta + \sum_{i=1}^{N} \theta_i \phi_i(\tau_i) \|x_{ri}\|^2,$$

where $\theta_i > 0$ and $\phi_i(\tau_i) > 0$. It is clear that $V(\xi) \geq 0$ and $V(\xi) = 0$ if and only if $\delta = x_r = 0$, so, $V(\xi)$ is an appropriate Lyapunov function candidate.

The objective is to analyze and show that the following inequality conditions hold, such that the system dynamics is asymptotically stable

$$\begin{aligned}
&\langle \nabla V(\xi), \mathscr{F}(\xi) \rangle \leq -\sigma_\delta(\delta) - \sum_{i=1}^{N} \sigma_r(x_{ri}), \text{ for } \xi \in \mathbb{F}, \\
&V(\xi^+) - V(\xi) \leq 0, \text{ for } \xi \in \mathbb{J},
\end{aligned}$$

where $\sigma_\delta(\cdot)$, $\sigma_r(\cdot)$ are class $\mathscr{K}_\infty$ functions. Then, the (δ, x_r) dynamics in ξ are UGAS. In other words, the consensus of the MAS is achieved.

First, consider the evolution of $V(\xi)$ along flow dynamics, for almost all $\tau \in \mathbb{R}^N_{\geq 0}$, $(\delta, x_r) \in \mathbb{R}^N \times \mathbb{R}^N$, we have

$$\begin{aligned}
&\langle \nabla V(\xi), \mathscr{F}(\xi) \rangle \\
=& -\delta^T((Z \otimes P)(\mathscr{L} \otimes K_P) + (\mathscr{L}^T \otimes K_P)(Z \otimes P))\delta \\
&- 2\delta^T(Z \otimes P)(\mathscr{L} \otimes K_I)x_r + \sum_{i=1}^{N} 2\theta_i \phi_i(\tau_i) \|x_{ri}\| \, \|\delta_i\| \qquad (9.18) \\
&- \sum_{i=1}^{N} s_i(\tau_i) \left(\theta_i^2 \phi_i^2(\tau_i) \|x_{ri}\|^2 + \theta_i^2 \|x_{ri}\|^2\right).
\end{aligned}$$

Note that $\delta^T \zeta = 0$, based on Lemma 9.1, we obtain

$$\begin{aligned}
&-\delta^T((Z \otimes P)(\mathscr{L} \otimes K_P) + (\mathscr{L}^T \otimes K_P)(Z \otimes P))\delta - 2\delta^T(Z \otimes P)(\mathscr{L} \otimes K_I)x_r \\
\leq& \delta^T(-(Z\mathscr{L} + \mathscr{L}^T Z) \otimes PK_P)\delta + x_r^T(I_N \otimes PK_I)x_r + \delta^T(Z\mathscr{L}\mathscr{L}^T Z \otimes PK_I)\delta \\
\leq& \delta^T(-(2\lambda_2(\mathscr{L}) - \alpha \left\|\mathscr{L}^2 Z\right\|)Z \otimes PK_P)\delta + x_r^T \|PK_I\| x_r \qquad (9.19) \\
=& -\delta^T(Z \otimes Q)\delta + x_r^T \|PK_I\| x_r.
\end{aligned}$$

Let the first matrix inequality in (9.16) left-multiplied by $[\delta^T \ x_r^T]$ and right-multiplied by $[\delta^T \ x_r^T]^T$, it can be derived that

$$
\begin{aligned}
& - \delta^T (Z \otimes Q)\delta + x_r^T \, \|PK_I\| \, x_r \\
\leq & - \sigma \delta^T (Z \otimes Q)\delta - \|\delta\|^2 + \frac{1}{2} x_r^T \Theta x_r - \epsilon \, \|x_r\|^2 + x_r^T \, \|PK_I\| \, x_r \\
\leq & - \sigma \delta^T (Z \otimes Q)\delta - \sum_{i=1}^{N} \left(\|\delta_i\|^2 - (\theta_i^2 - \epsilon) \, \|x_{ri}\|^2 \right).
\end{aligned}
\tag{9.20}
$$

Based on (9.18), (9.19), (9.20) and (9.14), when $\tau_i < \rho_i$, we have

$$
\begin{aligned}
& \langle \nabla V(\xi), \mathscr{F}(\xi) \rangle \\
\leq & - \sigma \delta^T (Z \otimes Q)\delta - \sum_{i=1}^{N} \left((\|\delta_i\| - \theta_i \phi_i(\tau_i) \, \|x_{ri}\|)^2 + \epsilon \, \|x_{ri}\|^2 \right) \\
\leq & - \sigma \delta^T (Z \otimes Q)\delta - \sum_{i=1}^{N} \epsilon \, \|x_{ri}\|^2 .
\end{aligned}
\tag{9.21}
$$

When $\tau_i \geq \rho_i$, $s_i(\tau_i) = 0$, in this case, we have

$$
\begin{aligned}
& \langle \nabla V(\xi), \mathscr{F}(\xi) \rangle \\
\leq & - \sigma \delta^T (Z \otimes Q)\delta - \sum\nolimits_{i=1}^{N} \epsilon \, \|x_{ri}\|^2 + \sum\nolimits_{i=1}^{N} \left(\theta_i^2 \phi_i^2(\tau_i) \, \|x_{ri}\|^2 + \theta_i^2 \, \|x_{ri}\|^2 \right) \\
\leq & - (\sigma - \sigma_\delta) \delta^T (Z \otimes Q)\delta - \sum\nolimits_{i=1}^{N} \epsilon \, \|x_{ri}\|^2 - \sum\nolimits_{i=1}^{N} \left(\sigma_\delta \zeta_i \delta_i^T Q \delta_i - \theta_i^2 (\phi_i^2(\tau_i) + 1) \, \|x_{ri}\|^2 \right) \\
\leq & - (\sigma - \sigma_\delta) \delta^T (Z \otimes Q)\delta - \sum\nolimits_{i=1}^{N} \epsilon \, \|x_{ri}\|^2 - \sum\nolimits_{i=1}^{N} \Psi_i(x_{ri}, \delta_i) \\
\leq & - (\sigma - \sigma_\delta) \delta^T (Z \otimes Q)\delta - \sum\nolimits_{i=1}^{N} \epsilon \, \|x_{ri}\|^2 .
\end{aligned}
\tag{9.22}
$$

The last inequality is derived according to (9.5) and (9.17).

Next, consider the evolution of $V(\xi)$ along system's jump dynamics. For each $i \in [1, N]$, if $(\delta_i, x_{ri}, \tau_i) \in \mathbb{J}_i$, based on the jump dynamics (9.11), there exist two cases.

Case 1: $\xi_i \in \mathbb{J}_i \ \wedge \ x_{ri}\delta_i \leq 0$.

From (9.12), the x_{ri} state jumps to zero, and

$$
\begin{aligned}
& V(\xi^+) - V(\xi) \\
= & \delta^+ (Z \otimes Q)\delta^+ + \sum\nolimits_{i=1}^{N} \theta_i \phi_i(\tau_i^+) \, \|x_{ri}^+\|^2 \\
& - \delta (Z \otimes Q)\delta - \sum\nolimits_{i=1}^{N} \theta_i \phi_i(\tau_i) \, \|x_{ri}\|^2 \overset{(9.4),(9.5)}{\leq} 0.
\end{aligned}
\tag{9.23}
$$

Case 2: $\xi_i \in \mathbb{J}_i \ \wedge \ \Psi_i(x_{ri}, \delta_i) \leq 0$.

Based on (9.12), the x_{ri} state jumps to $p_i x_{ri}$, and

$$
\begin{aligned}
&V(\xi^+) - V(\xi) \\
&= \delta^+ (Z \otimes Q)\delta^+ + \sum\nolimits_{i=1}^{N} \theta_i \phi_i(\tau_i^+) \left\| x_{ri}^+ \right\|^2 - \delta(Z \otimes Q)\delta - \sum\nolimits_{i=1}^{N} \theta_i \phi_i(\tau_i) \left\| x_{ri} \right\|^2 \quad (9.24) \\
&\overset{(9.11)}{\leq} \theta_i \phi_i(0) p_i^2 \left\| x_{ri} \right\|^2 - \theta_i \phi_i(\rho_i) \left\| x_{ri} \right\|^2 \overset{(9.16)}{\leq} 0.
\end{aligned}
$$

Then, based on (9.21), (9.22), (9.23) and (9.24), we can conclude that the required stability conditions in (9.18) are satisfied. Therefore, the (δ, x_r) dynamics is UGAS and consensus is achieved. □

Remark 9.2 As discussed in [15], the regularity of the reset control systems is important. Here, we adopt a simple but effective strategy (different from [15]) to guarantee the regularity, namely the time interval between two reset instants is enforced to be upper bounded by ρ_{iM}. Based on this idea, the reset element in each agent will experience jump dynamics.

Remark 9.3 The role of parameter ρ_i is to exclude Zeno solutions, thus it can be selected to be any positive value that satisfies the conditions in (9.15), and each agent can select its own ρ_i independent of each other. Generally, ρ_i is selected to be small (e.g., 10^{-3}), but according to (9.15), larger ρ_i is also allowed. In particular, if $\rho_i = \rho_{iM}$, agent i will experience resets periodically.

Remark 9.4 Based on the conditions $p_i^2 \phi_i(0) \leq \phi_i(\rho_i)$, small ρ_i implies relative large p_i and vice versa, but small p_i may weaken the role of reset control for improving transient performance. Thus, there exists tradeoff between ρ_i and p_i.

Remark 9.5 The developed Lyapunov-based consensus analysis implies that the results can be extended to robustness analysis for MAS with reset control protocol, for instance, consider the agent i with dynamics $\dot{x}_i = u_i + w_i$, where w_i denotes external disturbance. However, it is not easy to analyze the robustness property to external disturbances with the methods in [15].

9.3.2 *Consensus Based on Asynchronous Sampling*

Note from (9.2) and (9.3) that, to implement the control protocol, continuous communication between neighboring agents is needed. In practical applications, agents in MAS usually communicate with each other via a shared network, however, continuous communication implies ideal network environment with unlimited bandwidth, which is a restrictive requirement and not easy to be realized. In this section, we extend the results in previous section to the case with asynchronous sampling communication between agents, such that the clock synchronization and continuous communication are not required.

Let $t_{k_i}^i$ denote the k_i-th sampling instant of agent i with sampling period $h_i = t_{k_i+1}^i - t_{k_i}^i$, and define

$$\hat{x}_i(t) = x_i(t^i_{k_i}), \quad \text{for } t \in [t^i_{k_i}, t^i_{k_i+1}),$$
$$\hat{\delta}_i(t) = \sum_{j \in \mathcal{N}_i} (\hat{x}_j(t) - \hat{x}_i(t)), \quad \text{for } t \in [t^i_{k_i}, t^i_{k_i+1}). \tag{9.25}$$

With the sampled-data instead of real-time information, the distributed PI+R control law is designed as

$$u_i(t) = K_P \hat{\delta}_i(t) + K_I x_{ri}(t), \tag{9.26}$$

which has the same form as (9.3), but with $\hat{\delta}_i(t)$ instead of $\delta_i(t)$. Owning to the asynchronous sampling, $\hat{\delta}_i(t)$ is piecewise constant, and it is updated at sampling instants $t^i_{k_i}$, and is also updated at $t^j_{k_j}$, such that continuous communication is avoided.

Define error variable

$$e_i(t) = \hat{x}_i(t) - x_i(t). \tag{9.27}$$

It is obvious that the error $e_i(t)$, $i \in [1, N]$ jumps at sampling instant $t^i_{k_i}$ and flows during $[t^i_{k_i}, t^i_{k_{i+1}})$. The error dynamics can be described as follows by introducing a timer variable τ_i.

$$\left.\begin{aligned} \dot{\tau}_i &= \frac{1}{h_i} \\ \dot{e}_i &= -u_i \end{aligned}\right\} \text{ If } \tau_i < 1,$$
$$\left.\begin{aligned} \tau_i^+ &= 0 \\ e_i^+ &= 0 \end{aligned}\right\} \text{ If } \tau_i = 1. \tag{9.28}$$

By introducing a new timer variable $\tau_{i_R}(t)$, we propose the following reset element

$$\left.\begin{aligned} \dot{\tau}_{i_R} &= 1 \\ \dot{x}_{ri} &= \hat{\delta}_i \end{aligned}\right\} \text{ If } (x_{ri}\hat{\delta}_i \geq 0 \ \wedge \ \tau_i = 1 \ \wedge \ \tau_{i_R} < \Delta_i h_i) \ \vee \ \tau_i < 1,$$
$$\left.\begin{aligned} \tau_{i_R}^+ &= 0 \\ x_{ri}^+ &= a_{r_R i} x_{ri} \end{aligned}\right\} \text{ If } (x_{ri}\hat{\delta}_i \leq 0 \ \wedge \ \tau_i = 1) \ \vee \ \tau_{i_R} \geq \Delta_i h_i, \tag{9.29}$$

where $a_{r_R i} = 0$ if $x_{ri}\hat{\delta}_i \leq 0$, otherwise, $a_{r_R i} = p_i$ with $0 < p_i < 1$. Δ_i is a positive integer.

Remark 9.6 According to (9.29), it can be seen that the reset only occurs at sampling instants, thus, it is naturally to ensure that Zeno reset behavior does not occur. Due to sampling mechanism, $\delta_i(t)$ is not continuously available for agent i, different from the case with continuous communication, here, we use the reset condition $\tau_{i_R} \geq \Delta_i h_i$ instead of the reset function $\Psi(\cdot)$, such that the reset element state x_{ri} of each agent is guaranteed to have jump dynamics.

Next, based on the sampled-data based reset control law, we construct a hybrid model for the MAS, let's define

$$e=[e_1\ e_2\ \cdots\ e_N]^T,\ \tau=[\tau_1\ \tau_2\ \cdots\ \tau_N]^T,\ \tau_R=[\tau_{1_R}\ \tau_{2_R}\ \cdots\ \tau_{N_R}]^T,\ \bar{\xi}=[\delta^T\ x_r^T\ e^T\ \tau^T\ \tau_R^T]^T. \quad (9.30)$$

Then, based on (9.1), (9.25)–(9.29), the dynamics of the closed-loop MAS with asynchronous sampling can be represented as the following hybrid systems form

$$\begin{aligned} &\dot{\bar{\xi}} = \mathscr{F}(\bar{\xi}),\ \text{If}\ \bar{\xi} \in \bar{\mathbb{F}}, \\ &\bar{\xi}^+ \in \bigcup_{i=1}^{N} \mathscr{J}_i(\bar{\xi}),\ \ \text{If}\ \bar{\xi} \in \bar{\mathbb{J}}, \end{aligned} \quad (9.31)$$

with $\bar{\mathbb{F}} := \bigcap_{i=1}^N \bar{\mathbb{F}}_i$, $\bar{\mathbb{J}} := \bigcup_{i=1}^N \bar{\mathbb{J}}_i$, and

$$\bar{\mathbb{F}}_i := \{\bar{\xi} \in \bar{\mathbb{X}} | \tau_i < 1\},\ \ \bar{\mathbb{J}}_i := \{\bar{\xi} \in \bar{\mathbb{X}} | \tau_i = 1\},$$

where $\bar{\mathbb{X}} := \{(\delta, x_r, e, \tau, \tau_R) \in \mathbb{R}^N \times \mathbb{R}^N \times \mathbb{R}^N \times \mathbb{R}^N_{\geq 0} \times \mathbb{R}^N_{\geq 0}\}$. Let $H = [1/h_1\ 1/h_2\ \cdots\ 1/h_N]^T$, we have

$$\mathscr{F}(\bar{\xi}) = \begin{bmatrix} -K_P\mathscr{L}\delta - K_I\mathscr{L}x_r + K_P\mathscr{L}^2 e \\ \delta - \mathscr{L}e \\ -K_P\delta - K_I x_r + K_P\mathscr{L} \\ H \\ 1_N \end{bmatrix},$$

$$\mathscr{J}_i(\bar{\xi}) := \begin{cases} \mathscr{J}_{1i}(\bar{\xi}), & \text{If}\ \bar{\xi} \in \bar{\mathbb{J}}_i \wedge x_{ri}\hat{\delta}_i \geq 0 \wedge \tau_{i_R} < \Delta_i h_i \\ \mathscr{J}_{2i}(\bar{\xi}), & \text{If}\ \bar{\xi} \in \bar{\mathbb{J}}_i \wedge x_{ri}\hat{\delta}_i \leq 0 \wedge \tau_{i_R} < \Delta_i h_i \\ \mathscr{J}_{3i}(\bar{\xi}), & \text{If}\ \bar{\xi} \in \bar{\mathbb{J}}_i \wedge \tau_{i_R} \geq \Delta_i h_i \end{cases} \quad (9.32)$$

where

$$\begin{aligned} \mathscr{J}_{1i}(\bar{\xi}) &= \left[\delta^T\ x_r^T\ \left[e^T\ \tau^T\right](I_N - \Gamma_i)\ \tau_R^T\right]^T, \\ \mathscr{J}_{2i}(\bar{\xi}) &= \left[\delta^T\ \left[x_r^T\ e^T\ \tau^T\ \tau_R^T\right](I_N - \Gamma_i)\right]^T, \\ \mathscr{J}_{3i}(\bar{\xi}) &= [\delta^T\ \ x_r^T(I_N - (1 - p_i)\Gamma_i)\left[e^T\ \tau^T\ \tau_R^T\right](I_N - \Gamma_i)]^T. \end{aligned}$$

Hence, in the case with asynchronous sampling, the MAS can also be represented as hybrid systems form. Similar as in Sect. 9.3.1, we introduce an internal function $\phi_{i_R}(\tau_{i_R}) : \mathbb{R}_{\geq 0} \to \mathbb{R}_{\geq 0}$, $i \in [1, N]$ with the dynamics

$$\dot{\phi}_{i_R}(\tau_{i_R}) = -\theta_{i_R}(\phi_{i_R}^2(\tau_{i_R}) + 1), \tag{9.33}$$

where $\theta_{i_R} > 0$ is a constant and the initial condition $\phi_{i_R}(0)$ is given to be positive. By means of the results in Lemma 9.2, the parameters h_i and Δ_i can be selected to satisfy $0 < h_i < \rho_M^{iR}$ and $1 \le \Delta_i \in \mathbb{N} < \rho_M^{iR}/h_i$ with $\rho_M^{iR} = 1/\theta_{i_R}\arctan\phi_{i_R}(0)$.

Theorem 9.2 *Consider the MAS with hybrid systems form (9.31). Let $K_P > 0$, $K_I > 0$, if there exists constants $0 < \sigma < 1$, $0 < \epsilon < \min\{\theta_i^2, \theta_{i_R}^2\}$, $P > 0$, $0 < p_i < 1$ such that the following inequality conditions hold*

$$\begin{aligned} &\begin{bmatrix} \Xi_{11} & \Xi_{12} & \Xi_{13} \\ * & \Xi_{22} & \Xi_{23} \\ * & * & \Xi_{33} \end{bmatrix} \le 0, \\ &p_i^2\phi_i(0) \le \phi_i(\Delta_i h_i), \end{aligned} \tag{9.34}$$

where

$$\begin{aligned}
\Xi_{11} &= -(1-\sigma)Z \otimes Q + \sum\nolimits_{i=1}^{N}(R_i^T R_i + \bar{R}_i^T \bar{R}_i), \\
\Xi_{12} &= -(Z \otimes P)K_I\mathscr{L} + \sum\nolimits_{i=1}^{N} \bar{R}_i^T \bar{R}_{i_R}, \\
\Xi_{13} &= (Z \otimes P)K_P\mathscr{L}^2 + \sum\nolimits_{i=1}^{N}(R_i^T R_{i_E} + \bar{R}_i^T \bar{R}_{i_E}), \\
\Xi_{22} &= -\Theta_i + \epsilon I + \sum\nolimits_{i=1}^{N} \bar{R}_{i_R}^T \bar{R}_{i_R}, \ \Xi_{23} = \sum\nolimits_{i=1}^{N} \bar{R}_{i_R}^T \bar{R}_{i_E}, \\
\Xi_{33} &= -\Theta_{i_R} + \epsilon I + \sum\nolimits_{i=1}^{N}(R_{i_E}^T R_{i_E} + \bar{R}_{i_E}^T \bar{R}_{i_E}),
\end{aligned}$$

where $Q = 2\lambda_2(\mathscr{L})PK_P$. $R_i \in \mathbb{R}^{1\times N}$ is a row vector with the i-th entry is one and the other entries are zeros. $R_{iE} = -\mathscr{L}_{row}^i$, $\bar{R}_i = -K_P R_i$, $\bar{R}_{i_E} = K_P\mathscr{L}_{row}^i$, $\bar{R}_{i_R} = -K_I R_i$. $\mathscr{L}_{row}^i$ denotes the i-th row of Laplacian matrix $\mathscr{L}$. Θ and Θ_{i_R} are diagonal matrices with $\Theta = diag\{\theta_1^2, \theta_2^2, \ldots, \theta_N^2\}$ and $\Theta_R = diag\{\theta_{1_R}^2, \theta_{2_R}^2, \ldots, \theta_{N_R}^2\}$. Then, the (δ, x_r, e) dynamics is UGAS.

Proof Choose the Lyapunov function

$$V(\xi) = \delta^T(Z \otimes P)\delta + \sum\nolimits_{i=1}^{N} \theta_i\phi_i(\tau_i)\,\|x_{ri}\|^2 + \sum\nolimits_{i=1}^{N} \theta_{i_R}\phi_{i_R}(\tau_{i_R})\,\|e_i\|^2 .$$

Similar as the proof of Theorem 9.1, if we can show that the following conditions hold

$$\begin{aligned}
&\langle \nabla V(\xi), \mathscr{F}(\xi)\rangle \le -\sigma_\delta(\delta) - \sum_{i=1}^{N}\sigma_r(x_{ri}) - \sum_{i=1}^{N}\sigma_e(e_i), \text{ for } \xi \in \bar{\mathbb{F}}, \\
&V(\xi^+) - V(\xi) \le 0, \text{ for } \xi \in \bar{\mathbb{J}},
\end{aligned}$$

where $\sigma_\delta(\cdot), \sigma_r(\cdot), \sigma_e(\cdot)$ are class $\mathcal{K}_\infty$ functions. Then, the (δ, x_r, e) dynamics in $\bar{\xi}$ are UGAS.

First, consider the evolution of $V(\bar{\xi})$ along system's flow dynamics, for $\bar{\xi} \in \bar{\mathbb{F}}$, we have

$$\begin{aligned}
&\langle \nabla V(\bar{\xi}), \mathcal{F}(\bar{\xi})\rangle \\
=&\delta^T(-(Z\mathcal{L} + \mathcal{L}^T Z) \otimes PK_P)\delta - 2\delta^T(Z \otimes P)K_I\mathcal{L}x_r + 2\delta^T(Z \otimes P)K_P\mathcal{L}^2 e \\
&+\sum_{i=1}^{N} 2\theta_i\phi_i(\tau_i)\|x_{ri}\|\left\|R_i\delta + R_{i_E}e\right\| + \sum_{i=1}^{N} 2\theta_{i_R}\phi_{i_R}(\tau_{i_R})\|e_i\|\left\|\bar{R}_i\delta + \bar{R}_{i_R}x_r + \bar{R}_{i_E}e\right\| \\
&-\sum_{i=1}^{N}\left(\theta_i^2\phi_i^2(\tau_i)\|x_{ri}\|^2 + \theta_i^2\|x_{ri}\|^2\right) - \sum_{i=1}^{N}\left(\theta_{i_R}^2\phi_{i_R}^2(\tau_{i_R})\|e_i\|^2 + \theta_{i_R}^2\|e_i\|^2\right).
\end{aligned} \tag{9.35}$$

Then, according to Lemma 9.1, and let the matrix inequality in (9.34) left-multiplied by $\left[\delta^T\ x_r^T\ e^T\right]$ and right-multiplied by $\left[\delta^T\ x_r^T\ e^T\right]^T$, it follows that

$$\begin{aligned}
&\delta^T(-(Z\mathcal{L} + \mathcal{L}^T Z) \otimes PK_P)\delta - 2\delta^T(Z \otimes P)K_I\mathcal{L}x_r + 2\delta^T(Z \otimes P)K_P\mathcal{L}^2 e \\
\leq& -\sigma\delta^T(Z \otimes Q)\delta - \sum_{i=1}^{N}\left(\|R_i\delta + R_{i_E}e\|^2\right) - \sum_{i=1}^{N}\left(\|\bar{R}_i\delta + \bar{R}_{i_R}x_r + \bar{R}_{i_E}e\|^2\right) \\
&+ x_r^T\Theta x_r + e^T\Theta_R e - \epsilon\|x_r\|^2 - \epsilon\|e\|^2.
\end{aligned} \tag{9.36}$$

Note that $\|x_r\|^2 = \sum_{i=1}^N \|x_{ri}\|^2$, $\|e\|^2 = \sum_{i=1}^N \|e_i\|^2$, $x_r^T\Theta x_r = \sum_{i=1}^N \theta_i^2\|x_{ri}\|^2$, $e^T\Theta_R e = \sum_{i=1}^N \theta_{i_R}^2\|e_i\|^2$, then, according to (9.35) and (9.36), we have

$$\begin{aligned}
&\langle \nabla V(\bar{\xi}), \mathcal{F}(\bar{\xi})\rangle \\
\leq& -\sigma\delta^T(Z \otimes Q)\delta - \sum_{i=1}^{N}\epsilon\|x_{ri}\|^2 - \sum_{i=1}^{N}\epsilon\|e_i\|^2 - \sum_{i=1}^{N}\left(\theta_i\phi_i(\tau_i)\|x_{ri}\| - \|R_i\delta + R_{i_E}e\|\right)^2 \\
&-\sum_{i=1}^{N}\left(\theta_{i_R}\phi_{i_R}(\tau_{i_R})\|e_i\| - \|\bar{R}_i\delta + \bar{R}_{i_R}x_r + \bar{R}_{i_E}e\|\right)^2 \\
\leq& -\sigma\delta^T(Z \otimes Q)\delta - \sum_{i=1}^{N}\epsilon\|x_{ri}\|^2 - \sum_{i=1}^{N}\epsilon\|e_i\|^2.
\end{aligned} \tag{9.37}$$

Next, consider the evolution of $V(\bar{\xi})$ along system's jump dynamics, there exist three cases.

Case 1: $\bar{\xi}_i \in \bar{\mathbb{J}}_i \wedge x_{ri}\hat{\delta}_i \geq 0 \wedge \tau_{i_R} < \Delta_i h_i$.

The error variable e_i jumps and the reset element state x_{ri} does not jump because reset conditions are not satisfied, thus it can be derived that

$$
\begin{aligned}
&V(\xi^+) - V(\xi) \\
=&\delta^+(Z \otimes Q)\delta^+ - \delta(Z \otimes Q)\delta + \sum_{i=1}^{N} \theta_i\phi_i(\tau_i^+) \left\|x_{ri}^+\right\|^2 + \sum_{i=1}^{N} \theta_{i_R}\phi_{i_R}(\tau_{i_R}^+) \left\|e_i^+\right\|^2 \\
&- \sum_{i=1}^{N} \theta_i\phi_i(\tau_i) \left\|x_{ri}\right\|^2 - \sum_{i=1}^{N} \theta_{i_R}\phi_{i_R}(\tau_{i_R}) \left\|e_i\right\|^2 \\
\le& \ 0.
\end{aligned}
\tag{9.38}
$$

The last inequality is based on (9.32).

Case 2: $\bar{\xi}_i \in \bar{\mathbb{J}}_i \ \wedge \ x_{ri}\hat{\delta}_i \le 0 \ \wedge \ \tau_{i_R} < \Delta_i h_i$.

The error variable e_i jumps and the state x_{ri} jumps to zero because reset condition for the reset element is satisfied, thus, it can be derived from the jump dynamics (9.32) that

$$
\begin{aligned}
&V(\xi^+) - V(\xi) \\
=&\theta_i\phi_i(\tau_i^+) \left\|x_{ri}^+\right\|^2 + \theta_{i_R}\phi_{i_R}(\tau_{i_R}^+) \left\|e_i^+\right\|^2 - \theta_i\phi_i(\tau_i) \left\|x_{ri}\right\|^2 - \theta_{i_R}\phi_{i_R}(\tau_{i_R}) \left\|e_i\right\|^2 \\
\le& \ 0.
\end{aligned}
\tag{9.39}
$$

Case 3: $\bar{\xi}_i \in \bar{\mathbb{J}}_i \ \wedge \ \tau_{i_R} \ge \Delta_i h_i$.

The error variable e_i jumps and the state x_{ri} jumps to $p_i x_{ri}$ according to (9.32), thus, we have

$$
\begin{aligned}
&V(\xi^+) - V(\xi) \\
=&\theta_i\phi_i(\tau_i^+) \left\|x_{ri}^+\right\|^2 + \theta_{i_R}\phi_{i_R}(\tau_{i_R}^+) \left\|e_i^+\right\|^2 - \theta_i\phi_i(\tau_i) \left\|x_{ri}\right\|^2 - \theta_{i_R}\phi_{i_R}(\tau_{i_R}) \left\|e_i\right\|^2 \\
\le&\theta_i\phi_i(0) p_i^2 \left\|x_{ri}\right\|^2 - \theta_i\phi_i(\Delta_i h_i) \left\|x_{ri}\right\|^2 \\
\le& \ 0.
\end{aligned}
\tag{9.40}
$$

The last inequality is based on (9.35). Then, based on (9.37), (9.38), (9.39) and (9.40), we can conclude that the required stability conditions in (9.35) are satisfied. Therefore, the (δ, x_r, e) dynamics are UGAS, in other words, the consensus of the closed-loop MAS is achieved. □

Remark 9.7 Based on (9.40), owing to $e_i^+ = 0$, the reset of x_{ri} can be further relaxed according to $\theta_{i_R}\phi_{i_R}(\tau_{i_R}^+) \left\|e_i^+\right\|^2 - \theta_{i_R}\phi_{i_R}(\tau_{i_R}) \left\|e_i\right\|^2 \le -\theta_{i_R}\phi_{i_R}(\tau_{i_R}) \left\|e_i\right\|^2$. In particular, we can select parameter p_i to satisfy the following condition

$$
p_i \le \sqrt{\frac{1}{\theta_i\phi_i(0)} \left(\frac{\theta_{i_R}\phi_{i_R}(\tau_{i_R}) \left\|e_i\right\|^2}{\left\|x_{ri}^2\right\|} + \theta_i\phi_i(\Delta_i h_i) \right)}
$$

such that the condition $V(\xi^+) - V(\xi) \le 0$ still holds when $\bar{\xi}_i \in \bar{\mathbb{J}}_i \ \wedge \ \tau_{i_R} \ge \Delta_i h_i$. Moreover, this does not change the framework for consensus analysis.

Remark 9.8 Regarding the solvability of the obtained LMIs in the theorems, first, consider the inequality conditions in (9.16), there always exists $\alpha > 0$ such that the second inequality in (9.16) is satisfied, which results in $Q > 0$ owing to $P > 0$ and $K_P > 0$. Due to $Z > 0$, and $K_P > 0$ can be arbitrarily chosen, and $0 < \epsilon < \theta_i^2$, the first inequality in (9.16) is always solvable. Note from (9.14) that the function $\phi_i(\tau_i)$ is strictly decreasing for $0 < \tau_i < \rho_i$, so, there exists $0 < p_i < 1$ such that the third inequality in (9.16) is solvable. Then, consider the inequality conditions in (9.34), since Z is positive definite, we can select $K_P > 0$ and $P > 0$ such that $\Xi_{11} < 0$ holds. For the diagonal matrices Ξ_{22} and Ξ_{33}, the parameters θ_i and θ_{i_R} can be sufficiently large such that $\Xi_{22} < 0$ and $\Xi_{33} < 0$ hold, moreover, due to Ξ_{12}, Ξ_{13}, Ξ_{23} do not depend on θ_i and θ_{i_R}, thus, there always exist parameters θ_i, θ_{i_R} and matrix $Q > 0$ such that the first matrix inequality is solvable. Similar as in (9.16), the solvability of the second inequality condition in (9.34) comes from (9.33). Therefore, based on the aforementioned discussion, the obtained LMI conditions in (9.16) and (9.34) are solvable.

9.4 Simulation Study

This section presents an example to illustrate the proposed method, suppose that the communication graph with respect to the MAS has the form as shown in Fig. 9.1. Corresponding to the given communication graph, the Laplacian matrix $\mathscr{L}$ can be easily obtained and the minimum positive eigenvalue of $\mathscr{L}$ can be calculated as $\lambda_2(\mathscr{L}) = 1$. In [15], it was shown that the reset control with zero-crossing reset conditions can achieve better transient performance than traditional static state feedback method [30], finite time consensus method [31], quasi-reset control method [17] and dynamic state feedback method. Thus, choose the same control gains $K_P = 1$ and $K_I = 0.5$ as in [15], we compare our results with [15].

First, consider the consensus of MAS based on continuous communication. Choose parameters $\alpha = 0.5$, $\sigma = 0.2$, $\sigma_\delta = 0.18$, $\epsilon = 0.01$, $\theta_i = 5$, $p_i = 0.86$, $\phi_i(0) = 3$ for $i \in [1, 6]$, $P = 22$, $\rho_i = 0.02$, $Z = 1/6 I_6$. According to Theorem 9.1, it can be verified that $PK_I = 11 \leq \frac{1}{2}\theta_i^2 = 12.5$, $2\lambda_2(\mathscr{L}) - \alpha \left\| \mathscr{L}^Z \right\| = 0.3762 > 0$, $p_i^2 \phi_i(0) = 2.22 \leq \phi_i(0.02) = 2.23$. Thus, all the conditions in Theorem 9.1 are satisfied with the chosen parameters. The initial conditions are selected as $x(0) = [-2\ 2\ -4\ 4\ -6\ 6]^T$, $\tau(0) = [0\ 0\ 0\ 0\ 0\ 0]^T$. The simulation results with

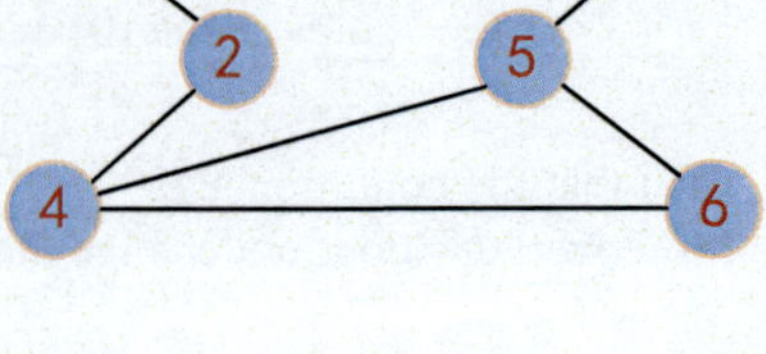

Fig. 9.1 Communication graph

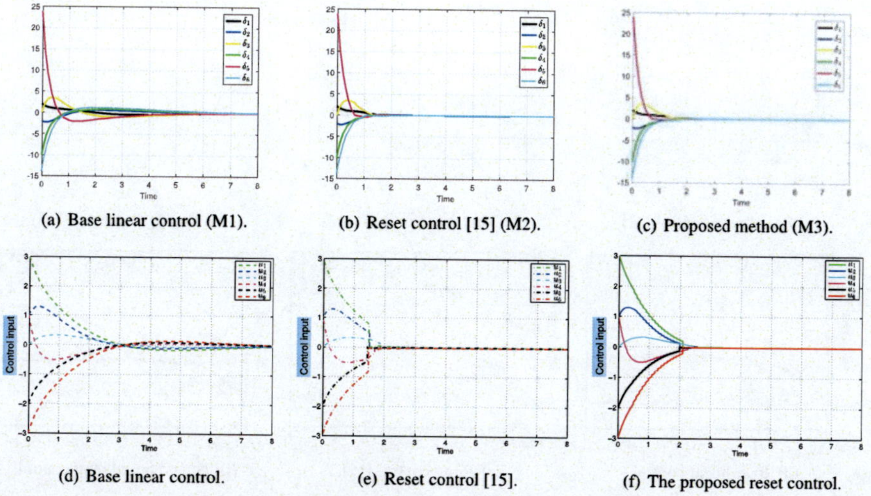

(a) Base linear control (M1). (b) Reset control [15] (M2). (c) Proposed method (M3).

(d) Base linear control. (e) Reset control [15]. (f) The proposed reset control.

Fig. 9.2 Trajectories of δ_i. (Zeno Initial State x_r)

$x_r(0) = [0\ 0\ 0\ 0\ 0\ 0]^T$ and $x_r(0) = [3\ 1\ 0\ 1\ 2\ 3]^T$ are shown in Figs. 9.2 and 9.3 respectively. Comparison results with different control methods are given in simulation results, including base linear dynamic state feedback method (**M1**), reset control with zero-crossing reset conditions [15] (**M2**), the proposed reset control method with continuous communication (**M3**) and the proposed reset control method with asynchronous sampling (**M4**). From Fig. 9.2, for the case with zero initial conditions of $x_r(0)$, both M2 and M3 are able to achieve better transient performance (e.g., settling time[1]) than M1. From Fig. 9.2b–c, the settling time is 7.305 with M2 and is 3.753 s with M3. Therefore, the methods M2 and M3 achieve almost the same consensus performance. However, centralized coordination of reset is required for M2, by contrast, the proposed method M3 is distributed and centralized coordination of reset is not needed. Moreover, the control inputs corresponding to different methods are given in Fig. 9.2d–f.

From Fig. 9.3, for the case with nonzero initial conditions of $x_r(0)$, both M2 and M3 still achieve better transient performance than M1. However, from Fig. 9.3b–c, the settling time is 2.235 s with M2 and is 2.245 s with the proposed M3, which increased by 226.8% and 67.2% respectively compared with the case with zero initial $x_r(0)$. It shows that the proposed method is more applicable to nonzero initial state, this is mainly because the flow set and jump set with the proposed reset conditions are more complicated than with zero-crossing reset conditions in [15]. Moreover, the control inputs corresponding to different methods are given in Fig. 9.3d–f.

---

[1] The settling time is defined as the time required for the combined measurement variable δ_i of all agents to reach and stay within a range of certain percentage of the origin and 2% is used here.

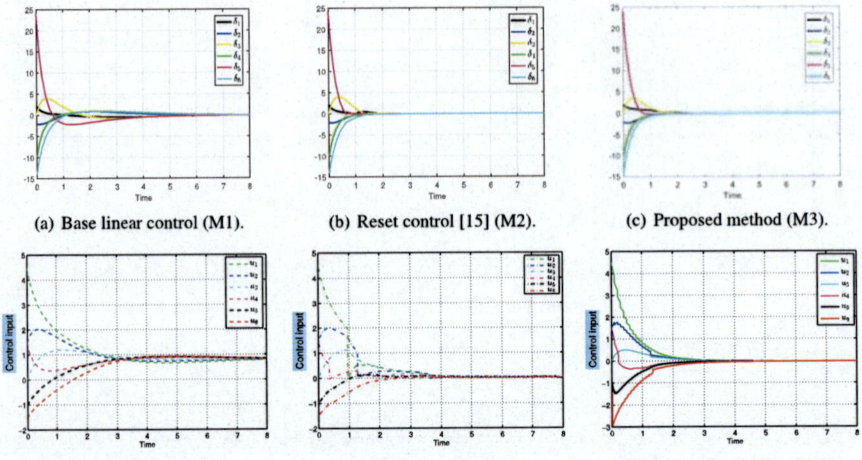

(a) Base linear control (M1). (b) Reset control [15] (M2). (c) Proposed method (M3).

(d) Base linear control. (e) Reset control [15]. (f) The proposed reset control.

Fig. 9.3 Trajectories of δ_i. (Non-Zeno Initial State x_r)

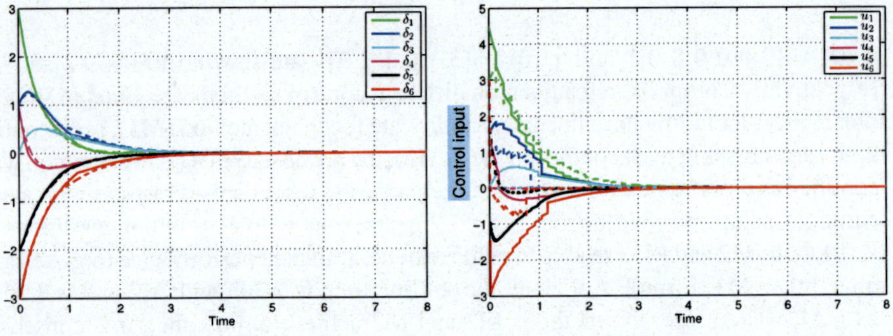

(a) (b)

Fig. 9.4 Comparison results with the proposed reset control method in the cases of continuous communication (solid line) and asynchronous sampling (dashed line)

Then, consider the consensus of MAS based on asynchronous sampling communication. Choose parameters $\theta_i = \theta_{i_R} = 8$ for $i \in [1, N]$, $h_1 = h_4 = 0.01$, $h_2 = h_5 = 0.02$, $h_3 = h_6 = 0.025$, $\Delta_1 = \Delta_4 = 10$, $\Delta_2 = \Delta_5 = 5$, $\Delta_3 = \Delta_6 = 4$, $p_i = 0.4$, $P = 50$, the rest parameters are chosen to be the same as in the case with continuous communication. Here, consider the initial conditions $x(0) = [-2\ 2\ -4\ 4\ -6\ 6]^T$, $\tau(0) = [0\ 0\ 0\ 0\ 0\ 0]^T$, $\tau_R(0) = [0\ 0\ 0\ 0\ 0\ 0]^T$, $x_r(0) = [3\ 1\ 0\ 1\ 2\ 3]^T$. The simulation results are shown in Fig. 9.4. Compared with the case based on continuous communication, similar consensus performance is achieved though only sampled-data can be obtained. Thus, it means that the proposed method M4 is effective, and it is more general than the case with continuous communication and can be implemented in practice.

Table 9.1 CE denotes control effort. ST denotes settling time. ZO/NZO denotes zero/nonzero initial x_r

		M1	M2	M3	M4
ZO	CE	13.874	12.928	12.037	10.493
	ST	7.403	2.235	2.245	4.252
NZO	CE	46.528	15.986	13.732	13.228
	ST	7.786	7.305	3.753	3.645

To further illustrate the effectiveness, some quantitative comparison results are given in Table 9.1, where the control effort (CE) denotes the integral index $\int_0^{t_f} u^T(t)u(t)dt$, where t_f denotes the terminal time. The results in Table 9.1 show that with zero initial $x_r(0)$, M2 achieves the best settling time performance. With nonzero initial $x_r(0)$, both M3 and M4 achieve better settling time performance than M2. Moreover, in contrast with M2, M3 and M4 are distributed and M4 is more general and practical.

9.5 Conclusion

In this chapter, a new consensus strategy consisting of traditional linear feedback control and reset element is proposed in order to achieve cooperative control of MAS and improve transient consensus performance. To make the presentation be clear and be easy to be understand for readers, first, the simple case with continuous communication is considered. Based on the assumption that neighbors' real-time information is available, a hybrid systems framework is proposed for stability analysis and reset control design. Then, the obtained results is extended to the case that only sampled information of neighbors is available, which is more general and more practical in application scenarios. A numerical example is finally given to demonstrate the effectiveness of the proposed reset control approach.

References

1. Yu, W., Chen, G., Cao, M., Kurths, J.: Second-order consensus for multiagent systems with directed topologies and nonlinear dynamics. IEEE Trans. Syst. Man, Cybern. B, Cybern., **40**(3), 881–891 (2010)
2. Tang, Y., Gao, H., Kurths, J., Fang, J.A.: Evolutionary pinning control and its application in UAV coordination. IEEE Trans. Ind. Informat., **8**(4), 828–838 (2012)
3. Liu, Q., Ye, M., Qin, J., Yu, C.: Event-triggered algorithms for leader–follower consensus of networked Euler-Lagrange agents. IEEE Trans. Syst., Man, Cybern., Syst., **49**(7), 1435–1447 (2019)

4. Ge, X., Han, Q.L., Wang, Z.: A dynamic event-triggered transmission scheme for distributed set-membership estimation over wireless sensor networks. IEEE Trans. Cybern., **49**(1), 171–183 (2019)
5. Ding, L., Han, Q.L., Wang, L.Y., Sindi, E.: Distributed cooperative optimal control of DC microgrids with communication delays. IEEE Trans. Ind. Informat., **14**(9), 3924–3935 (2018)
6. Wen, G., Yu, X., Liu, Z.W., Yu, W.: Adaptive consensus-based robust strategy for economic dispatch of smart grids subject to communication uncertainties. IEEE Trans. Ind. Informat., **14**(6), 2484–2496 (2018)
7. Du, S., Liu, T., Ho, D.W.C.: Dynamic event-triggered control for leader-following consensus of multiagent systems. IEEE Trans. Syst., Man, Cybern., Syst., **48**(6), 602–619 (2018)
8. Wu, Z.G., Xu, Y., Lu, R., Wu, Y., Huang, T.: Event-triggered control for consensus of multiagent systems with fixed/switching topologies. IEEE Trans. Syst., Man, Cybern., Syst., **48**(10), 1736–1746 (2018)
9. Almeida, J., Silvestre, C., Pascoal, A.: Synchronization of multiagent systems using event-triggered and self-triggered broadcasts. IEEE Trans. Autom. Control, **62**(9), 4741–4746 (2017)
10. Meng, X., Xie, L., Soh, Y.C.: Asynchronous periodic event-triggered consensus for multi-agent systems. Automatica, **84**, 214–220 (2017)
11. Yi, X., Liu, K., Dimarogonas, D.V., Hohansson, K.H.: Dynamic event-triggered and self-triggered control for multi-agent systems. IEEE Trans. Autom. Control, **64**(8), 3300–3307 (2019)
12. Ding, L., Han, Q.L., Ge, X., Zhang, X.M.: An overview of recent advances in event-triggered consensus of multiagent systems. IEEE Trans. Cybern., **48**(4), 1110–1123 (2018)
13. Zhang, X.M., Han, Q.L., Zhang, B.L.: An overview and deep investigation on sampled-data-based event-triggered control and filtering for networked systems. IEEE Trans. Ind. Informat., **13**(1), 4–16 (2017)
14. Qin, J., Ma, Q., Shi, Y., Wang, L.: Recent advances in consensus of multi-agent systems: A brief survey. IEEE Trans. Ind. Electron., **64**(6), 4972–4983 (2017)
15. Meng, X., Xie, L., Soh, Y.C.: Reset control for synchronization of multi-agent systems. Automatica, **104**, 189–195 (2019)
16. Bragagnolo, M.C., Morarescu, I.C., Daafouz, J., Riedinger, P.: Reset strategy for consensus in networks of clusters. Automatica, **65**, 53–63 (2016)
17. Yucelen, T., Haddad, W.M.: Consensus protocols for networked multi-agent systems with a uniformly continuous quasi-resetting architecture. Int. J. Control, **87**(8), 1716–1727 (2014)
18. Seron, M., Braslavsky, J., Goodwin, G.: Fundamental limitations in filtering and control. Springer London Ltd, (1997)
19. Fernandez, A., Barreiro, A., Carrasco, J.: Reset control for passive bilateral teleoperation. IEEE Trans. Ind. Electron., **58**(7), 3037–3045 (2011)
20. Vidal, A., Banos, A.: Reset compensation for temperature control: Experimental application on heat exchangers. Chemical Engineering Journal, **159**(1), 170–181 (2010)
21. Li, Y., Guo, G., Wang, Y.: Reset control for mid-frequency narrowband disturbance rejection with an application in hard disk drives. IEEE Trans. Control Syst. Tech., **19**(6), 1339–1348 (2011)
22. Clegg, J.C.: A nonlinear integrator for servomechanisms. Trans. Amer. Inst. Elect. Eng. II Appl. Ind. **77**(1), 41–42 (1958)
23. Beker, O., Hollot, C.V., Chait, Y., Han, H.: Fundamental properties of reset control systems. Automatica, **40**(6), 905–915 (2004)
24. Zaccarian, L., Nesic, D., Teel, A.R.: First order reset elements and the Clegg integrator revisited. 51st IEEE Conf. Decis. Control, 6825–6830 (2005)
25. Beker, O., Hollot, C.V., Chait, Y.: Plant with an integrator: An example of reset control overcoming limitations of linear feedback. IEEE Trans. Autom. Control, **46**(11), 1797–1799 (2001)
26. Hunnekens, B., Wouw, N., Nesic, D.: Overcoming a fundamental time-domain performance limitation by nonlinear control. Automatica, **67**, 277–281 (2016)
27. Zhao, G., Nesic, D., Tan, Y., Wang, J.: Open problems in reset control. IEEE Conf. Decis. Control, 3326–3331 (2013)

28. Zhao, G., Nesic, D., Tan, Y., Hua, C.: Overcoming overshoot performance limitations of linear systems with reset control. Automatica, **101**, 27–35 (2019)
29. Dolk, V.S., Borgers, D.P., Heemels, W.P.M.H.: Output-based and decentralized dynamic event-triggered control with guaranteed Lp-gain performance and Zeno-freeness. IEEE Trans. Autom. Control. **62**(1), 34–49 (2017)
30. Olfati-Saber, R., Murray, R.M.: Consensus problems in networks of agents with switching topology and time-delays. IEEE Trans. Autom. Control. **49**(9), 1520–1533 (2004)
31. Cortes, J.: Finite-time convergent gradient flows with applications to network consensus. Automatica. **42**(11), 1993–2000 (2006)

Chapter 10
Event-Based Reset Control of MAS

10.1 Introduction

In previous Chap. 9, we showed how to design reset based consensus control strategy for MAS under asynchronous sampling communication. The proposed reset control design approach has great potential to improve transient cooperative control performance, and has good application prospects in the field of mobile robots [1], unmanned air vehicles [2], sensor networks [3], and so on. However, with asynchronous sampling communication, although clock synchronization among agents is not required, each agent should sample required data at its own sampling instants, even if the sampled-data is not required. That is, sampled-data control periodically/aperiodically samples agent's information and broadcasts the information to neighbors, it does not take the system's current state into consideration. Thus, even if the MAS is working well and it is not necessary to transmit information and update controllers, the agents still sample data and exchange data mutually. Consider that the multiple agents in practical applications usually work under wireless network environment, it is well known that communication resources under wireless network environment are usually limited, such as communication bandwidth and node energy. Thus, on the basis of achieving consensus, it is expected that the information exchange among agents can be reduced as much as possible, such that the communication resources can be saved efficiently. Fortunately, event-triggered control, as introduced in some previous chapters, can be used to improve the efficiency of communication resource utilization, while achieving consensus control objective. Event-triggered control was proposed [4] and has been extensively applied to consensus of MAS, see [5–8], recent survey papers [9–11] and references therein.

In addition, similar as mentioned in Chap. 9, we note that most of the results in existing literature related to event-triggered consensus control of MAS only focus on asymptotic behavior of closed-loop system, but the same important transient performance was rarely studied, except for some recent works [12–14]. How to improve transient consensus performance for MAS with event-triggered communication is an interesting and important topic that worth to be investigated. Similar as in Chap. 9, we

G. Zhao et al., *Cooperative Control of Multi-agent Systems*,
https://doi.org/10.1007/978-981-97-0968-7_10

adopt reset control approach, and combine reset control and event-triggered control together in order to improve transient performance. Reset control was proposed in [15], has been studied by many researchers in recent years in different aspects, for example, zero-crossing reset model and reset conditions were proposed and analyzed in [16], to overcome limitations of zero-crossing reset conditions, sector reset conditions based on reset circuit analysis was developed in [17] under hybrid systems framework. Based on these pioneering works, reset control method has been widely studied in the following years and many outstanding results have been obtained [18–21], that clearly show the superiority of reset control in the aspects of performance improvement or even overcoming fundamental limitations of linear systems. So, inspired by the superiority of reset control, this chapter considers reset based consensus control of MAS with event-triggered communication.

Note that reset action leads to jump dynamics of controller state, event-triggering mechanism will also induce jump dynamics of some error signals, these two types of jump dynamics mutually interact and involve with each other, that brings great challenges for system stability analysis and design. In the existing results related to reset control for consensus of MAS, a quasi-resetting approach was proposed in [14], and backward Euler method was used to implement the integral behavior. In [13], quasi-periodic reset mechanism was applied to leaders of MAS in centralized manner. Reset control with zero-crossing reset conditions was proposed in [12]. However, how to apply reset control to consensus of MAS with event-triggered communications is still an open issue.

In this chapter, inspired by the superiority of reset control in the aspect of improving transient performance, and the superiority of event-triggered control in the aspect of saving usage of communication resources, we introduce how to design reset mechanism (RM) and event-triggering mechanism (ETM) for MAS. The RM determines when reset controller state has jump dynamics, and the ETM determines when to transmit information to neighbors and when to update controller state. In order to avoid Zeno reset behavior and Zeno triggering behavior, time regularization technique is utilized to impose lower bound of time intervals of reset event and triggering event. And, in order to deal with two different types of jump dynamics and the time regularization in a unified framework, we decompose the state space into multiple flow sets and jump sets that are related to reset and event-trigger. Based on the state space decomposition, a hybrid model is constructed for the closed-loop MAS by introducing some auxiliary variables. Then, a hybrid systems approach is presented for the consensus analysis and co-design of RM and ETM. The advantages of the proposed reset control method are shown via a simulation example.

10.2 System Model and Event-Based Reset Control Law

Since reset control of multi-input multi-output systems is still an open research topic, similar as in Chap. 9, we only consider reset control design for single-input single-output systems. Thus, the MAS with first-order dynamics is considered in this chapter. Next, we first give the MAS model with first-order dynamics.

The dynamics of N agents is described by

$$\dot{x}_i(t) = u_i(t), \quad i \in [1, N] \tag{10.1}$$

where $x_i(t) \in \mathbb{R}$ is the state and $u_i(t) \in \mathbb{R}$ is the control input. To be convenient, define the combined measurement variable $\delta_i(t)$ as follows

$$\delta_i(t) = \sum_{j \in \mathcal{N}_i} (x_j(t) - x_i(t)), \quad \text{for } i \in [1, N] \tag{10.2}$$

The neighboring agents exchange information via communication network, due to event-triggered communication is considered, it means that the agents only exchange information at triggering instants, i.e., the state $x_i(t)$ is sampled and transmitted to agent i's neighbors at triggering instants t_{k_i}, $k_i \in \mathbb{N}$. Each agent i receives its neighbor's information at t_{k_j}, and updates the control protocol at both t_{k_i} and t_{k_j}. So, it can be seen that, with event-triggered communication mechanism, only sampled information at triggering instants is available for the purpose of consensus control. This is different from the case with continuous communication mechanism, with which real-time neighbors' information is available. In contrast with continuous communication mechanism, ETM induced error signals bring new difficulties for system analysis. To construct the control protocol, define a combinational variable using sampled information at triggering instants,

$$\delta_{ik_i} = \delta_i(t) = \sum_{j \in \mathcal{N}_i} (x_j(t_{k_j}) - x_i(t_{k_i})), \ t \in (t_{k_i}, t_{k_i+1}] \tag{10.3}$$

where $x_i(t_{k_i})$ keeps unchanged for $t \in (t_{k_i}, t_{k_i+1}]$. Then, δ_{ik_i} can be viewed as a feedback signal and is used in the Reset Element, which is proposed in the following form

$$\begin{aligned} &\left.\begin{aligned} \dot{\tau}_{i_R}(t) &= 1 \\ \dot{x}_{ri}(t) &= \delta_{ik_i} \end{aligned}\right\} \text{If } (x_{ri}(t), \delta_{ik_i}, \tau_{i_R}(t)) \in \mathscr{R}_F \\ &\left.\begin{aligned} \tau_{i_R}(t^+) &= 0 \\ x_{ri}(t^+) &= \lambda_i(o_i(t)) x_{ri}(t) \end{aligned}\right\} \text{If } (x_{ri}(t), \delta_{ik_i}, \tau_{i_R}(t)) \in \mathscr{R}_J \end{aligned} \tag{10.4}$$

where $\mathscr{R}_F$ and $\mathscr{R}_J$ denote the flow set and jump set, $\lambda_i(o_i(t))$ is the reset map function which depends on $o_i(t)$ ($o_i(t)$ denotes the available local variables for agent i). $\tau_{i_R}(t)$ is an auxiliary timer variable that can record the time elapsed since the last reset instant, the role of the variable $\tau_{i_R}(t)$ is to achieve time regularization in order to avoid Zeno behavior. That is, reset only occurs when $\tau_{i_R}(t) \geq \rho_i > 0$, and the time interval between reset instants is enforced to be lower bounded by $\rho_i > 0$. The design of $\mathscr{R}_F$, $\mathscr{R}_J$ and $\lambda_i(o_i(t))$ will be given later.

Then, according to (10.3) and (10.4), to achieve the objective of state consensus of multiple agents while improving transient performance, compared with traditional consensus control protocols, and extra term with respect to reset element is added into the consensus control protocol, and the event-triggered reset consensus control protocol is proposed as follows

$$u_i(t) = K_P \delta_{ik_i} + K_I x_{ri}(t), \tag{10.5}$$

where (K_P, K_I) are constant gains. Hereafter, we call the control protocol (10.5) as PIR (Proportional Integral Reset) protocol.

To design the ETM, define the error variable with respect to real-time information $x_i(t)$ and sampled data $x_i(t_{k_i})$

$$e_i(t) = x_i(t_{k_i}) - x_i(t),\ t \in (t_{k_i}, t_{k_i+1}] \tag{10.6}$$

Then, we can design the ETM in the following form

$$t_{k_i+1} := \inf\{t \geq t_{k_i} + \tau_{MIET} | (e_i(t), x_{ri}(t), \delta_{ik_i}) \in \mathscr{E}_J\} \tag{10.7}$$

where $\tau_{MIET} > 0$ is an imposed lower bound of minimum inter-event time for the ETM, $\mathscr{E}_J$ denotes the jump set for the ETM. $x_i(t_{k_i})$ is updated to be $x_i(t)$ at the triggering instant t_{k_i+1}, that is, $e_i(t)$ has jump dynamics and $e_i(t^+_{k_i+1}) = 0$. How to design the jump set $\mathscr{E}_J$ and parameter τ_{MIET} will be given in the main results. The schematic diagram of MAS is shown in Fig. 10.1.

Remark 10.1 The proposed PIR protocol (10.5) is in distributed form, since from (10.7) and (10.4), it can be seen that only agent's own and neighbors' information is used in the event-triggered reset consensus protocol. Besides, in order to exclude Zeno reset phenomenon and Zeno triggering phenomenon, time regularization technique is adopted and lower bounds of reset intervals and triggering intervals are enforced.

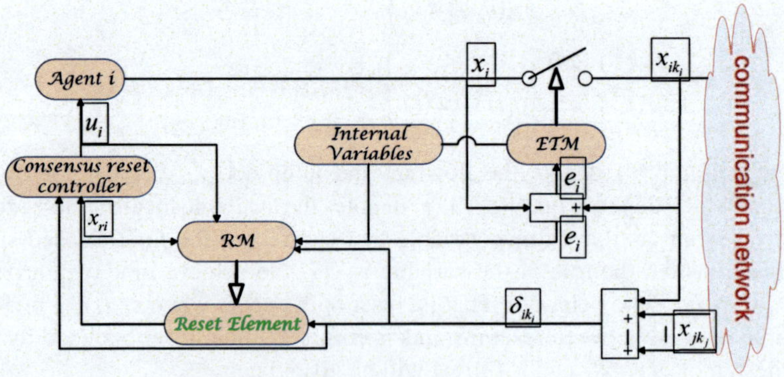

Fig. 10.1 Schematic diagram of MAS with RM and ETM

However, it is worth to be mentioned that, due to the hybrid ETM (10.7) and RM (10.4) induce hybrid dynamics in the closed-loop system, it leads to challenges for consensus analysis and control design.

In this chapter, we assume that the communication graph is undirected and connected. Then, from [22], it is known that the Laplacian matrix $\mathscr{L}$ is symmetric, and $\mathbf{1}_N$ is the left eigenvector of $\mathscr{L}$ with $\mathbf{1}_N \mathscr{L} = 0$. Moreover, the condition $\lambda_2(\mathscr{L}) = \min_{x^T \zeta = 0, x \neq 0} \frac{x^T \hat{\mathscr{L}} x}{x^T Z x} > 0$ holds, where $\lambda_2(\mathscr{L})$ is the minimum positive eigenvalue of $\mathscr{L}$. By means of the graph theory, the objective of this chapter is to design the hybrid ETM and RM, such that the proposed event-triggered PIR protocol (10.5) not only achieves consensus of the MAS, but also achieves better transient performance (e.g., settling time, overshoot) than traditional static and dynamic state feedback control protocols.

10.3 Hybrid Model Construction

For the purpose of consensus analysis and control design, we need to firstly construct a closed-loop model to describe the system dynamics. Notice that there exist different types of jump dynamics and flow dynamics in the closed-loop system, to completely describe the closed-loop system dynamics, we construct a hybrid model for the MAS under hybrid systems framework [23], and laying the foundation for stability analysis.

The objective of model construction is to describe the system dynamics in the following form

$$\mathbb{H} : \begin{cases} \dot{\xi} = F(\xi), & \text{If } \xi \in \mathbb{F}, \\ \xi^+ \in J(\xi), & \text{If } \xi \in \mathbb{J}. \end{cases} \tag{10.8}$$

where $F(\xi)/J(\xi)$ describes the flow/jump dynamics, $\mathbb{F}$ and $\mathbb{J}$ denote the flow set and jump set respectively.

To describe the MAS with PIR protocol into the form of (10.8), we introduce an auxiliary variable $\tau_{i_E}(t) \in \mathbb{R}_{\geq 0}$, for $i \in [1, N]$. $\dot{\tau}_{i_E}(t) = 1$ when triggering event does not occur, and $\tau_{i_E}(t^+) = 0$ at the triggering instant. The timer variable $\tau_{i_E}(t)$ has the same role as $\tau_{i_R}(t)$.

To construct the hybrid model, the overall state space should be decomposed into appropriate flow sets and jump sets, we need to find out which state has flow/jump dynamics at which flow/jump set, and distinguish different cases clearly. First, on the basis of the properties of τ_{i_R} and τ_{i_E}, define the following sets

$$
\begin{aligned}
\mathscr{C}_{R1}^{i} &:= \{\tau_{i_R} \in \mathbb{R}_{\geq 0} | \tau_{i_R} < \rho_i\} \\
\mathscr{C}_{R2}^{i} &:= \{\tau_{i_R} \in \mathbb{R}_{\geq 0} | \rho_i \leq \tau_{i_R} < \rho_{iM}\} \\
\mathscr{D}_{R}^{i} &:= \{\tau_{i_R} \in \mathbb{R}_{\geq 0} | \tau_{i_R} \geq \rho_{iM}\} \\
\mathscr{C}_{E}^{i} &:= \{\tau_{i_E} \in \mathbb{R}_{\geq 0} | \tau_{i_E} < \tau_{MIET}\} \\
\mathscr{D}_{E}^{i} &:= \{\tau_{i_E} \in \mathbb{R}_{\geq 0} | \tau_{i_E} \geq \tau_{MIET}\}
\end{aligned}
\tag{10.9}
$$

where the subscript 'R' corresponds to 'reset', and the subscript 'E' corresponds to 'event-trigger'. Based on the above defined sets, we can further define appropriate flow sets and jump sets for the closed-loop system. Based on the role of the timer variables τ_{i_R} and τ_{i_E}, it is known that the reset state x_{ri} must flow when $\tau_{i_R} < \rho_i$ and the error e_i must flow when $\tau_{i_E} < \tau_{MIET}$. However, x_{ri} may flow or jump when $\tau_{i_R} \geq \rho_i$ according to the RM in (10.4), and e_i may flow or jump when $\tau_{i_E} \geq \tau_{MIET}$ according to the ETM in (10.7). Therefore, based on whether the (x_{ri}, e_i) dynamics jump or not, we have the following cases to describe the flow sets and jump sets

$$
\begin{aligned}
\mathscr{F}_{1i} &:= \mathscr{C}_{R1}^{i} \wedge \mathscr{C}_{E}^{i} \\
\mathscr{J}_{1i} &:= \mathscr{C}_{E}^{i} \wedge (\mathscr{C}_{R2}^{i} \vee \mathscr{D}_{R}^{i}) \wedge \delta_{ik_i} x_{ri} \leq 0 \\
\mathscr{F}_{2i} &:= \mathscr{C}_{E}^{i} \wedge \mathscr{C}_{R2}^{i} \wedge \delta_{ik_i} x_{ri} \geq 0 \\
\mathscr{F}_{3i} &:= \mathscr{C}_{E}^{i} \wedge \mathscr{D}_{R}^{i} \wedge \delta_{ik_i} x_{ri} \geq 0 \wedge \phi_{1i}(o_i) \geq 0 \\
\mathscr{J}_{2i} &:= \mathscr{C}_{E}^{i} \wedge \mathscr{D}_{R}^{i} \wedge \delta_{ik_i} x_{ri} \geq 0 \wedge \phi_{1i}(o_i) \leq 0 \\
\mathscr{F}_{4i} &:= \mathscr{D}_{E}^{i} \wedge (\mathscr{C}_{R1}^{i} \vee (\mathscr{C}_{R2}^{i} \wedge \delta_{ik_i} x_{ri} \geq 0)) \wedge \phi_{2i}(o_i) \geq 0 \\
\mathscr{J}_{3i} &:= \mathscr{D}_{E}^{i} \wedge (\mathscr{C}_{R1}^{i} \vee (\mathscr{C}_{R2}^{i} \wedge \delta_{ik_i} x_{ri} \geq 0)) \wedge \phi_{2i}(o_i) \leq 0 \\
\mathscr{J}_{4i} &:= \mathscr{D}_{E}^{i} \wedge \mathscr{C}_{R2}^{i} \wedge \delta_{ik_i} x_{ri} \leq 0 \wedge \phi_{2i}(o_i) \geq 0 \\
\mathscr{J}_{5i} &:= \mathscr{D}_{E}^{i} \wedge \mathscr{C}_{R2}^{i} \wedge \delta_{ik_i} x_{ri} \leq 0 \wedge \phi_{2i}(o_i) \leq 0 \\
\mathscr{F}_{5i} &:= \mathscr{D}_{E}^{i} \wedge \mathscr{D}_{R}^{i} \wedge \delta_{ik_i} x_{ri} \geq 0 \wedge \phi_{3i}(o_i) \geq 0 \\
\mathscr{J}_{6i} &:= \mathscr{D}_{E}^{i} \wedge \mathscr{D}_{R}^{i} \wedge \delta_{ik_i} x_{ri} \geq 0 \wedge \phi_{3i}(o_i) \leq 0 \\
\mathscr{J}_{7i} &:= \mathscr{D}_{E}^{i} \wedge \mathscr{D}_{R}^{i} \wedge \delta_{ik_i} x_{ri} \leq 0 \wedge \phi_{3i}(o_i) \geq 0 \\
\mathscr{J}_{8i} &:= \mathscr{D}_{E}^{i} \wedge \mathscr{D}_{R}^{i} \wedge \delta_{ik_i} x_{ri} \leq 0 \wedge \phi_{3i}(o_i) \leq 0
\end{aligned}
\tag{10.10}
$$

where $\{\mathscr{F}_{1i}, \ldots, \mathscr{F}_{5i}\}$ denote the flow sets in which (x_{ri}, e_i) flow. $\{\mathscr{J}_{1i}, \ldots, \mathscr{J}_{8i}\}$ denote the jump sets in which x_{ri} or e_i or both of them jump. Base on the above representations, the states (x_{ri}, e_i) flow when $(\delta_{ik_i}, x_{ri}, e_i, \tau_{i_R}, \tau_{i_E}) \in \mathscr{F}_{1i} \vee \mathscr{F}_{2i} \vee \mathscr{F}_{3i} \vee \mathscr{F}_{4i} \vee \mathscr{F}_{5i}$, and x_{ri} or e_i jumps when $(\delta_{ik_i}, x_{ri}, e_i, \tau_{i_R}, \tau_{i_E}) \in \mathscr{J}_{1i} \vee \mathscr{J}_{2i} \vee \mathscr{J}_{3i} \vee \mathscr{J}_{4i} \vee \mathscr{J}_{5i} \vee \mathscr{J}_{6i} \vee \mathscr{J}_{7i} \vee \mathscr{J}_{8i}$. Specifically, the relationship between the jump dynamics of (x_{ri}, e_i) and the jump sets $\mathscr{J}_{ni}, n \in [1, 8]$ is given by

$$\begin{aligned}
&\mathscr{J}_{1i} \mapsto e_i^+ = e_i, x_{ri}^+ = 0; \quad \mathscr{J}_{2i} \mapsto e_i^+ = e_i, x_{ri}^+ = \lambda_i(o_i)x_{ri},\\
&\mathscr{J}_{3i} \mapsto e_i^+ = 0, x_{ri}^+ = x_{ri}; \quad \mathscr{J}_{4i} \mapsto e_i^+ = e_i, x_{ri}^+ = 0,\\
&\mathscr{J}_{5i} \mapsto e_i^+ = 0, x_{ri}^+ = 0; \quad \mathscr{J}_{6i} \mapsto e_i^+ = 0, x_{ri}^+ = \lambda_i(o_i)x_{ri},\\
&\mathscr{J}_{7i} \mapsto e_i^+ = e_i, x_{ri}^+ = 0; \quad \mathscr{J}_{8i} \mapsto e_i^+ = 0, x_{ri}^+ = 0.
\end{aligned} \tag{10.11}$$

The definitions of flow sets and jump sets in (10.10) is based on the values of timer variables τ_{i_R}, τ_{i_E} and the sign of product $\delta_{ik_i}x_{ri}$, where $\delta_{ik_i}x_{ri} \leq 0$ denotes the traditional reset condition as developed in [17, 24]. From (10.10) and (10.11), the reset state x_{ri} jumps to zero when $\tau_{i_R} \in \mathscr{C}_{R2}^i \vee \mathscr{D}_R^i$. In other cases, how the state x_{ri} is reset and the error e_i is updated are determined by the triggering functions $\phi_{1i}(o_i)$, $\phi_{2i}(o_i)$, $\phi_{3i}(o_i)$, and the state x_{ri} is reset to a non-zero value that depends on $\lambda_i(o_i)$ (here, $x_{ri}^+ = 0$ can be viewed as a special case with $\lambda_i(o_i) = 0$). Therefore, how to design the triggering functions $\phi_{1i}(o_i)$, $\phi_{2i}(o_i)$, $\phi_{3i}(o_i)$ and the function $\lambda_i(o_i)$ is needed to be considered. From (10.10) and (10.11), it seems that there exist many cases and how to judge which set the system states belong to and how to reset x_{ri} is complicated. Actually, we decompose the state space into these different cases, on the one hand, for certain system states, we can rigorously determine which sets they belong to, and the definitions of these sets are useful for stability analysis. On the other hand, consider implementation of the proposed reset consensus controller, it is very easy to program the logical relationship, such that it is not difficult to determine how to reset x_{ri}.

According to (10.2), (10.4) and (10.6), we define

$$\begin{aligned}
\delta(t) &= \left[\delta_1^T(t)\ \delta_2^T(t) \cdots \delta_N^T(t)\right]^T,\\
x_r(t) &= \left[x_{r1}^T(t)\ x_{r2}^T(t) \cdots x_{rN}^T(t)\right]^T,\\
e(t) &= \left[e_1^T(t)\ e_2^T(t) \cdots e_N^T(t)\right]^T.
\end{aligned} \tag{10.12}$$

Then, if the reset conditions in (10.4) and the event-triggered conditions in (10.7) are not satisfied, the flow dynamics of the closed-loop system can be described by

$$\begin{bmatrix} \dot{\delta}(t) \\ \dot{x}_r(t) \\ \dot{e}(t) \end{bmatrix} = \begin{bmatrix} -K_P\mathscr{L}\delta - K_I\mathscr{L}x_r + K_P\mathscr{L}^2 e \\ \delta - \mathscr{L}e \\ -K_P\delta - K_I x_r + K_P\mathscr{L}e \end{bmatrix}, \tag{10.13}$$

however, in contrast with most of the existing results in the literature, the closed-loop system states $(x_r(t), e(t))$ also have jump dynamics due to the reset mechanism and the ETM, which brings difficulties for consensus analysis.

Next, we construct a hybrid model for the MAS based on the decomposition of state space and the relation between the jump dynamics of (x_{ri}, e_i) and the jump sets $\mathscr{J}_{ni}$, $n \in [1, 8]$. Define

$$\tau_R = \left[\tau_{1_R}^T\ \tau_{2_R}^T \cdots \tau_{N_R}^T\right]^T,\ \tau_E = \left[\tau_{1_E}^T\ \tau_{2_E}^T \cdots \tau_{N_E}^T\right]^T,\ \xi = \left[\delta^T\ x_r^T\ e^T\ \tau_R^T\ \tau_E^T\right]^T. \tag{10.14}$$

Then, based on (10.3)–(10.7) and (10.10) and (10.11), the dynamics of the closed-loop MAS with PIR protocol can be represented as the following hybrid systems form

$$\begin{aligned} \dot{\xi} &= F(\xi), \text{ If } \xi \in \mathbb{F}, \\ \xi^+ &\in \bigcup_{i=1}^{N} J_i(\xi), \text{ If } \xi \in \mathbb{J}, \end{aligned} \tag{10.15}$$

with $\mathbb{F} := \bigcap_{i=1}^{N} \mathbb{F}_i$, $\mathbb{J} := \bigcup_{i=1}^{N} \mathbb{J}_i$, and

$$\begin{aligned} \mathbb{F}_i &:= \{\xi \in \mathbb{X} | \mathscr{F}_{1i} \vee \mathscr{F}_{2i} \vee \mathscr{F}_{3i} \vee \mathscr{F}_{4i} \vee \mathscr{F}_{5i}\} \\ \mathbb{J}_i &:= \{\xi \in \mathbb{X} | \mathscr{J}_{1i} \vee \mathscr{J}_{2i} \vee \mathscr{J}_{3i} \vee \mathscr{J}_{4i} \vee \mathscr{J}_{5i} \vee \mathscr{J}_{6i} \vee \mathscr{J}_{7i} \vee \mathscr{J}_{8i}\} \end{aligned}$$

where $\mathbb{X} := \{(\delta, x_r, e, \tau_R, \tau_E) \in \mathbb{R}^N \times \mathbb{R}^N \times \mathbb{R}^N \times \mathbb{R}^N_{\geq 0} \times \mathbb{R}^N_{\geq 0}\}$.

$$F(\xi) = \begin{bmatrix} -K_P \mathscr{L} \delta - K_I \mathscr{L} x_r + K_P \mathscr{L}^2 e \\ \delta - \mathscr{L} e \\ -K_P \delta - K_I x_r + K_P \mathscr{L} e \\ \mathbf{1}_N \\ \mathbf{1}_N \end{bmatrix} \tag{10.16}$$

$$J_i(\xi) := \begin{cases} J_{1i}(\xi), & \text{If } \xi \in \mathbb{J}_i \wedge (\mathscr{J}_{1i} \vee \mathscr{J}_{4i} \vee \mathscr{J}_{7i}) \\ J_{2i}(\xi), & \text{If } \xi \in \mathbb{J}_i \wedge (\mathscr{J}_{5i} \vee \mathscr{J}_{8i}) \\ J_{3i}(\xi), & \text{If } \xi \in \mathbb{J}_i \wedge \mathscr{J}_{3i} \\ J_{4i}(\xi), & \text{If } \xi \in \mathbb{J}_i \wedge \mathscr{J}_{2i} \\ J_{5i}(\xi), & \text{If } \xi \in \mathbb{J}_i \wedge \mathscr{J}_{6i} \end{cases}, \tag{10.17}$$

where

$$J_{1i}(\xi) = \begin{bmatrix} \delta \\ (I_N - \Lambda_i) x_r \\ e \\ (I_N - \Lambda_i) \tau_R \\ \tau_E \end{bmatrix}, \; J_{2i}(\xi) = \begin{bmatrix} \delta \\ (I_N - \Lambda_i) \begin{bmatrix} x_r \\ e \\ \tau_R \\ \tau_E \end{bmatrix} \end{bmatrix} \; J_{3i}(\xi) = \begin{bmatrix} \delta \\ x_r \\ (I_N - \Lambda_i) e \\ \tau_R \\ (I_N - \Lambda_i) \tau_E \end{bmatrix},$$

$$J_{4i}(\xi) = \begin{bmatrix} \delta \\ (I_N - (1 - \lambda_i(o_i)) \Lambda_i) x_r \\ e \\ (I_N - \Lambda_i) \tau_R \\ \tau_E \end{bmatrix}, \; J_{5i}(\xi) = \begin{bmatrix} \delta \\ (I_N - (1 - \lambda_i(o_i)) \Lambda_i) x_r \\ (I_N - \Lambda_i) \begin{bmatrix} e \\ \tau_R \\ \tau_E \end{bmatrix} \end{bmatrix}.$$

With the help of auxiliary variables and appropriate state space decomposition, the MAS with PIR protocol and event-triggered communication mechanism are constructed as hybrid model (10.15). From (10.2), the asymptotic stability of $\delta(t)$ indicates that the agent states $x_i(t)$ achieve consensus. Thus, we give the stability property of the MAS as defined in the following.

Definition 10.1 For the constructed hybrid model of the MAS as described by (10.15). The set given by $\{\xi \in \mathbb{X} | \delta = 0_N, x_r = 0_N, e = 0_N\}$ is said to be uniformly globally asymptotically stable (UGAS) if there exists a function $\beta \in \mathcal{KL}$ such that, for any initial condition $\xi(0) \in \mathbb{X}$, the following condition holds

$$\|(\delta(t), x_r(t), e(t))\| \leq \beta(\|(\delta(0), x_r(0), e(0))\|, t). \tag{10.18}$$

Remark 10.2 In the above definition, the UGAS is defined for $(\delta(t), x_r(t), e(t))$ instead of $x(t)$, because the goal is to achieve consensus of $x_i(t)$, $i \in [1, N]$, so $x(t)$ does not necessarily converge to zero. Therefore, we only need to show that the $(\delta(t), x_r(t), e(t))$ dynamics is UGAS, which will be given in the following section.

10.4 Co-Design of Hybrid ETM and RM

In this section, the aim is to introduce a co-design method of hybrid ETM and RM, while presenting sufficient conditions for closed-loop system stability. Specifically, we give conditions such that the (δ, x_r, e) dynamics is UGAS (in other words, the consensus is achieved), meanwhile, the explicit form of the triggering functions $\phi_{1i}(o_i)$, $\phi_{2i}(o_i)$, $\phi_{3i}(o_i)$ and the reset map function $\lambda_i(o_i)$ are introduced, and how to choose parameters ρ_i, ρ_{iM}, τ_{MIET} is also given.

Before presenting the co-design results of hybrid ETM and RM, we first introduce the following internal auxiliary functions with decreasing dynamics, such function is useful for determining minimum inter-event times of ETM and design control parameters in RM, and it is also useful for constructing Lyapunov function and stability analysis of closed-loop system. Introduce two auxiliary function for hybrid ETM and RM respectively,

$$\begin{aligned} \dot{\phi}_{i_R}(\tau_{i_R}) &= s_1(\tau_{i_R})(-\theta_{i_R}(\phi_{i_R}^2(\tau_{i_R}) + 1)), \\ s_1(\tau_{i_R}) &= \max\{\text{sign}(\rho_{iM} - \tau_{i_R}), 0\}, \end{aligned} \tag{10.19}$$

$$\begin{aligned} \dot{\phi}_{i_E}(\tau_{i_E}) &= s_2(\tau_{i_E})(-\theta_{i_E}(\phi_{i_E}^2(\tau_{i_E}) + 1)), \\ s_2(\tau_{i_E}) &= \max\{\text{sign}(\tau_{MIET} - \tau_{i_E}), 0\}, \end{aligned} \tag{10.20}$$

where $\theta_{i_R} > 0$, $\theta_{i_E} > 0$ are constants. The initial conditions $\phi_{i_R}(0) > 0$, $\phi_{i_E}(0) > 0$. The dynamics of ϕ_{i_R} and ϕ_{i_E} have similar form, take the dynamics of ϕ_{i_R} as example, when $\tau_{i_R} < \rho_{iM}$, according to (10.19), ϕ_{i_R} strictly decreases along τ_{i_R}. When $\tau_{i_R} \geq \rho_{iM}$, according to (10.19), $s_1(\tau_{i_R}) = 0$, and $\dot{\phi}_{i_R}(\tau_{i_R}) = 0$, so, ϕ_{i_R} remains unchanged until τ_{i_R} is reset to zero. The function $\phi_{i_R}(\tau_{i_R})$ and $\phi_{i_E}(\tau_{i_E})$ are useful for consensus analysis of MAS with hybrid model, and how to select parameters ρ_i, ρ_{iM}, τ_{MIET} is given in the following lemma.

Lemma 10.1 *For pre-specified initial conditions $\phi_{i_R}(0) > 0$, $\phi_{i_E}(0) > 0$, parameters $\theta_{i_R} > 0$, $\theta_{i_E} > 0$, $i \in [1, N]$, if ρ_i, ρ_{iM}, and τ_{MIET} satisfy the following conditions*

$$0 < \rho_i < \rho_{iM} < \rho^i_{MAX} \tag{10.21}$$

$$0 < \tau_{MIET} < \tau^i_{MAX} \tag{10.22}$$

where

$$\rho^i_{MAX} = \frac{1}{\theta_{i_R}} \arctan(\phi_{i_R}(0)),\ \tau^i_{MAX} = \frac{1}{\theta_{i_E}} \arctan(\phi_{i_E}(0))$$

Then, $\phi_{i_R}(\tau_{i_R}) > 0$ and $\phi_{i_E}(\tau_{i_E}) > 0$ are ensured for any $\tau_{i_R} > 0$ and $\tau_{i_E} > 0$.

Proof According to the dynamics of ϕ_{i_R} in (10.19), we have

$$\frac{d\phi_{i_R}(\tau_{i_R})}{dt} = -\theta_{i_R}(\phi^2_{i_R}(\tau_{i_R}) + 1)$$

and

$$\frac{d\phi_{i_R}(\tau_{i_R})}{\theta_{i_R}(\phi^2_{i_R}(\tau_{i_R}) + 1)} = -dt$$

when $\tau_{i_R} < \rho_{iM}$. Since it is required that ϕ_{i_R} should be always nonnegative and ϕ_{i_R} has decreasing dynamics, thus, there exists a ρ^i_{MAX} such that $\phi_{i_R}(\rho^i_{MAX}) = 0$. That is

$$\int_{\phi_{i_R}(0)}^{0} \frac{d\phi_{i_R}(\tau_{i_R})}{\theta_{i_R}(\phi^2_{i_R}(\tau_{i_R}) + 1)} = -\int_0^{\rho^i_{MAX}} dt$$

Then, the conclusion in this lemma can be easily obtained by integral operation, the calculation process is normal and is omitted here. □

Now, we are ready to present the co-design results of hybrid ETM and RM. The following theorem gives sufficient conditions for system stability, and the explicit form of reset functions and triggering functions are given.

Theorem 10.1 *Consider the MAS with hybrid systems form (10.15). Let $K_P > 0$, $K_I > 0$, if there exists constants $0 < \sigma_1 < \sigma < 1$, $0 < \epsilon < \min\{\theta^2_{i_R}, \theta^2_{i_R}\}$, $P > 0$, $0 < \eta_i < 1$ such that the following inequality conditions hold*

$$\begin{bmatrix} \Xi_{11} & \Xi_{12} & \Xi_{13} \\ * & \Xi_{22} & \Xi_{23} \\ * & * & \Xi_{33} \end{bmatrix} \leq 0 \tag{10.23}$$

$$\eta_i^2 \phi_{i_R}(0) \leq \phi_{i_R}(\rho_{iM}) \tag{10.24}$$

where

$$\begin{aligned}
\Xi_{11} &= -(1-\sigma) I_N \otimes Q + \sum\nolimits_{i=1}^{N} (R_i^T R_i + \bar{R}_i^T \bar{R}_i), \\
\Xi_{12} &= -(I_N \otimes P) K_I \mathscr{L} + \sum\nolimits_{i=1}^{N} \bar{R}_i^T \bar{R}_{i_R}, \\
\Xi_{13} &= (I_N \otimes P) K_P \mathscr{L}^2 + \sum\nolimits_{i=1}^{N} (R_i^T R_{i_E} + \bar{R}_i^T \bar{R}_{i_E}), \\
\Xi_{22} &= -\Theta_i + \epsilon I + \sum\nolimits_{i=1}^{N} \bar{R}_{i_R}^T \bar{R}_{i_R}, \; \Xi_{23} = \sum\nolimits_{i=1}^{N} \bar{R}_{i_R}^T \bar{R}_{i_E}, \\
\Xi_{33} &= -\Theta_{i_R} + \epsilon I + \sum\nolimits_{i=1}^{N} (R_{i_E}^T R_{i_E} + \bar{R}_{i_E}^T \bar{R}_{i_E}),
\end{aligned}$$

where $Q = 2\lambda_2(\mathscr{L}) P K_P$. $R_i \in \mathbb{R}^{1\times N}$ *is a row vector with the i-th entry is one and the other entries are zeros.* $R_{iE} = -\mathscr{L}_{row}^i$, $\bar{R}_i = -K_P R_i$, $\bar{R}_{i_E} = K_P \mathscr{L}_{row}^i$, $\bar{R}_{i_R} = -K_I R_i$. $\mathscr{L}_{row}^i$ *denotes the i-th row of Laplacian matrix* $\mathscr{L}$. Θ *and* Θ_{i_R} *are diagonal matrices with* $\Theta = \mathrm{diag}\{\theta_1^2, \theta_2^2, \ldots, \theta_N^2\}$ *and* $\Theta_R = \mathrm{diag}\{\theta_{1_R}^2, \theta_{2_R}^2, \ldots, \theta_{N_R}^2\}$. *The triggering functions and reset map function are designed as follows*

$$\begin{aligned}
\phi_{1i}(o_i) &= (u_i + \theta_{e_E}\phi_{i_E} e_i)^2 - 2\theta_{i_R}\phi_{i_R} x_{ri}\delta_{ik_i} + \delta_{ik_i}^2 + \theta_{i_R}^2 x_{ri}^2 \\
\phi_{2i}(o_i) &= (\delta_{ik_i}^2 - \theta_{i_R}\phi_{i_R} x_{ri})^2 + 2\theta_{i_E}\phi_{i_E} e_i u_i + u_i^2 + \theta_{i_E}^2 e_i^2 \\
\phi_{3i}(o_i) &= \delta_{ik_i}^2 + u_i^2 + 2\theta_{i_R}\phi_{i_R}(\tau_{i_R}) x_{ri}\delta_{ik_i} - 2\theta_{i_E}\phi_{i_E}(\tau_{i_E}) e_i u_i - \theta_{i_R}^2 x_{ri}^2 - \theta_{i_E}^2 e_i^2
\end{aligned} \tag{10.25}$$

$$\lambda_i(o_i) = \begin{cases} \eta_i, & \text{If } \xi \in \mathbb{J}_i \wedge \mathscr{J}_{2i}, \\ \psi_i & \text{If } \xi \in \mathbb{J}_i \wedge \mathscr{J}_{6i}. \end{cases} \tag{10.26}$$

where

$$\psi_i = \sqrt{\frac{\theta_{i_R}\phi_{i_R}(\rho_{iM}) x_{ri}^2 + \theta_{i_E}\phi_{i_E}(\tau_{MIET}) e_i^2}{\theta_{i_R}\phi_{i_R}(0) x_{ri}^2}} \tag{10.27}$$

Then, the (δ, x_r, e) *dynamics is UGAS, which implies that the consensus of the MAS is achieved.*

Proof Choose the Lyapunov function

$$V(\xi) = \delta^T (I_N \otimes P)\delta + \sum_{i=1}^{N} \theta_{i_R}\phi_{i_R}(\tau_{i_R}) x_{ri}^2 + \sum_{i=1}^{N} \theta_{i_E}\phi_{i_E}(\tau_{i_E}) e_i^2$$

According to Lemma 10.1, it is known that by selecting appropriate ρ_{iM} and τ_{MIET}, it is guaranteed that $\phi_{i_R}(\tau_{i_R})$ and $\phi_{i_E}(\tau_{i_E})$ are both always positive, besides, the Lyapunov function is radially unbounded, thus, it is an appropriate Lyapunov function for stability analysis. Due to the closed-loop system is a hybrid system, under hybrid systems framework [23], for the MAS with hybrid model (10.15), if the following conditions are satisfied

$$\begin{aligned} &\langle \nabla V(\xi), F(\xi)\rangle \leq -\sigma_\delta(\delta) - \sum_{i=1}^{N}\sigma_r(x_{ri}) - \sum_{i=1}^{N}\sigma_e(e_i), \text{ for } \xi \in \mathbb{F} \\ &V(\xi^+) - V(\xi) \leq 0, \text{ for } \xi \in \mathbb{J} \end{aligned} \tag{10.28}$$

for some positive definite functions $\sigma_\delta(\cdot), \sigma_r(\cdot), \sigma_e(\cdot)$, then, the (δ, x_r, e) dynamics in ξ converge to zeros asymptotically, namely the consensus is achieved and $\delta_i \to 0$.

First, consider the evolution of the Lyapunov function along the flow dynamics of the MAS, for all (τ_{i_R}, τ_{i_E}) and almost all (δ, x_r, e)

$$\begin{aligned} &\langle \nabla V(\xi), F(\xi)\rangle \\ =&\delta^T(-(\mathscr{L}+\mathscr{L}^T)\otimes PK_P)\delta - 2\delta^T(I_N\otimes P)K_I\mathscr{L}x_r + 2\delta^T(I_N\otimes P)K_P\mathscr{L}^2 e \\ &+\sum_{i=1}^{N}2\theta_{i_R}\phi_{i_R}(\tau_{i_R})x_{ri}\delta_{ik_i} - \sum_{i=1}^{N}s_1(\tau_{i_R})\theta_{i_R}^2(\phi_{i_R}^2(\tau_{i_R})+1)x_{ri}^2 \\ &+\sum_{i=1}^{N}2\theta_{i_E}\phi_{i_E}(\tau_{i_E})e_i(-u_i) - \sum_{i=1}^{N}s_2(\tau_{i_E})\theta_{i_E}^2(\phi_{i_E}^2(\tau_{i_E})+1)e_i^2 \end{aligned} \tag{10.29}$$

From [22, Definition 6], it is known that $\delta^T(-(\mathscr{L}+\mathscr{L}^T)\otimes PK_P)\delta \leq -\lambda_2(\mathscr{L})$ $\delta^T(I_N\otimes PK_P)\delta = -\delta^T(I_N\otimes Q)\delta$. Let the matrix inequality in (10.23) left-multiplied by $\left[\delta^T\ x_r^T\ e^T\right]$ and right-multiplied by $\left[\delta^T\ x_r^T\ e^T\right]^T$, and note that $\delta_{ik_i} = R_i\delta + R_{i_E}e$ and $-u_i = \bar{R}_i\delta + \bar{R}_{i_R}x_r + \bar{R}_{i_E}e$, we have

$$\begin{aligned} &\delta^T(-(\mathscr{L}+\mathscr{L}^T)\otimes PK_P)\delta - 2\delta^T(I_N\otimes P)K_I\mathscr{L}x_r + 2\delta^T(I_N\otimes P)K_P\mathscr{L}^2 e \\ \leq& -\sigma\delta^T(I_N\otimes Q)\delta - \sum_{i=1}^{N}(\delta_{ik_i}^2 + u_i^2 - \theta_{i_R}^2x_{ri}^2 - \theta_{i_E}^2e_i^2) - \epsilon\,\|x_r\|^2 - \epsilon\,\|e\|^2 \end{aligned} \tag{10.30}$$

Then, from (10.10), (10.19) and (10.20), according to the ranges of τ_{i_R} and τ_{i_E}, there exist several different cases.

Case 1: $\{\xi \in \mathbb{X} | \mathscr{F}_{1i} \vee \mathscr{F}_{2i}\}$

At this time, $\dot{\phi}_{i_R}(\tau_{i_R}) = -\theta_{i_R}(\phi_{i_R}^2(\tau_{i_R})+1)$, $\dot{\phi}_{i_E}(\tau_{i_E}) = -\theta_{i_E}(\phi_{i_E}^2(\tau_{i_E})+1)$. Hence, based on (10.29) and (10.30), it can be derived that

$$
\begin{aligned}
&\langle \nabla V(\xi), F(\xi)\rangle \\
&\leq -\sigma\delta^T(I_N \otimes Q)\delta - \sum_{i=1}^{N}(\theta_{i_R}\phi_{i_R}(\tau_{i_R})x_{ri} - \delta_{ik_i})^2 - \sum_{i=1}^{N}(\theta_{i_E}\phi_{i_E}(\tau_{i_E})e_i + u_i)^2 - \sum_{i=1}^{N}\epsilon(x_{ri}^2 + e_i^2) \\
&\leq -\sigma\delta^T(I_N \otimes Q)\delta - \sum_{i=1}^{N}\epsilon(x_{ri}^2 + e_i^2)
\end{aligned}
\tag{10.31}
$$

Case 2: $\{\xi \in \mathbb{X} | \mathscr{F}_{3i}\}$

At this time, $\dot{\phi}_{i_R}(\tau_{i_R}) = 0$, $\dot{\phi}_{i_E}(\tau_{i_E}) = -\theta_{i_E}(\phi_{i_E}^2(\tau_{i_E}) + 1)$. Thus, according to (10.29) and (10.30), the following conditions are derived

$$
\begin{aligned}
&\langle \nabla V(\xi), F(\xi)\rangle \\
&\leq -\sigma\delta^T(I_N \otimes Q)\delta - \sum_{i=1}^{N}(\theta_{i_E}\phi_{i_E}(\tau_{i_E})e_i + u_i)^2 - \sum_{i=1}^{N}\left(\epsilon(x_{ri}^2 + e_i^2) - 2\theta_{i_R}\phi_{i_R}(\tau_{i_R})x_{ri}\delta_{ik_i} + \delta_{ik_i}^2\right) \\
&\leq -\sigma\delta^T(I_N \otimes Q)\delta - \sum_{i=1}^{N}\epsilon(x_{ri}^2 + e_i^2) - \sum_{i=1}^{N}\phi_{1i}(o_i) \\
&\overset{(10.10)}{\leq} -\sigma\delta^T(I_N \otimes Q)\delta - \sum_{i=1}^{N}\epsilon(x_{ri}^2 + e_i^2)
\end{aligned}
\tag{10.32}
$$

Case 3: $\{\xi \in \mathbb{X} | \mathscr{F}_{4i}\}$

At this time, $\dot{\phi}_{i_R}(\tau_{i_R}) = -\theta_{i_R}(\phi_{i_R}^2(\tau_{i_R}) + 1)$, $\dot{\phi}_{i_E}(\tau_{i_E}) = 0$. Thus, from (10.29) and (10.30), it follows that

$$
\begin{aligned}
&\langle \nabla V(\xi), F(\xi)\rangle \\
&\leq -\sigma\delta^T(I_N \otimes Q)\delta - \sum_{i=1}^{N}(\theta_{i_R}\phi_{i_R}(\tau_{i_R})x_{ri} - \delta_{ik_i})^2 - \sum_{i=1}^{N}\left(\epsilon(x_{ri}^2 + e_i^2) + 2\theta_{i_E}\phi_{i_E}(\tau_{i_E})e_iu_i + u_i^2\right) \\
&\leq -\sigma\delta^T(I_N \otimes Q)\delta - \sum_{i=1}^{N}\epsilon(x_{ri}^2 + e_i^2) - \sum_{i=1}^{N}\phi_{2i}(o_i) \\
&\overset{(10.10)}{\leq} -\sigma\delta^T(I_N \otimes Q)\delta - \sum_{i=1}^{N}\epsilon(x_{ri}^2 + e_i^2)
\end{aligned}
\tag{10.33}
$$

Case 4: $\{\xi \in \mathbb{X} | \mathscr{F}_{5i}\}$

At this time, $\dot{\phi}_{i_R}(\tau_{i_R}) = 0$, $\dot{\phi}_{i_E}(\tau_{i_E}) = 0$. Therefore, according to (10.29) and (10.30), it follows that

$$
\begin{aligned}
&\langle \nabla V(\xi), F(\xi)\rangle \\
&\leq -\sigma\delta^T(I_N \otimes Q)\delta + \sum_{i=1}^{N}(2\theta_{i_R}\phi_{i_R}(\tau_{i_R})x_{ri}\delta_{ik_i} - \delta_{ik_i}^2) \\
&\quad - \sum_{i=1}^{N}\left(\epsilon(x_{ri}^2 + e_i^2) + 2\theta_{i_E}\phi_{i_E}(\tau_{i_E})e_iu_i + u_i^2\right) + \sum_{i=1}^{N}(\theta_{i_R}^2x_{ri}^2 + \theta_{i_E}e_i^2) \\
&\leq -\sigma\delta^T(I_N \otimes Q)\delta - \sum_{i=1}^{N}\epsilon(x_{ri}^2 + e_i^2) - \sum_{i=1}^{N}\phi_{3i}(o_i) \\
&\overset{(10.10)}{\leq} -\sigma\delta^T(I_N \otimes Q)\delta - \sum_{i=1}^{N}\epsilon(x_{ri}^2 + e_i^2)
\end{aligned}
\tag{10.34}
$$

Next, we show that the Lyapunov function $V(\xi)$ does not increase when it evolves along the jump dynamics, there exist several cases as follows

Case 1: $\{\xi \in \mathbb{X} | \mathscr{J}_{1i} \vee \mathscr{J}_{3i} \vee \mathscr{J}_{4i} \vee \mathscr{J}_{5i} \vee \mathscr{J}_{7i} \vee \mathscr{J}_{8i}\}$

Based on (10.11) and (10.18), it is possible that $e_i = 0$ or $x_{ri} = 0$ or both of them equal zeros when jump dynamics occurs. It can be easily seen that the Lyapunov function $V(\xi)$ is non-increasing according to the form of $V(\xi)$. Hence, we have $V(\xi^+) - V(\xi) \leq 0$ in this case.

Case 2: $\{\xi \in \mathbb{X} | \mathscr{J}_{2i}\}$

From (10.11) and (10.26), the reset state x_{ri} jumps to $\eta_i x_{ri}$ in this case, it follows that

$$\begin{aligned} &V(\xi^+) - V(\xi) \\ =&\delta^+(I_N \otimes Q)\delta^+ + \sum_{i=1}^{N} \theta_{i_R}\phi_{i_R}(\tau_{i_R}^+)x_{ri}^{+2} + \sum_{i=1}^{N} \theta_{i_E}\phi_{i_E}(\tau_{i_E})e_i^{+2} \\ &- \delta(I_N \otimes Q)\delta - \sum_{i=1}^{N} \theta_{i_R}\phi_{i_R}(\tau_{i_R}){x_{ri}}^2 - \sum_{i=1}^{N} \theta_{i_E}\phi_{i_E}(\tau_{i_E})e_i^2 \\ \leq&\theta_{i_R}\phi_{i_R}(0)\eta_i^2{x_{ri}}^2 - \theta_{i_R}\phi_{i_R}(\rho_{iM}){x_{ri}}^2 \overset{(10.24)}{\leq} 0 \end{aligned} \tag{10.35}$$

Case 3: $\{\xi \in \mathbb{X} | \mathscr{J}_{6i}\}$

From (10.11) and (10.26), the state x_{ri} jumps to $\psi_i x_{ri}$ with ψ as defined in (10.26). Similar as (10.35) in Case 2 and note that $\dot{\phi}_{i_R}(\tau_{i_R}) = 0$ when $\tau_{i_R} \geq \rho_{iM}$ and $\dot{\phi}_{i_E}(\tau_{i_E}) = 0$ when $\tau_{i_E} \geq \tau_{MIET}$, we have

$$\begin{aligned} &V(\xi^+) - V(\xi) \\ &\leq\theta_{i_R}\phi_{i_R}(0)\psi_i^2{x_{ri}}^2 - \theta_{i_R}\phi_{i_R}(\rho_{iM}){x_{ri}}^2 - \theta_{i_E}\phi_{i_E}(\tau_{MIET})e_i^2 \\ &\overset{(10.27)}{\leq} 0 \end{aligned} \tag{10.36}$$

Based on (10.31)–(10.34), (10.35) and (10.36), we can conclude that the conditions in (10.28) are satisfied. Therefore, the (δ, x_r, e) dynamics are UGAS and the consensus is achieved. □

Remark 10.3 Based on Lemma 10.1, the minimum inter-event time τ_{MIET} is required to be less than the minimum ρ_{MAX}^i, $i \in [1, N]$, which seems to be a little conservative. However, this condition can be relaxed by selecting independent τ_{MIET}^i for each agent i, that does not change the framework for consensus analysis. Here, we use a common τ_{MIET} just for brevity.

Remark 10.4 It is known that both reset mechanism and ETM may lead to Zeno behavior, however, Zeno-freeness analysis is not necessary for the proposed hybrid systems approach, because time regularization (based on timer variables τ_{i_R} and τ_{i_E}) imposes lower bounds for the reset time intervals and event-triggering intervals, therefore, Zeno behavior is naturally excluded. Moreover, the proposed event-triggered

PIR protocol is distributed, because the triggering functions $\phi_{1i}(o_i)$, $\phi_{2i}(o_i)$, $\phi_{3i}(o_i)$, the map function $\lambda_i(o_i)$, and the PIR protocol (10.5) only use local available information and continuous communication between neighboring agents is not needed, thus, the proposed control protocol is appropriate for practical applications.

Remark 10.5 Based on the state space decomposition (10.10) and jump dynamics (10.18), it seems that the implementation of the proposed event-triggered PIR protocol is complex, however, it is not like this. In particular, from (10.10) and (10.11), the sets $\mathscr{R}_F$, $\mathscr{R}_J$ in the reset element (10.4) and the set $\mathscr{E}$ in ETM (10.7) have the following form

$$\begin{aligned}
\mathscr{R}_F &= \mathscr{C}^i_{R1} \vee (\mathscr{C}^i_{R2} \wedge \delta_{ik}x_{ri} \geq 0) \vee (\mathscr{D}^i_R \wedge \delta_{ik}x_{ri} \geq 0 \wedge \phi_{1i}(o_i) \geq 0) \\
\mathscr{R}_J &= ((\mathscr{C}^i_{R2} \vee \mathscr{D}^i_R) \wedge \delta_{ik}x_{ri} \leq 0) \vee (\mathscr{D}^i_R \wedge \phi_{1i}(o_i) \leq 0) \\
\mathscr{E} &= \mathscr{D}^i_E \wedge (\phi_{2i}(o_i) \leq 0 \vee \phi_{3i}(o_i) \leq 0)
\end{aligned}$$

The reset element only needs to check if the local available variables belong to the jump set $\mathscr{R}_J$, and the ETM only needs to check if the local available variables belong to the set $\mathscr{E}$, then, the error e_i and state x_{ri} will be reset according to (10.11). Therefore, the implementation of the proposed event-triggered PIR protocol is not complex.

Remark 10.6 In this chapter, in order to design reset based consensus protocol for MAS, we only consider MAS with single integrator dynamics, i.e., such MAS is a class of single-input single-output system, this is mainly due to how to design proper reset conditions for multi-input multi-output systems is still an open problem that is not resolved yet. In addition, if we want to design reset controller with multiple inputs and multiple outputs, how to reset controller states according to values of controller input and controller output is not easy, since they are possibly involved with each other. Moreover, there exist both reset induced jump dynamics and ETM induced jump dynamics in the closed-loop system, the reset conditions design for MAS with multiple inputs and multiple outputs will become more complex, which will cause more difficulty for the state space decomposition, closed-loop system model construction, and stability analysis. Therefore, though the reset control approach has potential to be extended to MASs with general linear dynamics, the extension is not direct and there is much future work that needs further investigation.

10.5 Simulation Study

This section gives simulation results to verify the effectiveness of the proposed reset consensus control method. Suppose that there exist 6 agents and the communication relationship among them is represented as a communication graph as shown in Fig. 10.2. Based on the communication graph, the Laplacian matrix $\mathscr{L}$ can be easily obtained, and the minimum positive eigenvalue of $\mathscr{L}$ can be calculated as

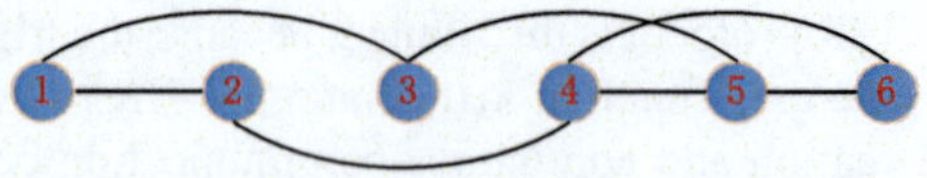

Fig. 10.2 Communication graph

$\lambda_2(\mathscr{L}) = 1$. The same control gains $K_P = 1$ and $K_I = 0.5$ are selected as in [12]. The other parameters are designed as $\sigma = 0.2$, $\sigma_1 = 0.18$, $\epsilon = 0.001$, $\theta_{i_R} = 8$, $\theta_{i_E} = 8$, $\eta_i = 0.5$, $\phi_{i_R}(0) = 3$, $\phi_{i_E}(0) = 3$ for $i \in [1, 6]$, $P = 20$, $\rho_i = 0.02$, $\rho_{iM} = 0.08$, $\tau_{MIET} = 0.1$, $\phi_{i_R}(0) = 3$, $\phi_{i_E}(0) = 3$. $\eta_i^2 \phi_{i_R}(0) = 0.75 \leq \phi_i(0.08) = 1.14$. The initial conditions are set to be $x(0) = [1\ 2\ 3\ 4\ 5\ 6]^T$, $\tau_{i_R}(0) = [0\ 0\ 0\ 0\ 0\ 0]^T$, $\tau_{i_E}(0) = [0\,0\,0\,0\,0\,0]^T$, $x_r(0) = [0\,0\,0\,0\,0\,0]^T$.

For the purpose of illustrating the proposed method, comparison results will be given. Take the static state feedback protocol $u_i(t) = \sum_{j \in \mathcal{N}_i}(x_j(t) - x_i(t))$ (**M1**) as a benchmark method. Since M1 requires continuous communication between neighboring agents, some recent results (e.g., [8, 26, 27], to just name a few) have been developed to avoid continuous communication. However, the consensus performance is usually degraded compared with M1 because only sampled or estimated information of neighbors can be obtained. Thus, it is reasonable to verify the effectiveness of consensus protocols by investigating to what extent the achieved consensus performance can approximate the obtained consensus performance with M1.

With the given design parameters and different control strategies, the simulation verification results are presented in Figs. 10.3, 10.4, 10.5 and 10.6. The results with base dynamic state feedback protocol[1] (**M2**), the consensus protocol with dynamic ETM in recent work [7] (**M3**), and the proposed PIR protocol (**M4**) are given for comparison. We consider the settling time[2] performance, which is an important transient performance index in many engineering applications. From Fig. 10.3, in contrast with M1 (the settling time is 4.77 s), the method M2 achieves less rising time but with larger overshoot and settling time (8.02 s), the method M3 with dynamic ETM and the proposed M4 achieve settling time 9.32 s and 5.81 s respectively. Thus, the proposed method M4 achieves less settling time compared with M2 and M3, but due to continuous neighbors' information cannot be obtained (induced by ETM), the consensus performance is a little worse compared with M1. It is worth to be noted that if we suppose that continuous communication is available, the PIR protocol $u_i(t) = K_P\delta_i(t) + K_I x_{ri}(t)$ can achieve better consensus performance than the benchmark method M1 (as shown in Fig. 10.4 with settling time 2.15 s), but it is not practical. Here, ETM-based control leads to performance degradation, but by means of reset control, the transient consensus performance can be improved. Figure 10.5 presents the corresponding control inputs.

As the ETM is also utilized in method M3 [7], the triggering instants as well as triggering intervals with M3 and M4 are shown in Fig. 10.6. From Fig. 10.6a, and together with Fig. 10.3c, we can see that though the settling time with M3 (dynamic

[1] The protocol (10.5) with $\delta_i(t)$ instead of δ_{ik_i} and without reset mechanism.

[2] The settling time is defined as the time required for all agents to reach and stay within a range of certain percentage of the consensus value and 2% is used here.

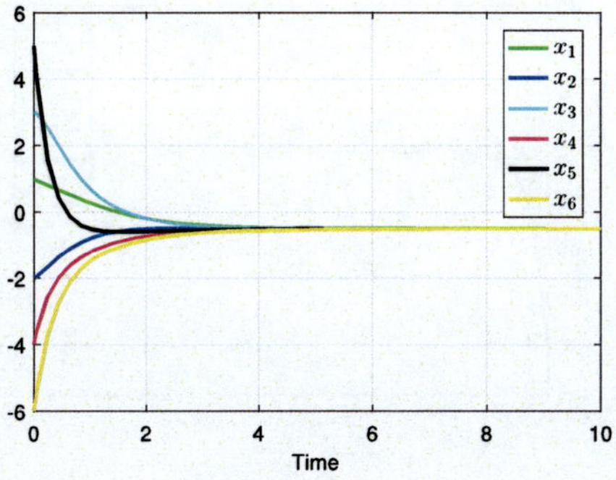

(a) Static Feedback (M1).

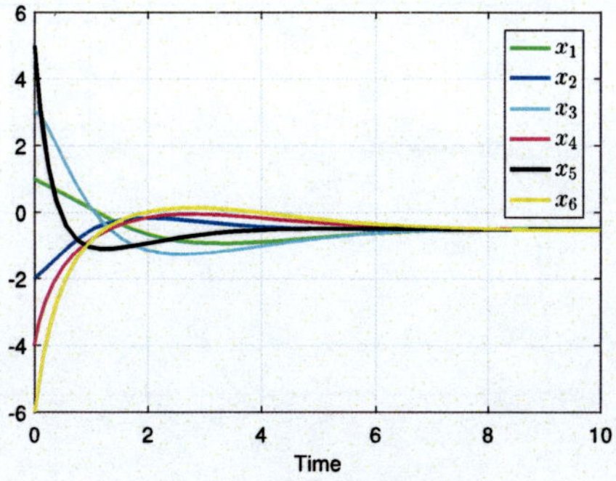

(b) Dynamic Feedback (M2).

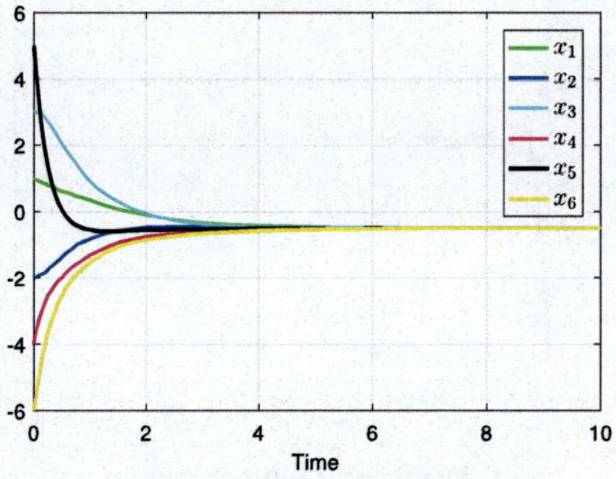

(c) The method in [7] (M3).

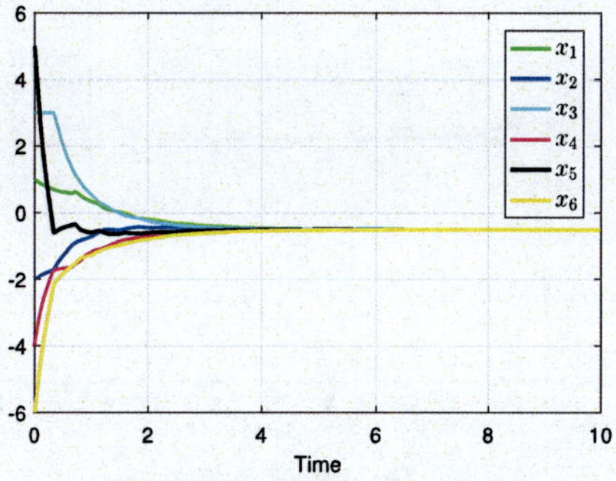

(d) The proposed PIR protocol (M4).

Fig. 10.3 Agent states x_i. **a** and **b** are based on continuous communication. **c** and **d** are based on event-triggered communication

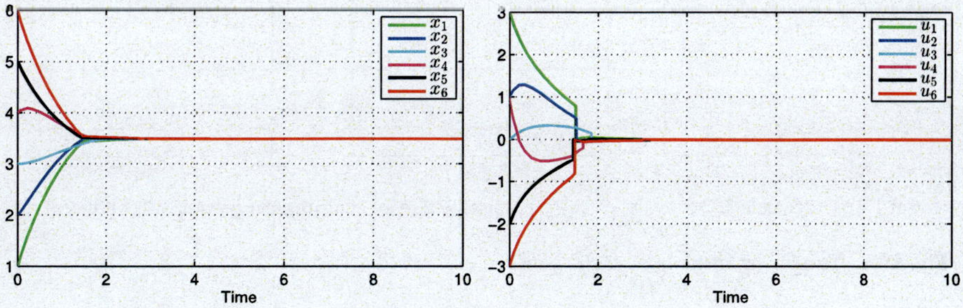

Fig. 10.4 Simulation results with $u_i(t) = K_P\delta_i(t) + K_I x_{ri}(t)$

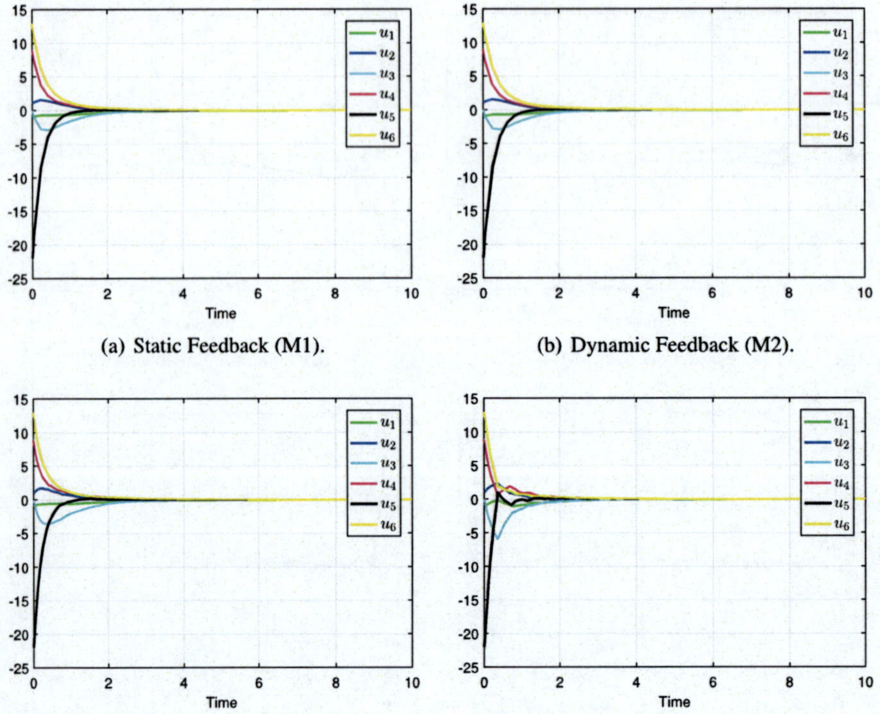

(a) Static Feedback (M1). (b) Dynamic Feedback (M2).

(c) The method in [7] (M3). (d) The proposed PIR protocol (M4).

Fig. 10.5 Control input u_i with different consensus protocols

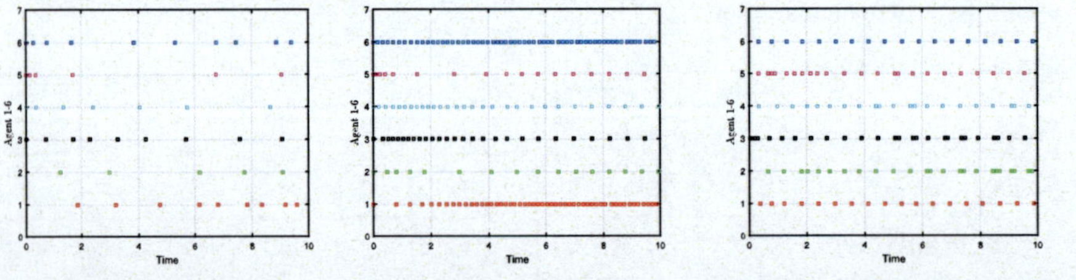

(a) [7] M3:dynamic ETM. (b) [7] M3:static ETM. (c) The proposed PIR protocol (M4).

Fig. 10.6 Triggering instants of each agent

ETM) is large, the number of triggers is less than with method M4. This is mainly because dynamic ETM is utilized, but at the cost of degraded consensus performance. To improve the consensus performance, the method M3 (static ETM) can be used to achieve less settling time (5.75 s, which is almost the same as with method M4 (5.81 s)), but with much more number of triggers as shown in Fig. 10.6b. From Fig. 10.6b and c, when the achieved settling time performance is almost the same (5.75 s versus 5.81 s), the proposed method M4 achieves less number of triggers (134

Table 10.1 Performance comparison. M3D/M3S denotes the method M3 with Dynamic/Static ETM. NoT denotes the number of triggers. ST denotes the settling time. (M3D is repeated)

	Agent1	Agent2	Agent3	Agent4	Agent5	Agent6	Total
M3D:NoT	27	9	8	9	10	28	91
M3D:ST	9.32	8.12	7.82	7.68	7.93	9.15	
M3S:NoT	61	19	15	16	20	62	193
M3S:ST	5.75	5.18	5.16	4.76	4.75	5.42	
M4:NoT	19	20	30	25	21	19	134
M4:ST	5.81	5.14	4.95	4.97	4.96	5.76	

versus 193). Moreover, the proposed method M4 can also be extended to the case with hybrid dynamic ETM without changing the analysis framework, and potential less number of triggers is expected, but may also at the cost of degraded consensus performance.

To further illustrate the effectiveness of the proposed results, performance comparison results are given in Table 10.1. The proposed method M4 achieves less number of triggers compared with M3 (static ETM), while achieving almost the same settling time performance. In addition, in contrast with M3 (dynamic ETM), the proposed method M4 achieves better consensus performance, but at the cost of more triggering events. However, we cannot say that the method M4 must be better than M3 [7], since both methods are effective for saving resources with the consensus performance as a tradeoff. The main contribution of this work is that reset control is utilized to improve transient consensus performance of MAS, and a novel hybrid systems approach is proposed for consensus analysis and ETM design with guaranteed Zeno-freeness. The advantage of reset method is that it provides a simple and easy way to improve the transient performance without fine tuning of parameters.

10.6 Conclusion

This chapter provides an event-triggered cooperative control method combining reset and consensus control, which is useful for improving transient consensus performance under event-triggered communication environment. Reset element with sector reset conditions and time regularization is introduced, and a hybrid ETM is proposed. Due to there exist two types of jump dynamics: state jump dynamics induced by reset and error jump dynamics induced by event-trigger, the state space is carefully decomposed into multiple flow sets and jump sets, based on which, and by means of auxiliary variables, a hybrid model is constructed to describe the closed-loop dynamics of the MAS. Then, Lyapunov-based consensus analysis and RM/ETM design results are derived.

References

1. Dimarogonas, D.V., Johansson, K.H.: Stability analysis for multiagent systems using the incidence matrix: Quantized communication and formation control. Automatica **46**(4), 695–700 (2010)
2. Tang, Y., Gao, H., Kruths, J., Fang, J.A.: Evolutionary pinning control and its application in UAV coordination. IEEE Trans. on Ind. Informat. **8**(4), 828–838 (2012)
3. Ge, X., Han, Q.L., Wang, Z.: A dynamic event-triggered transmission scheme for distributed set-membership estimation over wireless sensor networks. IEEE Trans. Cybern. **49**(1), 171–183 (2019)
4. Tabuada, P.: Event-triggered real-time scheduling of stabilizing control tasks. IEEE Trans. Autom. Control **52**(9), 1680–1685 (2007)
5. Hu, W., Liu, L., Feng, G.: Consensus of linear multiagent systems by distributed event-triggered strategy. IEEE Trans. Cybern. **46**(1), 148–157 (2016)
6. Garcia, E., Cao, Y., Casbeer, D.W.: Periodic event-triggered synchronization of linear multiagent systems with communication delays. IEEE Trans. Autom. Control **62**(1), 366–371 (2017)
7. Yi, X., Liu, K., Dimarogonas, D.V., Hohansson, K.H.: Dynamic event-triggered and self-triggered control for multiagent systems. IEEE Trans. Autom. Control **64**(8), 3300–3307 (2019)
8. Selivanov, A., Fridman, E.,: Synchronization of multiagent systems using event-triggered and self-triggered broadcasts. IEEE Trans. Autom. Control **62**(9), 4741–4746 (2017)
9. Ding, L., Han, Q.L., Ge, X., Zhang, X.: An overview of recent advances in event-triggered consensus of multiagent systems. IEEE Trans. Cybern. **48**(4), 1110–1123 (2018)
10. Qin, J., Ma, Q., Shi, Y., Wang, L.: Recent advances in consensus of multi-agent systems: A brief survey. IEEE Trans. Ind. Electron. **64**(6), 4972–4983 (2017)
11. Nowzari, C., Carcia, E., Cortes, J.: Event-triggered communication and control of networked systems for multiagent consensus. Automatica **105**, 1–27 (2019)
12. Meng, X., Xie, L., Soh, Y.C.: Reset control for synchronization of multi-agent systems, Automatica **104**, 189–195 (2019)
13. Bragagnolo, M.C., Morarescu, I.C., Daafouz, J., Riedinger, P.: Reset strategy for consensus in networks of clusters, Automatica **65**, 53–63 (2016)
14. Yucelen, T., Haddad, W.M.: Consensus protocols for networked multi-agent systems with a uniformly continuous quasi-resetting architecture, Int. J. Control **87**(8), 1716–1727 (2014)
15. Clegg, J.C.: A nonlinear integrator for servomechanisms. Trans. Amer. Inst. Elect. Eng. II Appl. Ind. **77**(1), 41–42 (1958)
16. Beker, O., Hollot, C.V., Chait, Y., Han, H.: Fundamental properties of reset control systems, Automatica **40**(6), 905–915 (2004)
17. Zaccarian, L., Nesic, D., Teel, A.R.: First order reset elements and the Clegg integrator revisited. 51st IEEE Conf. Decis. Control 6825–6830 (2005)
18. Hunnekens, B., Wouw, N., Nesic, D.: Overcoming a fundamental time-domain performance limitation by nonlinear control. Automatica **67**, 277–281 (2016)
19. Zhao, G., Nesic, D., Tan, Y., Wang, J.: Open problems in reset control. IEEE Conf. Decis. Control 3326–3331 (2013)
20. Zhao, G., Wang, J.: On L2 gain performance improvement of linear systems with Lyapunov-based reset control. Nonlinear Anal.-Hybrid Syst. **21**, 105–117 (2016)
21. Zhao, G., Nesic, D., Tan, Y., Hua. C.: Overcoming overshoot performance limitations of linear systems with reset control. Automatica **101**, 27–35 (2019)
22. Yu, W., Chen, G., Cao, M., Kurths, J.: Second-order consensus for multiagent systems with directed topologies and nonlinear dynamics. IEEE Trans. Syst. Man, Cybern. B, Cybern. **40**(3), 881–891 (2010)
23. Goebel, R., Sanfelice, R.G., Teel, A.R..: Hybrid Dynamical Systems: Modeling, Stability, Robustness. Princeton, NJ, USA: Princeton Univ. Press, (2012).
24. Nesic, D., Zaccarian, L., Teel, A.R.: Stability properities of reset systems. Automatica **44**(8), 2019–2026 (2008)

25. Nesic, D., Teel, A.R.: Stability and performance of SISO control systems with first-order reset elements. IEEE Trans. Autom. Control **56**(11), 2567–2582 (2011)
26. Cheng, T., Kan, Z., Klotz, J.R., Shea, J.M., Dixon, W.E.: Event-triggered control of multiagent systems for fixed and time-varying network topologies. IEEE Trans. Autom. Control **62**(10), 5365–5371 (2017)
27. Yang, D., Ren, W., Liu, X., Chen, W.: Decentralized eventtriggered consensus for linear multi-agent systems under general directed graphs. Automatica **69**, 242–249 (2016)

Chapter 11
Reset Observer Based Event-Triggered Control of MAS

11.1 Introduction

In Chaps. 9 and 10, sampled-data based reset control approach and event-based reset control approach are respectively proposed to tackle the cooperative control problem of MAS. As demonstrated by the obtained results in these two chapters, the reset control approach is able to effectively improve transient consensus control performance, such as the settling time performance. However, one common feature of the two different reset control approaches is that full state information is required to be available. It is widely known that the requirement of full state information in application scenarios is usually not easy to be realized, since some state information is not measurable or dedicated and expensive sensors are required to measure the state, that causes difficulties in implementing control strategies or high control cost. To relax the requirement of full state information, state observers are proposed and applied to estimate real state information by utilizing input and output of a system, with the help of state observers, the requirement of full state information is relaxed by slightly sacrificing control performance. Thus, in this chapter, we consider observer-based consensus control problem of MAS, and focus on designing distributed controllers and dynamic event-triggering mechanism by using observer states rather than real states.

Note that the observer-based event-triggered consensus control problems have been studied in some recent works, such as [1–3]. Specifically, in [1], for general linear MAS, the authors proposed observer-based event-triggered consensus protocol to reduce the frequency of controller updates and save computational burden, however, to realize the proposed control strategy in practice, continuous communication between agents is required, which causes unnecessary and excessive usage of network communication bandwidth. To overcome the limitation of relying on continuous communication among agents, by only using sampled data, an observer-based consensus control strategy was proposed in [2]. Moreover, [3] took the usability of global topology information into consideration and employed adaptive distributed observer-based event-triggered cooperative control strategy for MAS with external

G. Zhao et al., *Cooperative Control of Multi-agent Systems*,
https://doi.org/10.1007/978-981-97-0968-7_11

disturbances, such that the developed control strategy can be applied to MAS with possible time-varying communication topology and unfixed agent number. However, it is worth to be noted that linear observers are considered in the above-mentioned results. Different from the existing observer-based event-triggered control method, in this chapter, we will introduce how to design reset observer to accelerate the convergence speed of observer state, and further improve transient consensus control performance, this is an interesting and worthwhile researching topic. To investigate this topic, reset control, which consists of linear controller and reset conditions [4], can provide us some inspirations. It is known that well-defined reset conditions endows reset control the ability to improve transient performance or even overcome limitations of linear control, such superiority has been confirmed in some recent results, e.g., [5–7]. The authors in [8, 9] proposed reset adaptive observer for nonlinear systems for the purpose of improving transient convergence performance, however, one limitation is that the defined flow set may overflow into the jump set when system is running, and an imposed assumption is required to guarantee the well-posedness of reset conditions. Similar issues can also be found in [10, 11]. In this chapter, we introduce refined reset conditions with appropriately defined flow set and jump set, such that the reset conditions is ensured to be well-defined.

In addition, note that in some recent works [12–15] that are relevant to event-triggered control of MAS, static ETMs were employed, which is able to reduce triggering number compared with sampled-data based control method. To further reduce the communication frequency between agents, the idea of dynamic event-triggered control is applied to cooperative control problem of MAS, such as [15–17], where dynamic ETM with internal dynamic variables were developed. Due to the dynamic event-triggering conditions are not only related to current system state, but also record past system information by means of internal dynamic variables, it makes the dynamic ETM have good potential to further reduce the communication frequency compared with its counterpart of static ETM. Recently, [18] proposed a novel distributed dynamic ETM, but the results could only be applied to MAS with single integrators. An adaptive dynamic event-triggered consensus protocol was proposed in [19], where the coupling strength could be adaptively adjusted. In [20], a dynamic ETM with exponential term was proposed to further reduce the triggering number, and [21] extended the dynamic ETM method to MAS with directed networks. In the aforementioned results, it has been shown that Zeno behavior can be excluded, however, most of them did not consider the influence of disturbances. In recent work [22], it was formally shown that even arbitrary small disturbances may lead to excessive triggering events or Zeno behavior for some popular ETMs. Thus, whether the existing static/dynamic ETMs can exclude Zeno behavior in the presence of disturbances is unknown. In this chapter, we consider reset observer-based dynamic event-triggered consensus control of MAS with disturbances. First, a reset observer is developed in order to improve the error convergence performance compared with linear observers. Second, a novel hybrid dynamic ETM is proposed to impose that the triggering intervals have strictly positive lower bound. However, due to reset observer states include jump dynamics, and internal timer variables (with jump dynamics) are introduced to describe the reset conditions and the triggering conditions, the existence of jump dynamics brings difficulties for stability analysis and design. To address the jump dynamics induced problem, a hybrid model (under hybrid systems framework [23]) will be constructed to describe the overall closed-loop MAS dynamics, and Lyapunov-based consensus analysis and hybrid dynamic ETM design are given. From this chapter, readers can learn how to design reset observer with well-defined reset conditions, and how to construct closed-loop hybrid model for MAS with reset observer-based consensus control strategy. Moreover, the readers can also learn how to address the considered reset observer-based consensus control problem under hybrid systems framework.

11.2 System Model and Reset Observer

Consider a class of linear MASs with disturbances. The dynamics of each agent i has the following form

$$\begin{aligned} \dot{x}_i(t) &= Ax_i(t) + Bu_i(t) + B_w w_i(t) \\ y_i(t) &= Cx_i(t) \end{aligned} \tag{11.1}$$

where $x_i(t) \in \mathbb{R}^n$, $i \in [1, N]$ denotes the state of agent i, $u_i(t) \in \mathbb{R}^m$ denotes the control input of agent i, $y_i(t) \in \mathbb{R}$ is the measured output, $w_i(t) \in \mathbb{R}^l$ denotes bounded disturbance with $\|w_i(t)\| \leq \bar{w}_i$, where $\bar{w}_i > 0$. The matrices A, B, B_w, C are system matrices with appropriate dimensions. In practice, the agents are usually communicate with each other via wireless network. With event-triggered control strategy, for agent i, when the triggering event in agent i is activated, it collects and sends information to its neighbors at triggering instant $t^i_{k_i}$, $0 \leq t^i_0 < t^i_1 < \cdots$.

Some mathematical symbols and preliminaries on communication graph have been introduced in Chap. 1, in this chapter, we do not introduce them repeatedly, readers can found them in Sect. 1.4 if necessary. For proceeding this chapter, we make the following assumptions.

Assumption 11.1 (A, B) is stabilizable. (C, A) is detectable.

Assumption 11.2 The graph $\mathcal{G}$ is strongly connected.

According to graph theory, the following Lemmas are introduced that are necessary for the following stability analysis and design results.

Lemma 11.1 ([21]) *If the graph $\mathcal{G}$ is strongly connected, there exists a vector $z = \begin{bmatrix} z_1 & z_2 & \cdots & z_N \end{bmatrix}^T$ with all the elements being positive such that $z^T \mathcal{L} = 0$ and $\sum_{i=1}^N z_i = 1$. Moreover, define $Z = \mathrm{diag}\{z_1, z_1, \ldots, z_N\}$. Then, $\hat{\mathcal{L}} = \frac{1}{2}(Z\mathcal{L} + \mathcal{L}^T Z)$ is a symmetric positive definite matrix and $\sum_{j=1}^N \hat{\mathcal{L}}_{ij} = \sum_{j=1}^N \hat{\mathcal{L}}_{ji} = 0$ for $i \in [1, N]$.*

Lemma 11.2 ([24]) *For a strongly connected graph $\mathcal{G}$ with the Laplacian matrix $\mathcal{L}$, the following condition holds*

$$a(\mathcal{L}) = \min_{x^T z=0, x\neq 0} \frac{x^T \hat{\mathcal{L}} x}{x^T Z x} > 0 \tag{11.2}$$

where $a(\mathcal{L})$ is called the general algebraic connectivity for a strongly connected graph. In this chapter, the communication topology among agents is assumed to be directed, which is more general than undirected communication topology. For directed communication topology, a certain agent at the end of a directed path only receives its in-neighbors' information, and transmits its own information to out-neighbors at its triggering instants. By contrast, for undirected communication topology [3, 13, 19], the agent i and its neighbors can broadcast and receive information from their neighbors at agent i's triggering instant $t^i_{k_i}$. In addition, assume that the disturbance signal $w_i(t)$ persistently exists but it is bounded, due to the influence of such bounded disturbances, asymptotic consensus of MAS cannot be realized. Similar as [3], the following consensus/stability definition is used.

Definition 11.1 [3] (Consensus) Consider the linear MAS (11.1) subject to external disturbances with a fixed directed graph $\mathcal{G}$. Given any initial conditions $x_i(0)$, $i \in [1, N]$, if the agents' states satisfy the following condition:

$$\lim_{t\to\infty} \left\| x_i(t) - x_j(t) \right\| \leq \epsilon_0, \quad i, j \in [1, N]$$

where $\epsilon_0 \geq 0$ is a constant parameter, then, the consensus is said to be achieved for system (11.1).

11.2.1 Reset Observer

In the following, a reset observer is developed, which has potential to improve transient performance of observation error, and further leads to better consensus performance compared with using linear observers [2, 3]. The proposed reset observer has the following form

$$\left.\begin{aligned} &\dot{\tau}_{i_R}(t) = 1 \\ &\dot{\hat{x}}_i(t) = A\hat{x}_i(t) + Bu_i(t) + K_r x_{ri}(t) + K_p \tilde{y}_i(t) \\ &\dot{x}_{ri}(t) = a_r x_{ri}(t) + b_r \tilde{y}_i(t) \end{aligned}\right\} \text{ If } (\tau_{i_R}(t), \tilde{y}_i(t), x_{ri}(t)) \in \mathcal{F}_{Ri}$$

$$\left.\begin{aligned} &\tau_{i_R}(t^+) = 0 \\ &\hat{x}_i(t^+) = \hat{x}_i(t) \\ &x_{ri}(t^+) = 0 \end{aligned}\right\} (\tau_{i_R}(t), \tilde{y}_i(t), x_{ri}(t)) \in \mathcal{J}_{Ri} \tag{11.3}$$

where $\hat{x}_i(t) \in \mathbb{R}^n$ is the observer state, $x_{ri}(t) \in \mathbb{R}$ is the reset state, $\tilde{y}_i(t) = y_i(t) - \hat{y}_i(t)$, $\hat{y}_i(t) = C\hat{x}_i(t)$, K_r, K_p, a_r, b_r are gain matrices to be determined. $\tau_{i_R}(t)$ is a timer variable. $\mathscr{F}_{Ri}$ and $\mathscr{J}_{Ri}$ denote the flow set and jump set respectively.

$$\begin{aligned} \mathscr{F}_{Ri} &= \{\tau_{i_R}(t), \tilde{y}_i(t), x_{ri}(t) | \varepsilon \tilde{y}_i^2(t) + 2\tilde{y}_i(t)x_{ri}(t) \geq 0 \text{ or } \tau_{i_R}(t) \leq \rho\}, \\ \mathscr{J}_{Ri} &= \{\tau_{i_R}(t), \tilde{y}_i(t), x_{ri}(t) | \varepsilon \tilde{y}_i^2(t) + 2\tilde{y}_i(t)x_{ri}(t) \leq 0 \text{ and } \tau_{i_R}(t) \geq \rho\}. \end{aligned} \tag{11.4}$$

where $\varepsilon > 0$ and $\rho > 0$.

Remark 11.1 To guarantee well-posedness of reset observer, sector reset conditions are designed in (11.4), the state space related to $(\tau_{i_R}(t), \tilde{y}_i(t), x_{ri}(t))$ is decomposed into two parts: flow set $\mathscr{F}_{Ri}$ and jump set $\mathscr{J}_{Ri}$. With the designed reset conditions, it is guaranteed that when $(\tau_{i_R}(t), \tilde{y}_i(t), x_{ri}(t)) \in \mathscr{J}_{Ri}$, we obtain $(\tau_{i_R}(t^+), \tilde{y}_i(t^+), x_{ri}(t^+)) \in \mathscr{F}_{Ri}$, this is different from [8], where the after reset condition is guaranteed by assumption. Besides, it is well known that reset action may lead to Zeno behavior, which is not desired in practice since it may lead to uncertain system dynamics. To ensure exclusion of Zeno behavior, a timer variable τ_{i_R} is added into the reset conditions, such that the time interval between two successive reset instants is imposed to be lower bounded by $\rho > 0$. So, Zeno reset behavior is explicitly excluded.

Next, before introducing the ETM and control law, define combined measurement variables

$$\begin{aligned} q_i(t) &= \sum_{j \in \mathscr{N}_i^{in}} (\hat{x}_j(t) - \hat{x}_i(t)), t \in [0, \infty), \\ \hat{q}_i(t) &= \sum_{j \in \mathscr{N}_i^{in}} \left(e^{A(t-t_{k_j}^j)} \hat{x}_j\left(t_{k_j}^j\right) - e^{A(t-t_{k_i}^i)} \hat{x}_i\left(t_{k_i}^i\right) \right), t \in (t_{k_i}^i, t_{k_i+1}^i], \end{aligned} \tag{11.5}$$

and error variables

$$\begin{aligned} \hat{e}_i(t) &= x_i(t) - \hat{x}_i(t), \\ e_i(t) &= \hat{x}_i(t) - e^{A(t-t_{k_i}^i)} \hat{x}_i\left(t_{k_i}^i\right), \end{aligned} \tag{11.6}$$

where $t_{k_i}^i$ denotes the triggering instant of agent i. Then, the reset observer-based control law is given as follows

$$u_i(t) = K\hat{q}_i(t), \tag{11.7}$$

where the matrix K is the controller gain to be determined.

As mentioned before, the disturbances $\omega_i(t)$ is bounded implies that asymptotic consensus cannot be achieved. Based on Definition 11.1, we can define a combined measurement variable $\delta_i(t) = \sum_{j \in \mathscr{N}_i^{in}} (x_j(t) - x_i(t))$ such that the boundedness of $\delta_i(t)$ implies the consensus of the MAS. Moreover, based on (11.5) and (11.6), we have $q(t) = \hat{q}(t) - (\mathscr{L} \otimes I_n)e(t)$ and $\delta(t) = q(t) - (\mathscr{L} \otimes I_n)\hat{e}(t)$ ($\delta(t) = \left[\delta_1^T(t)\ \delta_2^T(t)\ \cdots\ \delta_N^T(t)\right]^T$), it indicates that if we can obtain the uniform ultimate boundedness of $q(t)$, $\hat{e}(t)$, it follows that the bounded consensus of MAS is achieved.

11.2.2 Hybrid Dynamic ETM with Guaranteed Zeno-Freeness

In this section, we develop a hybrid dynamic ETM with guaranteed Zeno-freeness, the triggering conditions are designed as follows

$$t^i_{k_i+1} = \inf\{t \geq t^i_{k_i} + \tau_{MIET} | \eta_i(t) \leq 0\}, \tag{11.8}$$

where $\tau_{MIET} > 0$ is a pre-specified constant value, whose role is mainly to avoid Zeno behavior. Based on the triggering conditions in (11.8), it is obviously that the minimum inter-event time (MIET) between adjacent triggering instants is lower bounded by τ_{MIET}.

The variable $\eta_i(t)$ is an internal dynamic variable, that has the following dynamics

$$\dot{\eta}_i(t) = \Psi_i(v^i_{loc}(t)),$$

where $\Psi_i()$ is a function with respect to $v^i_{loc}(t)$, and $v^i_{loc}(t)$ denotes the locally available information of agent i, such as $\eta_i(t)$, $e_i(t)$, $\hat{q}_i(t)$. It can be seen that the proposed hybrid ETM (11.8) is dynamic rather than static, and the form of the developed ETM is in distributed manner. Hence, the ETM in each agent is triggered asynchronously. Here, only the form and idea of the design of ETM are given, the specific form of $\Psi_i(v^i_{loc}(t))$ and how to determine τ_{MIET} will be given later.

With the proposed reset observer, distributed control law and hybrid dynamic ETM, the overall control structure can be represented as shown in Fig. 11.1.

In this chapter, we mainly focus on designing reset observer and hybrid dynamic ETM, and aim to improve the transient consensus control performance by utilizing the advantages of reset observer, the specific problem can be formulated as follows.

Problem: Consider the MAS (11.1) with external disturbances, design reset observer (11.3) to estimate agents' states, and design the controller $u_i(t)$ and a hybrid dynamic ETM (11.8) based on the estimated states, such that the consensus is achieved. Meanwhile, the inter-event times $t^i_{k_i+1} - t^i_{k_i} \geq \tau_{MIET}$ hold for $i \in [1, N]$, $k_i \in \mathbb{N}$, where the lower bound τ_{MIET} is independent of disturbances, namely Zeno-freeness can be guaranteed even in the presence of disturbances.

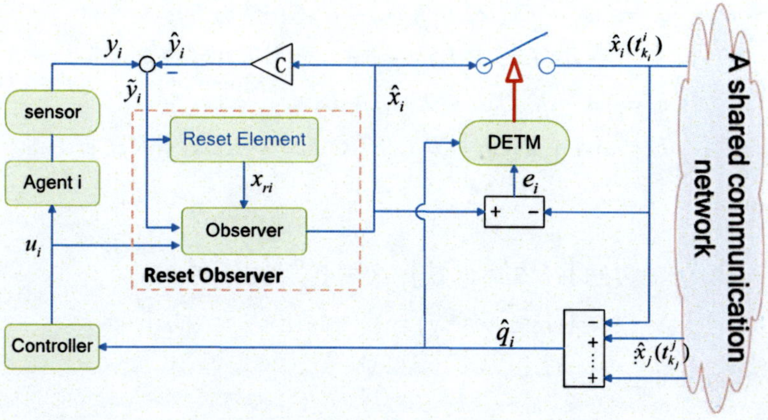

Fig. 11.1 Control structure with reset observer

11.3 Reformulation of MAS's Dynamics as Hybrid Form

Note that due to reset behavior in reset observer and the discontinuous controller update induced by event-triggered communication, there exist both flow dynamics and jump dynamics in the closed-loop system, such as $x_{ri}(t)$, $\hat{q}_i(t)$ that have jump dynamics when reset conditions are satisfied or triggering conditions are satisfied. Thus, in order to appropriately describe the closed-loop dynamics with both flow dynamics and jump dynamics, we utilize the hybrid systems framework introduced in [23] and construct a hybrid model to completely describe the closed-loop dynamics with reset observer and hybrid ETM.

First, a timer variable $\tau_{i_E}(t)$ is introduced with flow dynamics $\dot{\tau}_{i_E}(t) = 1$ for $t \in (t^i_{k_i}, t^i_{k_i+1}]$ and jump dynamics $\tau_{i_E}(t^+) = 0$ for $t = t^i_{k_i}$ (triggering instant). $\tau_{i_E}(t)$ can record the time elapsed since the last triggering instant, based on $\tau_{i_E}(t)$, we define the following sets

$$\begin{aligned}\mathscr{F}_{Ei} &= \{(\eta_i, e_i, \hat{q}_i) \in \mathbb{R} \times \mathbb{R}^n \times \mathbb{R}^n | \tau_{i_E} < \tau_{MIET} \vee \eta_i > 0\} \\ \mathscr{J}_{Ei} &= \{(\eta_i, e_i, \hat{q}_i) \in \mathbb{R} \times \mathbb{R}^n \times \mathbb{R}^n | \tau_{i_E} \geq \tau_{MIET} \wedge \eta_i \leq 0\}\end{aligned} \tag{11.9}$$

where the conditions in the definition of the jump set $\mathscr{J}_{Ei}$ is equivalent to the conditions in the ETM (11.8). Thus, by means of $\tau_{i_E}(t)$, we have reformulated the hybrid ETM (11.8) into a form of set, which is useful for constructing the closed-loop system model.

For agent i, it follows from (11.1), (11.3), (11.5) and (11.6) that

$$
\begin{aligned}
\dot{q}_i &= Aq_i - K_r(\mathscr{L}_r^i \otimes I_n)x_r - K_pC(\mathscr{L}_r^i \otimes I_n)\hat{e} \\
&\quad - BK(\mathscr{L}_r^i \otimes I_n)q - BK(\mathscr{L}_r^i \otimes I_n)(\mathscr{L} \otimes I_n)e \\
\dot{\hat{e}}_i &= (A - K_pC)\hat{e}_i - K_rx_{ri} + B_ww_i \\
\dot{e}_i &= Ae_i + BKq_i - BK(\mathscr{L}_r^i \otimes I_n)e + K_rx_{ri} + K_pC\hat{e}_i
\end{aligned}
\tag{11.10}
$$

where

$$
\begin{aligned}
q &= \left[q_1^T\ q_2^T\ \cdots\ q_N^T\right]^T, e = \left[e_1^T\ e_2^T\ \cdots\ e_N^T\right]^T, \\
\hat{e} &= \left[\hat{e}_1^T\ \hat{e}_2^T\ \cdots\ \hat{e}_N^T\right]^T, x_r = \left[x_{r1}^T\ x_{r2}^T\ \cdots\ x_{rN}^T\right]^T.
\end{aligned}
$$

Moreover, define

$$
\begin{aligned}
\tau_R &= \left[\tau_{1_R}^T\ \tau_{2_R}^T\ \cdots\ \tau_{N_R}^T\right]^T,\ \tau_E = \left[\tau_{1_E}^T\ \tau_{2_E}^T\ \cdots\ \tau_{N_E}^T\right]^T, \\
w &= \left[w_1^T\ w_2^T\ \cdots\ w_N^T\right]^T,\ \xi = \left[q^T\ e^T\ \hat{e}^T\ x_r^T\ \tau_R^T\ \tau_E^T\right].
\end{aligned}
$$

Then, based on the above definitions and the previous derivations, the closed-loop system including reset observer, agent's model, distributed controller and hybrid ETM can be formulated as a hybrid system. Specifially, the flow dynamics is described by

$$
\dot{\xi} = F(\xi, w),\ \text{If } \xi \in \mathbb{F}, \tag{11.11}
$$

where

$$
F(\xi, w) = \begin{bmatrix} A_{11}q + A_{12}\hat{e} + A_{13}x_r + A_{14}e \\ A_{21}q + A_{22}\hat{e} + A_{23}x_r + A_{24}e \\ A_{31}\hat{e} + A_{32}x_r + \bar{B}_ww \\ A_{41}\hat{e} + A_{42}x_r \\ 1_N \\ 1_N \end{bmatrix},
$$

$$
\begin{aligned}
A_{11} &= (I_N \otimes A - \mathscr{L} \otimes BK), A_{12} = -(\mathscr{L} \otimes K_pC), \\
A_{13} &= -(\mathscr{L} \otimes K_r), A_{14} = -(\mathscr{L}^2 \otimes BK), \\
A_{21} &= (I_N \otimes BK), A_{22} = (I_N \otimes K_pC), \\
A_{23} &= I_N \otimes K_r, A_{24} = (I_N \otimes A - \mathscr{L} \otimes BK), \\
A_{31} &= (I_N \otimes (A - K_pC)), A_{32} = -(I_N \otimes K_r), \\
A_{41} &= (I_N \otimes b_rC), A_{42} = (I_N \otimes a_r), \bar{B}_w = (I_N \otimes B_w).
\end{aligned}
$$

and the flow set $\mathbb{F}$ is defined as

$$\mathbb{F} := \bigcap_{i\in[1,N]} \left(\mathscr{F}_{Ri} \bigcap \mathscr{F}_{Ei}\right),$$

where $\mathscr{F}_{Ri}$ and $\mathscr{F}_{Ei}$ are defined in (11.4) and (11.9) respectively.

At reset instants $t^i_{r_i}$, the reset observer state x_{ri} and the timer variable τ_{i_R} are reset to zeros. At triggering instants $t^i_{k_i}$, the error variable e_i and the timer variable τ_{i_E} jump to zeros. Thus, the jump dynamics of the closed-loop MAS can be described by

$$\xi^+ \in \bigcup_{i\in[1,N]} J_i(\xi), \;\; \text{If } \xi \in \mathbb{J}, \tag{11.12}$$

where

$$J_i(\xi) := \begin{cases} J_{1i}(\xi), & \text{If } \xi \in \mathbb{J} \wedge \xi \in \mathscr{J}_{Ri} \\ J_{2i}(\xi), & \text{If } \xi \in \mathbb{J} \wedge \xi \in \mathscr{J}_{Ei} \end{cases} \tag{11.13}$$

$$J_{1i}(\xi) = \begin{bmatrix} q \\ e \\ \hat{e} \\ (I_N - \Gamma_i)x_r \\ (I_N - \Gamma_i)\tau_R \\ \tau_E \end{bmatrix}, \; J_{2i}(\xi) \;\; = \begin{bmatrix} q \\ (I_N - \Gamma_i)e \\ \hat{e} \\ x_r \\ \tau_R \\ (I_N - \Gamma_i)\tau_E \end{bmatrix},$$

where $\Gamma_i \in \mathbb{R}^{N\times N}$ is a diagonal matrix with ii-th element being 1 and all the other elements being zeros. The jump set $\mathbb{J}$ is given as

$$\mathbb{J} := \bigcup_{i\in[1,N]} \left(\mathscr{J}_{Ri} \bigcup \mathscr{J}_{Ei}\right),$$

Note that the states $q(t)$, $\hat{e}(t)$, $e(t)$ belong to the closed-loop state $\xi(t)$, thus, we give stability property of the closed-loop system as follows.

Definition 11.2 For the constructed hybrid model of the MAS as described by (11.11) and (11.12). The MAS is said to be uniformly globally bounded stable if there exists $\Delta_w > 0$ such that, for any initial condition $\xi(0)$, the following condition holds

$$\left\|(q(t), e(t), \hat{e}(t), x_r(t))\right\| \leq \Delta_w. \tag{11.14}$$

where Δ_w is a constant depending on $\bar{w}$.

11.4 Stability Analysis and ETM Design

There exist two type of jump dynamics in the closed-loop system: reset induced jump dynamics of observer state, and triggering event induced jump dynamics of controller. These two types of jump dynamics are implicitly correlated with each other and bring challenges for stability analysis. In this section, we present how to construct appropriate Lyapunov function for consensus analysis, while designing the hybrid dynamic ETM for saving communication resources. Some discussions about the superiority of the proposed method compared with some recent works are also provided.

11.4.1 Consensus Analysis and ETM Design

First, design auxiliary function $\phi_i(\tau_{i_E}) : \mathbb{R}_{\geq 0} \to \mathbb{R}_{\geq 0}$ with respect to τ_{i_E}

$$\begin{aligned} \dot{\phi}_i(\tau_{i_E}) &= s(\tau_{i_E})(-(\phi_i(\tau_{i_E}) + \kappa_i^2)), \\ s(\tau_{i_E}) &= \max\{\text{sign}(\tau_{MIET} - \tau_{i_E}), 0\}, \end{aligned} \tag{11.15}$$

where $\kappa_i > 0$ is a constant. sign$(\cdot)$ denotes sign function. The initial conditions $\phi_i(0) > 0$ can be chosen by designer, and $\phi_i(\tau_{i_E}) > 0$ can be always ensured by choosing τ_{MIET} to satisfy the condition in the following Lemma.

Lemma 11.3 *For given initial conditions $\phi_i(0) > 0$, $\kappa_i > 0$, $i \in [1, N]$, if τ_{MIET} satisfies the following condition*

$$0 < \tau_{MIET} < \tau_M^i \tag{11.16}$$

with $\tau_M^i = \arctan(\phi_i(0)/\kappa_i^2)$. Then, $\phi_i(\tau_{i_E}) > 0$ is ensured for $\tau_{i_E} \leq \tau_{MIET}$.

Proof The result can be easily obtained from (11.15) with initial condition $\phi_i(0) > 0$, so it is omitted here. □

According to the distributed control stratey (11.7) and the ETM (11.8), we are now in a position to present the main results.

Theorem 11.1 *Consider the MAS (11.1) with reset observer-based control protocol (11.7) under Assumptions 1–2. Let $K = \mu B^T P$, where $P = P^T > 0$ is the solution of the following algebraic Riccati equation*

$$PA + A^T P - 2\mu a(\mathscr{L}) PBB^T P + Q = 0. \tag{11.17}$$

If there exist symmetric positive definite matrices P_ζ, Q_ζ, constants $\tau_F > 0$, $\tau_J > 0$, such that the following linear matrix inequalities hold

$$\begin{aligned} P_\zeta A_\zeta + A_\zeta^T P_\zeta + \tau_F M &\leq -Q_\zeta, \\ A_R^T P_\zeta A_R - P_\zeta - \tau_J M &\leq 0, \end{aligned} \tag{11.18}$$

with

$$A_\zeta = \begin{bmatrix} A - K_p C & -K_r \\ b_r C & a_r \end{bmatrix}, A_R = \begin{bmatrix} I & 0 \\ 0 & 0 \end{bmatrix}, M = \begin{bmatrix} \varepsilon C^T C & C^T \\ C & 0 \end{bmatrix},$$

where $K_r > 0$, $b_r > 0$, $a_r < 0$, K_p is chosen such that $A - K_p C$ is Hurwitz.
For each agent i, the triggering function $\Psi_i(v_{loc}^i)$ is designed as

$$\Psi_i(v_{loc}^i) := \begin{cases} -\pi_i \eta_i + \gamma_{0i} \|e_i\|^2, & \text{for } 0 \leq \tau_{i_E} < \tau_{MIET} \\ -\pi_i \eta_i - \gamma_{1i} \|e_i\|^2 + \gamma_{2i} \|\hat{q}_i\|^2, & \text{for } \tau_{i_E} \geq \tau_{MIET} \end{cases} \tag{11.19}$$

where $\gamma_{0i} > 0$, and

$$\begin{aligned} \gamma_{1i} &= \mathscr{M}_{1i} + \lambda_{1i}\left(\frac{1}{\alpha} - 1\right)\left\|\mathscr{L}^T \mathscr{L} \otimes I_n\right\|, \\ \gamma_{2i} &= \lambda_{1i}(1-\alpha), \\ \lambda_{1i} &= \lambda_{\min}(\alpha_Z Z \otimes Q - 3\alpha I - \phi_i \|BK\|^2), \\ \mathscr{M}_{1i} &= \phi_i\left(\frac{1}{\alpha} + 2\|A\|\right) + \max_{i \subset [1,N]} \left\|BK(\mathscr{L}_r^i \otimes I_n)\right\|^2 N + \frac{\mu^2}{\alpha}\left\|Z\mathscr{L}^2 \otimes PBB^T P\right\|^2, \\ & 0 < \alpha < 1, 0 < \alpha_Z < 1, \pi_i > 0, \mu > 0 \end{aligned} \tag{11.20}$$

and choose $\kappa_i > \sqrt{\mathscr{M}_{1i} + \gamma_{0i} - \phi_i}$. Then, the combined measurement variable $\delta(t)$ converges to a neighborhood of the origin, namely the consensus of the MAS under external disturbances is achieved.

Proof Denote $\zeta_i = \left[\hat{e}_i^T \; x_{ri}^T\right]^T$, we obtain

$$\begin{aligned} \dot{\zeta}_i &= \underbrace{\begin{bmatrix} A - K_p C & -K_r \\ b_r C & a_r \end{bmatrix}}_{A_\zeta} \zeta_i + \underbrace{\begin{bmatrix} B_w \\ 0 \end{bmatrix}}_{B_\zeta} w_i, \\ \zeta_i^+ &= A_R \zeta_i. \end{aligned} \tag{11.21}$$

Define $\zeta = \left[\zeta_1^T \; \zeta_2^T \; \cdots \; \zeta_N^T\right]^T$, and construct Lyapunov function

$$V_1(\zeta) = \sum\nolimits_{i=1}^N \zeta_i^T P_\zeta \zeta_i. \tag{11.22}$$

From the definition of ζ and based on the hybrid model (11.11) and (11.12), ζ_i has jump dynamics when $\zeta_i \in \mathscr{J}_{Ri}$, and the dynamics of ζ_i flows when $\zeta_i \in \mathscr{F}_{Ri}$. Hence, the evolution of $V_1(\zeta)$ along the trajectory of ζ_i satisfies

$$
\begin{aligned}
\dot{V}_1(\zeta) &= \sum_{i=1}^{N}\Big(\zeta_i^T(P_\zeta A_\zeta + A_\zeta^T P_\zeta)\zeta_i + 2\zeta_i^T P_\zeta B_\zeta w_i\Big) \\
&\le \sum_{i=1}^{N}\Big(-\zeta_i^T(Q_\zeta - \tau_F M)\zeta_i + \alpha_\zeta \zeta_i^T\zeta_i + \frac{1}{\alpha_\zeta} w_i^T(P_\zeta B_\zeta)^T(P_\zeta B_\zeta)w_i\Big) \\
&\le \sum_{i=1}^{N}\Big(-\zeta_i^T(Q_\zeta - \alpha_\zeta I)\zeta_i + \frac{1}{\alpha_\zeta}\left\|P_\zeta B_\zeta\right\|^2 \bar{w}_i^2\Big) \\
&\le -\lambda_{\min}(Q_\zeta - \alpha_\zeta I)\left\|\zeta\right\|^2 + \frac{1}{\alpha_\zeta}\left\|P_\zeta B_\zeta\right\|^2 \bar{w},
\end{aligned}
\tag{11.23}
$$

where $\alpha_\zeta > 0$, $\bar{w} = \bar{w}_1^2 + \cdots + \bar{w}_N^2$, and the Young's inequality $2x^T y \le ax^T x + (1/a)y^T y$ is used. When $\zeta_i \in \mathscr{J}_{Ri}$, consider the jump dynamics of ζ_i, we have

$$
\begin{aligned}
V_1(\zeta^+) - V_1(\zeta) &= \sum_{j=1, j\neq i}^{N} \zeta_j^T P_\zeta \zeta_j + \zeta_i^{+T} P_\zeta \zeta_i^+ - \sum_{j=1, j\neq i}^{N} \zeta_j^T P_\zeta \zeta_j - \zeta_i^T P_\zeta \zeta_i \\
&= \zeta_i^T(A_R^T P_\zeta A_R - P_\zeta)\zeta_i \\
&\le \tau_J \zeta_i^T M \zeta_i \le 0.
\end{aligned}
\tag{11.24}
$$

Second, we consider the dynamics of q and e, choose the following Lyapunov function candidate

$$
V_2(q, e, \eta) = q^T(Z \otimes P)q + \sum\nolimits_{i=1}^{N} \phi_i e_i^T e_i + \sum\nolimits_{i=1}^{N} \eta_i. \tag{11.25}
$$

Take the derivative of $V_2(q, e, \eta)$ along the flow dynamics (11.11) and $\dot{\eta}_i = \Psi_i(v_{loc}^i)$, we have

$$
\begin{aligned}
&\dot{V}_2(q, e, \eta) \\
=&q^T((Z \otimes P)A_{11} + A_{11}^T(Z \otimes P))q + 2q^T(Z \otimes P)A_{12}\hat{e} \\
&+ 2q^T(Z \otimes P)A_{13}x_r + 2q^T(Z \otimes P)A_{14}e + \sum_{i=1}^{N}\dot{\eta}_i \\
&+ \sum_{i=1}^{N}(-s(\tau_{i_E})(\phi_i + \kappa_i^2)\left\|e_i\right\|^2 + 2\phi_i e_i^T \dot{e}_i) \\
=&q^T(Z \otimes (PA + A^T P) - \mu(Z\mathscr{L} + \mathscr{L}^T Z) \otimes PBB^T P)q \\
&- 2q^T(\mu Z\mathscr{L}^2 \otimes PBB^T P)e - 2q^T(Z\mathscr{L} \otimes PK_pC)\hat{e} \\
&- 2q^T(Z\mathscr{L} \otimes PK_r)x_r + \sum_{i=1}^{N}\Psi_i(v_{loc}^i) + \sum_{i=1}^{N}(-s(\tau_{i_E})(\phi_i + \kappa_i^2)\left\|e_i\right\|^2 + 2\phi_i e_i^T \dot{e}_i)
\end{aligned}
\tag{11.26}
$$

Notice that $q^T(\zeta \otimes I_n) = 0$, based on Lemma 11.2, we obtain

$$
q^T(-\mu(Z\mathscr{L} + \mathscr{L}^T Z) \otimes PBB^T P)q \le -2\mu a(\mathscr{L})q^T(Z \otimes PBB^T P)q, \tag{11.27}
$$

substitute (11.27) and (11.17) into (11.26), and note that $\dot{e}_i = Ae_i + BKq_i - BK(\mathscr{L}_r^i \otimes I_n)e + K_r x_{ri} + K_p C\hat{e}_i$, then, according to Young's inequality, it leads to

$$\begin{aligned}
&\dot{V}_2(q, e, \eta) \\
&\leq q^T(Z \otimes (PA + A^T P - 2\mu a(\mathscr{L})PBB^T P))q + \alpha q^T q + \frac{1}{\alpha} e^T (Z\mathscr{L}^2 \otimes PBB^T P)^T (Z\mathscr{L}^2 \otimes PBB^T P)e \\
&+ \alpha q^T q + \frac{\mu^2}{\alpha} \hat{e}^T (Z\mathscr{L} \otimes PK_pC)^T (Z\mathscr{L} \otimes PK_pC)\hat{e} + \alpha q^T q + \frac{1}{\alpha} x_r^T (Z\mathscr{L} \otimes PK_r)^T (Z\mathscr{L} \otimes PK_r)x_r \\
&+ \sum_{i=1}^{N} \phi_i (2\|A\| \|e_i\|^2 + \frac{1}{\alpha}\|e_i\|^2 + \alpha \|BK\|^2 \|q_i\|^2 + \frac{1}{\alpha}\|e_i\|^2 + \alpha \left\|BK(\mathscr{L}_r^i \otimes I_n)\right\|^2 \|e\|^2 + \frac{1}{\alpha}\|e_i\|^2 \\
&+ \alpha \|K_r\|^2 \|x_{ri}\|^2 + \frac{1}{\alpha}\|e_i\|^2 + \alpha \left\|K_pC\right\|^2 \|\hat{e}_i\|^2) + \sum_{i=1}^{N} (\Psi_i(v_{loc}^i) - s(\tau_{i_E})(\phi_i + \kappa_i^2)\|e_i\|^2) \\
&\leq q^T(-Z \otimes Q + 3\alpha I + \phi_i \|BK\|^2)q + \sum_{i=1}^{N} (\phi_i(\frac{4}{\alpha} + 2\|A\|) + \max_i\{\left\|BK(\mathscr{L}_r^i \otimes I_n)\right\|^2\}N \\
&+ \frac{\mu^2}{\alpha}\left\|Z\mathscr{L}^2 \otimes PBB^T P\right\|^2)\|e_i\|^2 + \frac{1}{\alpha}\max\{\alpha^2\|K_r\|^2, \left\|Z\mathscr{L} \otimes PK_pC\right\|^2, \left\|Z\mathscr{L} \otimes PK_r\right\|^2, \alpha^2\left\|K_pC\right\|^2\}\|\zeta\|^2 \\
&+ \sum_{i=1}^{N} (\Psi_i(v_{loc}^i) - s(\tau_{i_E})(\phi_i + \kappa_i^2)\|e_i\|^2) \\
&= q^T(-Z \otimes Q + 3\alpha I + \phi_i \|BK\|^2)q + \sum_{i=1}^{N} \mathscr{M}_{1i}\|e_i\|^2 + \sum_{i=1}^{N} (\Psi_i(v_{loc}^i) - s(\tau_{i_E})(\phi_i + \kappa_i^2)\|e_i\|^2) + \mathscr{M}_2\|\zeta\|^2,
\end{aligned} \tag{11.28}$$

where $\mathscr{M}_{1i}$ is defined in (11.20), and $\mathscr{M}_2 = \frac{1}{\alpha}\max\{\alpha^2\|K_r\|^2, \left\|Z\mathscr{L} \otimes PK_pC\right\|^2, \|Z\mathscr{L} \otimes PK_r\|^2, \alpha^2\left\|K_pC\right\|^2\}$.

Then, construct the Lyapunov function for the whole closed-loop MAS as

$$V(q, \eta, e, \hat{e}, x_r) = \theta_1 V_1(\hat{e}, x_r) + V_2(q, e, \eta). \tag{11.29}$$

According to (11.23) and (11.28), choose $\theta_1 \geq \mathscr{M}_2/\lambda_{\min}(Q_\zeta - \alpha_\zeta I)$, and consider the value of $\Psi_i(v_{loc}^i) - s(\tau_{i_E})(\phi_i + \kappa_i^2)\|e_i\|^2$ in (11.28), according to the range of τ_{i_E}, there exist two cases.

CASE 1: $0 \leq \tau_{i_E} < \tau_{MIET}$

$$\Psi_i(v_{loc}^i) - s(\tau_{i_E})(\phi_i + \kappa_i^2)\|e_i\|^2 = -\pi_i\eta_i + \gamma_{0i}\|e_i\|^2 - (\phi_i + \kappa_i^2)\|e_i\|^2. \tag{11.30}$$

Based on (11.28) and (11.20), we have

$$\dot{V}(q, \eta, e, \hat{e}, x_r) \leq (1 - \alpha_z)q^T(-Z \otimes Q)q + \frac{\theta_1}{\alpha_\zeta}\|P_\zeta B_\zeta\|^2 \bar{w} - \sum_{i=1}^{N} \pi_i\eta_i. \tag{11.31}$$

CASE 2: $\tau_{i_E} \geq \tau_{MIET}$

$$\Psi_i(v_{loc}^i) - s(\tau_{i_E})(\phi_i + \kappa_i^2)\|e_i\|^2 = -\pi_i\eta_i - \gamma_{1i}\|e_i\|^2 + \gamma_{2i}\left\|\hat{q}_i\right\|^2. \tag{11.32}$$

Based on (11.28) and (11.20), we have

$$\begin{aligned}
&\dot{V}(q, e, \eta, \hat{e}, x_r) \\
&\leq -(1-\alpha_z)q^T(Z \otimes Q)q - q^T(\alpha_z Z \otimes Q - 3\alpha I - \phi_i \|BK\|^2)q + \frac{\theta_1}{\alpha_\zeta} \|P_\zeta B_\zeta\|^2 \bar{w} \\
&+ \sum_{i=1}^{N}((\mathscr{M}_{1i} - \gamma_{1i}) \|e_i\|^2 + \gamma_{2i} \|\hat{q}_i\|^2) - \sum_{i=1}^{N} \pi_i \eta_i.
\end{aligned} \tag{11.33}$$

Let $\lambda_{1i} = \lambda_{\min}(\alpha_z Z \otimes Q - 3\alpha I - \phi_i \|BK\|^2)$, and using $q = \hat{q} - (\mathscr{L} \otimes I_n)e$, one has

$$\begin{aligned}
&-q^T(\alpha_z Z \otimes Q - 3\alpha I - \phi_i \|BK\|^2)q \\
&\leq -\lambda_{1i}(\hat{q} - (\mathscr{L} \otimes I_n)e)^T(\hat{q} - (\mathscr{L} \otimes I_n)e) \\
&\leq -\lambda_{1i}(\hat{q}^T\hat{q} - 2\hat{q}^T(\mathscr{L} \otimes I_n)e + e^T(\mathscr{L}^T\mathscr{L} \otimes I_n)e) \\
&\leq -\lambda_{1i}(1-\alpha)\hat{q}^T\hat{q} - \lambda_{1i}(1 - \frac{1}{\alpha}) \left\|\mathscr{L}^T\mathscr{L} \otimes I_n\right\| \|e\|^2.
\end{aligned} \tag{11.34}$$

According to (11.20), and based on (11.33) and (11.34), we derive that

$$\dot{V}(q, \eta, e, \hat{e}, x_r) \leq (1-\alpha_z)q^T(-Z \otimes Q)q + \frac{\theta_1}{\alpha_\zeta} \|P_\zeta B_\zeta\|^2 \bar{w} - \sum_{i=1}^{N} \pi_i \eta_i. \tag{11.35}$$

From (11.31) and (11.35), we can get

$$\dot{V}(q, e, \eta, \hat{e}, x_r) \leq (\alpha_z - 1)q^T(Z \otimes Q)q + \Delta \tag{11.36}$$

where $\Delta = \frac{\theta_1}{\alpha_\zeta} \|P_\zeta B_\zeta\|^2 \bar{w}$.

Moreover, consider the jump dynamics of $V_2(q, e, \eta)$ at triggering instants $t^i_{k_i}$

$$\begin{aligned}
&V_2(q^+, e^+, \eta^+) \\
&= q^{+T}(Z \otimes P)q^+ + \sum_{j=1, j \neq i}^{N} \phi_j e_j^{+T} e_j^+ + \sum_{i=1}^{N} \eta_i^+ - q^T(Z \otimes P)q \\
&- \sum_{j=1, j \neq i}^{N} \phi_j e_j^T e_j - \sum_{i=1}^{N} \eta_i + \phi_i(0) e_i^{+T} e_i^+ - \phi_i(\tau_{MIET}) e_i^T e_i \\
&\leq 0.
\end{aligned} \tag{11.37}$$

Thus, from (11.24) and (11.37), it is clear that no matter reset event occurs or triggering event occurs, the Lyapunov function $V(q, \eta, e, \hat{e}, x_r)$ can be ensured to be decreasing. Then, according to (11.36), (11.37) and (11.23), we have $\|q(t)\|^2 \leq \Delta/(1-\alpha_z)\lambda_{\min}(Z \otimes Q)$, and $\|\hat{e}(t)\|^2 + \|x_r(t)\|^2 \leq \frac{\|P_\zeta B_\zeta\|^2 \bar{w}}{\alpha_\zeta \lambda_{\min}(Q_\zeta - \alpha_\zeta I)}$ respectively. Due to $\delta(t) = q(t) - (\mathscr{L} \otimes I_n)\hat{e}(t)$, it implies that $\delta(t)$ converges to a neighborhood of the origin, namely the consensus of the MAS is achieved. □

The combined measurement variable $\delta(t)$ can be viewed as the consensus error, it has been shown in [3, 21, 25] that the convergence of $\delta(t)$ implies consensus of the MAS. In addition, the consensus error $\delta(t)$ is related to agent dynamics via $\delta(t) = -(\mathscr{L} \otimes I_n)x(t)$, and from $\delta(t) = q(t) - (\mathscr{L} \otimes I_n)\hat{e}(t)$, we have $\|\delta(t)\| \leq \sqrt{2}(\|q(t)\| + \|\mathscr{L} \otimes I_n\| \, \|\hat{e}(t)\|) \leq \sqrt{2\Delta/(1-\alpha_z)\lambda_{\min}(Z \otimes Q)} + \sqrt{\frac{2\|P_\zeta B_\zeta\|^2 \bar{w}}{\alpha_\zeta \lambda_{\min}(Q_\zeta - \alpha_\zeta I)}}$. Since Δ depends on $\bar{w}$, and $P_\zeta, Q_\zeta, B_\zeta, \alpha_\zeta$ and Q are implicitly related to control parameters (e.g., K, K_p, K_r, etc.) according to (11.17) and (11.18), thus, the upper bound of consensus error is related to disturbances and control parameters.

Remark 11.2 From (11.5)–(11.8), (11.19), and $\dot{\eta}_i(t) = \Psi_i(v^i_{loc}(t))$, it indicates that the implementation of reset observer-based consensus control strategy (11.7) and (11.8) only requires agent's locally available information. In addition, to realize the hybrid dynamic ETM, based on (11.8) and (11.19), the variables $e_i(t)$ and $\hat{q}_i(t)$ are required to be computed. Besides, note that some parameters (such as π_i, γ_{1i}, γ_{2i}, K, K_r, etc.) can be pre-given, thus, the online computational burden is not heavy. In summary, the physical feasibility of the proposed reset observer-based event-triggered control strategy can be guaranteed.

Remark 11.3 The MAS with disturbances were also considered in [3, 14, 26]. The eigenvalues of system matrix A in [26] was required to have nonpositive real parts, which limits the application of the proposed approach. Compared with [3], a reset observer with sector reset conditions is developed that can improve the convergence performance of observation error. In addition, notice that sliding-mode technique proposed in [14] can also be employed to handle matched disturbances, by contrast, we unmatched disturbances and directed network are taken into consideration in this part. Thus, the results in this work is more general than [14], but it is worth to be mentioned that the consensus control strategy in [14] is fully distributed.

Remark 11.4 One limitation of the proposed reset observer-based control method is that some global information, such as the whole communication topology, is required in applications. This means that the practicability of the obtained results is restricted to some extent, since the communication topology in application scenarios is not always fixed, for example, some agents depart from the network due to communication/hardward network or some new agents are added into the network. Thus, if the whole communication topology information is needed in control strategy, the dynamic variation of topology may lead to performance degradation or even instability. To further improve the practicability of the proposed methods, how to design fully distributed consensus protocol and hybrid dynamic ETM is an interesting topic, some notable results [14, 27, 28] regarding fully distributed ETMs can provide us some ideas and inspirations.

11.4.2 Further Discussions

In the proof of the main results, it can be seen that the variable $\eta_i(t)$ in triggering conditions is also a part of the Lyapunov function, based on Lyapunov function stability theory, $\eta_i(t) \geq 0$ should be always held. From the design of dynamics of $\eta_i(t)$, it is not explicitly whether $\eta_i(t) \geq 0$ can always be ensured. To ensure this point, the following proposition is given.

Proposition 11.1 *For given initial condition $\eta_i(0) \geq 0$, parameters $\pi_i > 0$, $\gamma_{1i} > 0$, $\gamma_{2i} > 0$, $\gamma_{0i} > 0$, the following conditions hold.*

$$\eta_i(t) \geq 0,\ i \in [1, N], t \in [0, \infty)$$

Proof Suppose that there exists time t_- such that $\eta_i(t_-) < 0$, due to $\eta_i(t)$ is continuous, there must exists a $t_z < t_-$ such that $\eta_i(t_z) = 0$ and $\dot{\eta}_i(t_z^+) < 0$. According to the ETM (11.8), (11.9), and (11.19), when $\eta_i(t_z) = 0$, if $\tau_{i_E} < \tau_{MIET}$, we have $\dot{\eta}_i(t_z^+) = -\pi_i \eta(t_{z^+}) + \gamma_{0i} \|e_i\|^2 \geq 0$, which contradicts with $\dot{\eta}_i(t_z^+) < 0$. If $\tau_{i_E} \geq \tau_{MIET}$, according to (11.9), we have $\dot{\eta}_i(t_z^+) = -\pi_i \eta(t_z^+) - \gamma_{1i} \left\|e_i(t_z^+)\right\|^2 + \gamma_{2i} \left\|\hat{q}_i(t_z^+)\right\|^2 \geq 0$, which also contradicts with $\dot{\eta}_i(t_z^+) < 0$. Therefore, we have $\eta_i(t) \geq 0$, for $t \geq 0$. □

Next, we show the advantage of the proposed hybrid dynamic ETM compared with its counterpart of static ETM and the existing dynamic ETM.

(1) *Static ETM*

For static ETM, the next triggering instant for agent i is determined by

$$t_{k_i+1}^i = \inf\{t > t_{k_i}^i | \gamma_{1i} \|e_i(t)\|^2 - \gamma_{2i} \left\|\hat{q}_i(t)\right\|^2 \geq 0\} \tag{11.38}$$

where $\gamma_{1i} > 0$, $\gamma_{2i} > 0$, this form of static ETM have been widely applied in [3, 13, 26] and references therein.

(2) *Existing Dynamic ETM*

For the existing dynamic ETM, the next triggering instant for agent i is determined by

$$t_{k_i+1}^i = \inf\{t > t_{k_i}^i | \eta_i(t) + \sigma_i(\gamma_{2i} \left\|\hat{q}_i(t)\right\|^2 - \gamma_{1i} \|e_i(t)\|^2 \leq 0\} \tag{11.39}$$

where $\gamma_{1i} > 0$, $\gamma_{2i} > 0$, and $\dot{\eta}_i(t) = -\pi_i \eta_i(t) - \gamma_{1i} \|e_i(t)\|^2 + \gamma_{2i} \left\|\hat{q}_i(t)\right\|^2$. This form of dynamic ETM was originally proposed in [16], and has been used in [19–21].

Let $t_{k_i+1}^{i(s)}, t_{k_i+1}^{i(d)}, t_{k_i+1}^{i(dm)}$ denote the next triggering instant determined by the ETMs (11.38), (11.39), and (11.8), respectively. The following results formally show that the next triggering instant $t_{k_i+1}^{i(dm)}$ determined by the dynamic ETM (11.8) is ensured to be not less than the triggering instant $t_{k_i+1}^{i(s)}$ and $t_{k_i+1}^{i(d)}$.

Proposition 11.2 *Given a triggering instant $t_{k_i}^i$, by applying the static ETM (11.38), the existing dynamic ETM (11.39), and the proposed hybrid dynamic ETM (11.8) to* *the MAS* (11.1) *separately, the next triggering instants determined by these ETMs satisfy* $t_{k_i+1}^{i(s)} \le t_{k_i+1}^{i(d)} \le t_{k_i+1}^{i(dm)}$.

Proof Let us assume that $t_{k_i+1}^{i(s)} > t_{k_i+1}^{i(d)}$, from the static event-triggering condition in (11.38), we have

$$-\gamma_{1i} \left\| e_i(t_{k_i+1}^{i(d)}) \right\|^2 + \gamma_{2i} \left\| \hat{q}_i(t_{k_i+1}^{i(d)}) \right\|^2 > 0 \tag{11.40}$$

However, based on the dynamic event-triggering condition in (11.39), the following condition holds

$$\sigma_i \left(\gamma_{2i} \left\| \hat{q}_i(t_{k_i+1}^{i(d)}) \right\|^2 - \gamma_{1i} \left\| e_i(t_{k_i+1}^{i(d)}) \right\|^2 \right) \le -\eta_i \left(t_{k_i+1}^{i(d)} \right) \le 0$$

which contradicts (11.40). Hence, $t_{k_i+1}^{i(s)} \le t_{k_i+1}^{i(d)}$.

Note that with the proposed hybrid dynamic ETM (11.8), $t_{k_i+1}^{i(dm)} - t_{k_i}^{i} \ge \tau_{MIET} > 0$ must be satisfied. Thus, if $t_{k_i+1}^{i(d)} - t_{k_i}^{i} \le \tau_{MIET}$, it is naturally ensured that $t_{k_i+1}^{i(d)} \le t_{k_i+1}^{i(dm)}$. If $t_{k_i+1}^{i(d)} - t_{k_i}^{i} > \tau_{MIET}$, we suppose that $t_{k_i+1}^{i(d)} > t_{k_i+1}^{i(dm)}$, then, based on the dynamic event-triggering condition in (11.39), we obtain

$$\sigma_i \left(\gamma_{2i} \left\| \hat{q}_i(t_{k_i+1}^{i(dm)}) \right\|^2 - \gamma_{1i} \left\| e_i(t_{k_i+1}^{i(dm)}) \right\|^2 \right) + \eta_i \left(t_{k_i+1}^{i(dm)} \right) > 0 \tag{11.41}$$

It follows from (11.8) that $\eta_i(t_{k_i+1}^{i(dm)}) \le 0$, which, combined with (11.41), leads to $\sigma_i(\gamma_{2i} \left\| \hat{q}_i(t_{k_i+1}^{i(dm-)}) \right\|^2 - \gamma_{1i} \left\| e_i(t_{k_i+1}^{i(dm-)}) \right\|^2) > 0$, and it further leads to $\dot{\eta}_i \left(t_{k_i+1}^{i(dm-)} \right) > 0$, which implies that $\eta_i \left(t_{k_i+1}^{i(dm)} \right) > 0$, that contradicts with $\eta_i(t_{k_i+1}^{i(dm)}) \le 0$. Hence, $t_{k_i+1}^{i(d)} \le t_{k_i+1}^{i(dm)}$. □

11.5 Simulation Study

A numerical example is provided to show the effectiveness of the proposed reset observer-based control method in this chapter. Suppose that there exist four agents in the MAS, the communication relationship among agents is represented by a strongly connected directed graph as shown in Fig. 11.2. The system matrices are given as follows

$$A = \begin{bmatrix} -1 & -2 & -2 \\ 0 & -1 & 1 \\ 1 & 0 & -1 \end{bmatrix}, B = \begin{bmatrix} 2 \\ 0 \\ 1 \end{bmatrix}, B_w = \begin{bmatrix} 1 \\ 0 \\ 1 \end{bmatrix}, C = \begin{bmatrix} 1 \\ 0 \\ 1 \end{bmatrix}^T. \tag{11.42}$$

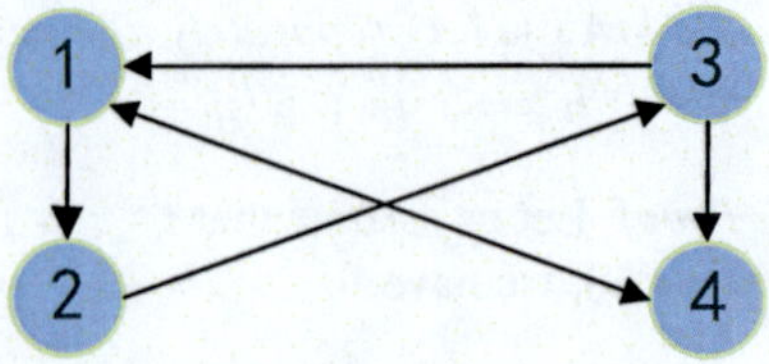

Fig. 11.2 Communication topology

It can be easily verified that the pair (A, B) is stabilizable and the pair (C, A) is observable. The simulation is performed using MATLAB/Simulink, in particular, the modules (e.g., Integrator, Subsystem, Gain, etc.) in Simulink are used to construct the the MAS with hybrid dynamics, MATLAB Function module is used to compute some intermediate signals (e.g., $q_i(t)$, $\Psi_i(v^i_{loc}(t))$), and the parameters and solution of conditions in the main results are programmed into '.m' file. Based on Lemma 11.1, it it not hard to calculate that

$$z = \begin{bmatrix}0.2222\ 0.3333\ 0.3333\ 0.1111\end{bmatrix}.$$

With respect to the parameter $a(\mathscr{L})$, according to the algorithm in [24], $a(\mathscr{L}) = 1.3485$ can be calculated. Then, select $Q = 8I$, $\mu = 0.005$, the gain matrix K can be obtained by solving the algebraic Riccati equation (11.17), and it follows that

$$K = \begin{bmatrix}0.0360\ 0.0104\ 0.0426\end{bmatrix}.$$

The rest parameters are set to be $a_r = -2$, $b_r = 2$, $\kappa_i = 7$, $\alpha = 0.1$, $\alpha_z = 0.9$, $\gamma_{0i} = 0.01$, $\pi = 0.5$, $\rho = 0.01$, $\varepsilon = 0.01$, $K_p = \begin{bmatrix}11.6489\ {-4.9677}\ {-5.2592}\end{bmatrix}^T$, $K_r = \begin{bmatrix}3.0754\ {-12.5712}\ {-8.6547}\end{bmatrix}^T$. Based on these parameters and the results in Theorem 11.1, it can be calculated that $\lambda_{1i} = 0.2915$, $\gamma_{1i} = 73.7297$, $\gamma_{2i} = 0.2623$, $\mathscr{M}_{1i} = 48.5329$. To determine the MIET τ_{MIET}, we first compute τ^i_M based on Lemma 11.3, and obtain $\tau^i_M = 0.0611$, so we select $\tau_{MIET} = 0.05$ in the simulation to guarantee that the triggering interval between any two triggering instants is at least 0.05 s, i.e., $t^i_{k_i+1} - t^i_{k_i} \geq 0.05$, which ensures that Zeno behavior is excluded no matter there exist disturbances or not. The initial conditions are chosen as $x_i(0) = i \times \begin{bmatrix}1\ 2\ 3\end{bmatrix}^T$, $\hat{x}_i(0) = \begin{bmatrix}0\ 0\ 0\end{bmatrix}^T$, $\tau_{i_R}(0) = 0$, $\tau_{i_E}(0) = 0$, $\phi_i(0) = 3$, $\eta_i(0) = 3$, for $i \in [1, 4]$. Suppose that each agent subjects to the bounded random disturbance signal as shown in Fig. 11.3.

To show the superiority of the proposed method in this chapter, comparison results are provided. Specifically, let **M1** denote the proposed reset observer-based control method in this chapter, let **M2** denote the method in existing results (e.g., dynamic ETM in [19–21] and linear continuous observers in [3, 29]). The comparison results are shown in Fig. 11.4, where the simulation results clearly demonstrates that the estimation errors with reset observer converge faster. This is mainly due to the role of reset behavior, specifically, before the estimation errors reach a small neighborhood of the origin, under the influence of the term $K_r x_{ri}$ in the reset observer, faster rising

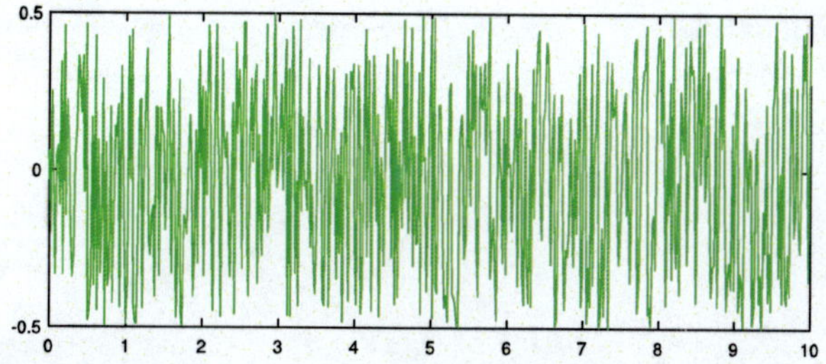

Fig. 11.3 Disturbance

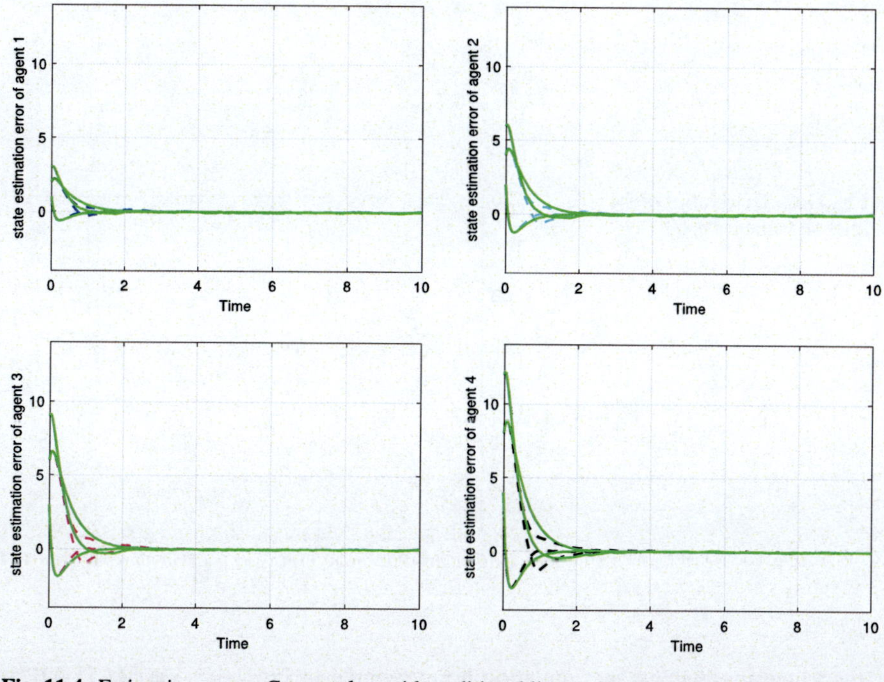

Fig. 11.4 Estimation errors. Green color: with traditional linear observer. Other colors: with reset observer

time is achieved compared to linear observer. Note that if there is no reset behavior, faster rising time usually leads to larger overshoot and longer setting time, this is a limitation of linear control. However, from Fig. 11.4, we can see that faster rising time and shorter setting time can be simultaneously achieved with the proposed reset observer.

The variation curve of combined variable $\delta(t)$ is given in Fig. 11.5, from which, it implies that the expected consensus performance under influence of disturbances is achieved. In addition, we have proved that the variables $\eta_i(t)$ are always nonnegative, to show this fact, the trajectories of $\eta_i(t)$ are given in Fig. 11.6, which confirms the theoretical result. To show the control effect of event-triggering mechanism,

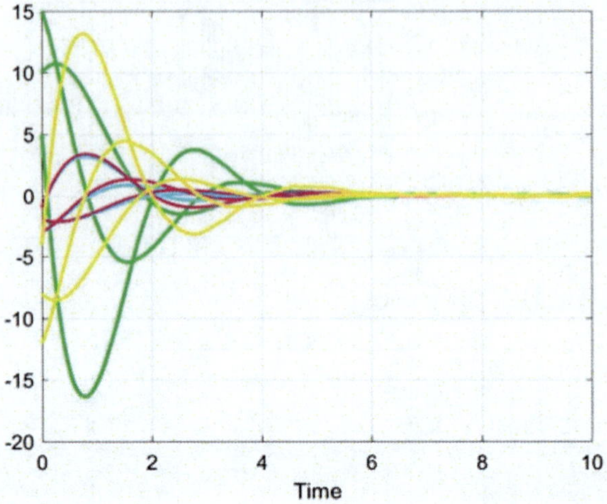

Fig. 11.5 Trajectories of $\delta(t) = -(\mathscr{L} \otimes I_n)x(t)$

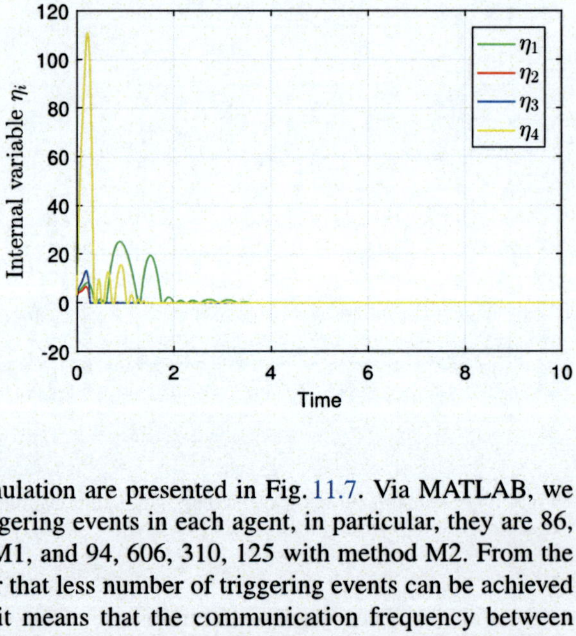

Fig. 11.6 Trajectories of the internal variables $\eta_i(t)$

the triggering instants in simulation are presented in Fig. 11.7. Via MATLAB, we can count the number of triggering events in each agent, in particular, they are 86, 189, 186, 120 with method M1, and 94, 606, 310, 125 with method M2. From the comparison results, it is clear that less number of triggering events can be achieved with the proposed method, it means that the communication frequency between agents is reduced efficiently, and it further means that more communication resources (e.g., node energy and network bandwidth) are saved, that is useful for engineering applications.

To further illustrate the effectiveness of the proposed hybrid dynamic ETM, the triggering numbers with different methods and different parameters are shown in Table 11.1. Based on the main results and Lemma 11.3, it is known that τ_{MIET} can be arbitrarily chosen as long as it satisfies $0 < \tau_{MIET} < \tau_M^i$. Based on the chosen parameters $\tau_M^i = 0.0611$, thus, we select $\tau_{MIET} = 0.01$, $\tau_{MIET} = 0.03$, $\tau_{MIET} = 0.05$

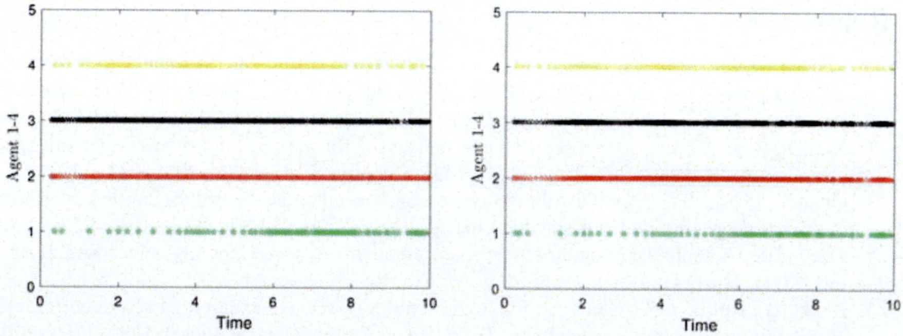

Fig. 11.7 Triggering intervals

Table 11.1 Triggering number. DU denotes the upper bound of the absolute value of disturbances

Parameters	Method	Agent1	Agent2	Agent3	Agent4	Total
DU = 1, MIET = 0.01	M1	88	433	255	111	887
DU = 1, MIET = 0.03	M1	77	289	221	100	687
DU = 1, MIET = 0.05	M1	86	189	186	120	581
DU = 1	M2	94	606	310	125	1135

respectively, and do simulations with these different values of MIET. The triggering number of each agent and total triggering number are shown in Table 11.1, from which, we can see that as τ_{MIET} increases, the triggering number becomes less and less. It implies that the communication resources can be further saved with relatively larger τ_{MIET}.

11.6 Conclusion

Differently from the results in the existing literature, reset observer rather than linear observer is design in this chapter, which has potential to improve transient consensus performance. With appropriate designed sector reset conditions, the well-posedness of reset observer is guaranteed. Then, in order to reduce communication frequency and save communication resources, a reset observer-based distributed control strategy with hybrid dynamic ETM is developed. Time regularization technique is used in the reset observer and dynamic ETM to exclude Zeno reset/triggering behavior, and Zeno-freeness can always be guaranteed even in the presence of disturbances. Stability analysis and ETM design are performed under hybrid systems framework. Finally, the effectiveness of the proposed methods is demonstrated via a simulation example.

References

1. Zhang, H., Feng, G., Yan, H., Chen, Q.: Observer-based output feedback event-triggered control for consensus of multi-agent systems. IEEE Trans. Ind. Electron. **61**(9), 4885–4894 (2014)
2. Xia, D., Ren, W., Sun, X.: Observer-based consensus for multiagent systems under stochastic sampling mechanism. IEEE Trans. Syst., Man, Cybern.: Syst. **48**(12), 2328–2338 (2018)
3. Qian, Y., Liu, L., Feng, G.: Distributed event-triggered adaptive control for consensus of linear multi-agent systems with external disturbances. IEEE Trans. Cybern. **50**(5), 2197–2208 (2020)
4. Clegg, J.C.: A nonlinear integrator for servomechanisms. Trans. Amer. Inst. Elect. Eng. II Appl. Ind. **77**(1), 41–42 (1958)
5. Beker, O., Hollot, C.V., Chait, Y.: Plant with an integrator: An example of reset control overcoming limitations of linear feedback. IEEE Trans. Autom. Control **46**(11), 1797–1799 (2001)
6. Hunnekens, B., Wouw, N., Nesic, D.: Overcoming a fundamental time-domain performance limitation by nonlinear control. Automatica **67**, 277–281 (2016)
7. Zhao, G., Nesic, D., Tan, Y., Hua, C.: Overcoming overshoot performance limitations of linear systems with reset control. Automatica **101**, 27–35 (2019)
8. Paesa, D., Franco, C., Llorente, S., Lopez-Nicolas, G., Sagues, C.: Reset adaptive observer for a class of nonlinear systems. IEEE Trans. Autom. Control **57**(2), 506–511 (2012)
9. Zhao, G., Wang, J.: On L2 gain performance improvement of linear systems with Lyapunov-based reset control. Nonlinear Anal. Hybrid Syst. **21**, 105–117 (2016)
10. Zhao, G., Wang, J.: Reset observers for linear time-varying delay systems: Delay-dependent approach. J. Franklin Inst. **351**(2), 5133–5147 (2014)
11. Paesa, D., Banos, A., Sagues, C.: Optimal reset adaptive observer design. Syst. Control Lett. **60**(2), 877–883 (2011)
12. Garcia, E., Cao, Y., Casbeer, D.W.: Periodic event-triggered synchronization of linear multi-agent systems with communication delays. IEEE Trans. Autom. Control **62**(1), 366–371 (2017)
13. Fan, Y., Liu, L., Feng, G., Wang, Y.: Self-triggered consensus for multi-agent systems with Zeno-free triggers. IEEE Trans. Autom. Control **60**(10), 2779–2784 (2015)
14. Cheng, B., Li, Z.: Designing fully distributed adaptive event-triggered controllers for networked linear systems with matched uncertainties. IEEE Trans. Neural Netw. Learn. Syst. **30**(12), 3645–3655 (2019)
15. Seyboth, G.S., Dimarogonas, D.V., Johansson, K.H.: Event-based broadcasting for multi-agent average consensus. Automatica **49**(1), 245–252 (2013)
16. Liu, T., Jiang, Z.: Event-based control of nonlinear systems with partial state and output feedback. Automatica **53**, 10–22 (2015)
17. Girard, A.: Dynamic triggering mechanisms for event-triggered control. IEEE Trans. Autom. Control **60**(7), 1992–1997 (2015)
18. Yi, X., Liu, K., Dimarogonas, D.V., Hohansson, K.H.: Dynamic event-triggered and self-triggered control for multi-agent systems. IEEE Trans. Autom. Control **64**(8), 3300–3307 (2019)
19. He, W., Xu, B., Han, Q.L., Qian, F.: Adaptive consensus control of linear multiagent systems with dynamic event-triggered strategies. IEEE Trans. Cybern. **50**(7), 2996–3008 (2020)
20. Du, S., Liu, T., Ho, D.: Dynamic event-triggered control for leaderfollowing consensus of multiagent systems. IEEE Trans. Syst., Man, Cybern.: Syst. **23**(6), 602–619 (2018)
21. Hu, W., Yang, C., Huang, T., Gui, W.: A distributed dynamic event-triggered control approach to consensus of linear multiagent systems with directed networks. IEEE Trans. Cybern. **50**(2), 2996–3008 (2020)
22. Borgers, D.P., Heemels, W.P.M.H.: Event-separation properties of event-triggered control systems. IEEE Trans. Autom. Control **59**(10), 2644–2656 (2014)
23. Goebel, R., Sanfelice, R.G., Teel, A.R.: Hybrid Dynamical Systems: Modeling, Stability, Robustness. Princeton, NJ, USA: Princeton Univ. Press, 2012.
24. Yu, W., Chen, G., Cao, M., Kurths, J.: Second-order consensus for multiagent systems with directed topologies and nonlinear dynamics. IEEE Trans. Syst. Man, Cybern. B, Cybern. **40**(3), 881–891 (2010)

25. Fan, Y., Feng, G., Wang, Y., Song, C.: Distributed event-triggered control of multi-agent systems with combinational measurements. Automatica **49**(2), 671–675 (2013)
26. Xing, L., Wen, C., Guo, F., Liu, Z., Su, H.: Event-based consensus for linear multiagent systems without continuous communication. IEEE Trans. Cybern. **47**(8), 2132–2142 (2017)
27. Cheng, B., Li, Z.: Coordinated tracking control with asynchronous edge-based event-triggered communications. IEEE Trans. Autom. Control **64**(10), 4321–4328 (2019)
28. Cheng, B., Li, Z.: Fully distributed event-triggered protocols for linear multiagent networks. IEEE Trans. Autom. Control **64**(4), 1655–1662 (2019)
29. Hu, W., Liu, L., Feng, G.: Output consensus of heterogeneous linear multi-agent systems by distributed event-triggered/self triggered strategy. IEEE Trans. Cybern. **47**(8), 1914–1924 (2017)